V WAS FOR
VICTORY

Also by John Morton Blum

Joe Tumulty and the Wilson Era

The Republican Roosevelt

Woodrow Wilson and the Politics of Morality

From the Morgenthau Diaries
   I. Years of Crisis, 1928–1938
   II. Years of Urgency, 1938–1941
   III. Years of War, 1941–1945

Yesterday's Children (Editor)

The National Experience:
  A History of the United States (Editor)

The Promise of America

Roosevelt and Morgenthau

The Price of Vision (Editor)

Handbook of American History (with Donald B. Cole)

The Progressive Presidents: Theodore Roosevelt,
  Woodrow Wilson, Franklin D. Roosevelt, Lyndon B. Johnson

Public Philosopher:
  Selected Letters of Walter Lippmann (Editor)

Years of Discord: American Politics and Society, 1961–1974

V WAS FOR VICTORY Politics and American Culture during World War II

JOHN MORTON BLUM

A HARVEST BOOK
HARCOURT BRACE & COMPANY
San Diego New York London

Requests for permission to make copies of any part of the work should
be mailed to: Permissions Department, Harcourt Brace & Company,
6277 Sea Harbor Drive, Orlando, Florida 32887-6777.

The lines by Randall Jarrell are from *The Complete Poems* by Randall Jarrell,
copyright renewed 1948 by Randall Jarrell, copyright © 1969 by Mrs. Randall Jarrell,
copyright renewed © 1976 by Mary von Schrader Jarrell, and are reprinted with
the permission of Farrar, Straus & Giroux, Inc. Lines by Marianne Moore
from her *Collected Poems*, copyright 1944 by Marianne Moore, renewed 1972
by Marianne Moore, are reprinted with permission of Macmillan Publishing Co.
Inc., and lines from *The Complete Poems of Marianne Moore* are reprinted with
permission of Faber and Faber Ltd. Pauli Murray's poem in *The Crisis* is reprinted
by permission of that magazine. The quotations from "Little Boxes," words and
music by Malvina Reynolds, copyright © 1962 by Schroder Music Company
(ASCAP), are used by permission.

Library of Congress Cataloging-in-Publication Data
Blum, John Morton, 1921–
V was for victory.
(A Harvest book)
Includes bibliography and index.
1. United States—Politics and government—1933-1945.
2. United States—Social life and customs—1918-1945.
3. World War, 1939–1945—United States.   I. Title
[E806.B58 1976b]   973.917   77-3426
ISBN 0-15-693628-3

Printed in the United States of America
First Harvest edition 1977
X   W   V   U   T   S

FOR E. S. M. AND C. V. W.
incomparable colleagues, indispensable friends

Contents

Preface

This book is not a history of the American people during the years of World War II, not even a history of the home front. There are such books, as the bibliography at the end of this one indicates, and there are still other books about American military and diplomatic history in the war years. This book examines only selected parts or facets of the history of American politics and culture in those years. There is a story within each part, and all the parts relate to each other. Considered separately or together, they suggest some of the ways in which American politics and American culture interacted. They describe some of the habits of thought, some of the hopes, some of the anxieties, some of the institutions that touched the war, often through politics, and that the war touched in turn. They were selected for examination because each in its own way, and especially all taken together, help to explain how the wartime experience of Americans, nurtured in their culture and expressed in their politics, shaped American expectations about the postwar period at home and abroad.

Preceding the substantive parts of the work there is a prologue that attempts to anticipate them all. It is by design an impressionistic prologue, an alternation—not chronologically arranged—of episodes and observations that may, as they are written, help those too young to recall World War II to sense the purpose and the themes of the rest of this book.

There is also an epilogue. Designed in the manner of the prologue, it goes beyond the war years to point out the connections between the wartime expectations of the American people and the kind of postwar society they created for themselves. Those connections are also implicit in the book. They call to mind an admonition that Theodore Roosevelt often gave to those for whom he cared. "God save you," T. R. warned them, "from the werewolf and from your heart's desire."

John Morton Blum

New Haven, Connecticut

V WAS FOR
VICTORY

Prologue: The President, the People, the War

"Yesterday, December 7, 1941—a date which will live in infamy—the United States of America was suddenly and deliberately attacked by naval and air forces of the Empire of Japan. . . . I ask the Congress declare that since the unprovoked and dastardly attack . . . a state of war has existed. . . ."

Forty months passed, each month forever.

". . . The once powerful, malignant Nazi state is crumbling. The Japanese war lords are receiving, in their own homeland, the retribution for which they asked when they attacked Pearl Harbor. . . ."

After those forty months, the ship pulled out of the ping-line where for several days, with others in the task unit, it had been echo-ranging around the perimeter of the staging area at Savo Island. That late in the war, in April 1945, there were few Japanese submarines near the Solomons, but the Navy was taking no chances. Anyhow, for the men on the ship—five officers, all under twenty-five; sixty crew, most of them in their teens—the ping-line provided relief from the boredom of the South Pacific. The motion of the ship, even at eight knots, stirred a breeze that cut the heavy heat. At sea it was bad only during the noon watch, when the sun converted the steel decks to skillets until the invariable daily shower first made the skillets steam and then left them merely warm again. It was a small ship, displacing about three hundred tons, marked by the sleek lines of a destroyer; an unimportant ship that sailed on unimportant missions; a patrol craft, the U.S.S. *PC 616*, thousands of miles from home and going nowhere except back to Tulagi, where at least there would be mail.

As the ship rounded Bungana Island, the captain, a lieutenant, took the deck to bring her in. The gunnery officer, whom he had relieved, moved to the bow to chat with the men standing by the lines. The executive officer, also a lieutenant, lounged against the railing near the captain on the flying bridge. He was chatting, too, now with the captain, now with the signalman who stood alongside the port light, his hand on the lever that controlled it, ready to receive and acknowledge directions from the shore about where to drop the anchor. The light from the shore began to beam its flashes. As the signalman replied, the captain and the executive officer could hear the clacking of the unoiled lever on the ship's light, the da-dit-da-dit that spoke as clearly as the flashes that emerged each time the lever opened the screening of the light. On so small a ship, officers and men had long since learned to play each other's parts. Every one on the bridge knew Morse code, every one read semaphore.

The directions clear, the captain ordered left rudder, slowed the engines, alerted the men at the starboard anchor. As he always did when those routines were completed, the signalman used his light to ask his counterpart ashore whether there was any news. The ship had had no news at sea. Now the executive officer turned his head to see the reply. Maybe, he thought, the Germans had surrendered; maybe the Russians had come in against the Japs; maybe Tokyo was ready to quit; maybe the war was over. His reveries stopped. The light from the shore was flashing out incredible letters: "Y-E-S-T-E-R-D-A-Y P-R-E-S-I-D-E-N-T R-O-O-S-E-V-E-L-T D-I-E-D O-V-E-R." Immediately the signalman flashed back short-long-short. Simultaneously the executive officer, rising on his toes so as better to be seen, thrust his arms out from his sides in a horizontal line parallel to the deck. They had sent back, respectively in Morse and in semaphore, the letter R, meaning "Repeat." Back came the message: ". . . D-A-Y . . . R-O-O-S . . . D-I-E-D . . ." They had hoped they had misunderstood, though they knew they had not. "Who the hell," the signalman said, half to himself, "is President now?" None of the boys or the young men on the ship really remembered any other President.

All ships, the Navy ordered, were to hold memorial services for Franklin Roosevelt the next Sunday. On the *PC 616* the executive officer, as he had for more than a year, followed the standard prescription for ships without chaplains while he presided over the services that Sunday. The rules led him to his texts, some lines from the Old, some from the New Testament. He adapted a short sermon from pages of Thomas Wolfe. Schmaltz, he suspected, but *You Can't Go Home Again* had seemed appropriate reading for so long. He made a silent count of the congregation on the afterdeck: three officers, almost forty men, more in all than gathered there on any Sunday, even after an air attack, even after a typhoon. They had recited the Lord's Prayer. They had sung the first and fourth stanzas of "America." They had mumbled the creed. Now they sang the closing Navy Hymn: "Eternal Father, strong to save . . ." The lieutenant, a secular man uncomfortable with his Sunday duty, felt the metaphor in his stomach. Who was eternal father now?

At sea and on Pacific islands, in France and Italy and England, in India and Burma and along the Persian Gulf, American soldiers—Americans away from home, from the home where they wanted to be and where their wives and sweethearts and parents wanted them—felt their guts twist any day when they were surprised by a metaphor or a memory, one they had almost suppressed. Suddenly there was the timbre of a remembered voice heard while the ear listened for a footstep, or the glimpse of a street or a face or a field in an eye that was looking for a bomber. The stanzas of "America" twisted few stomachs. "Sweet land of liberty," if it meant anything, meant liberty from the Army, not for a few hours, but forever and ever. "Freedom's holy light" was neon.

Inadvertently, because he represented his constituency and expressed its culture, Franklin Roosevelt planned it that way. His "missus," the President often implied, worried about realizing ideals like freedom, but no more than any husband, so he said, could he control his "missus." Now and then he talked about freedom, but he worried constantly about winning the war as fast and as thoroughly as possible. To achieve that end he did what he considered necessary. Only victory, he believed, would bring the boys home and keep them there. "The great white father," his critics, intending slander, sometimes called him. So he really was to the armies of his political children caught up in the war. The soldier boys were always glad to see him. When, to their astonishment, he appeared, smiling his wonderful smile and waving his wonderful wave from a seat in an open car in North Africa, they were delighted. He was part of home. "Oh my god," a GI said, so Roosevelt pretended to recall, adding to the story his own apocryphal reply: "Just for one more term, son." Not an eternal father, then, he was also not a crusader, not a messiah, not in his own eyes and not in theirs. But he was strong to save the lives he thought he could save, strong to fight the war he believed his country had decisively to win, strong for the victory Americans wanted. Physically he was not quite strong enough, by only four months not strong enough, to see that victory. It has never been entirely clear, it was not wholly clear to him, what he would have done with that victory after he had it.

He would not have had a free hand any more than he had had a free hand during or before the war, any more than any President or any other politician ever has a free hand. During the war the possibilities of politics were circumscribed by the limitations of the culture. There was some elasticity in the culture, of course, but there was also a constraining weight of long-standing habits of mind and behavior, of changing expectations, some mutually contradictory, of powerful institutions, of history and of hope. That weight settled on those in Washington who had to manage the war but could not much manipulate the culture. They had not the time for that, and often not the will.

Steward of the people, manager of the managers, the President, the most sensitive of politicians, felt the weight just as he felt the burden of his terrible responsibilities. Some things he could and did decide, especially matters of strategy and diplomacy that fell for decision to heads of state and their appointed agents. Other questions, especially those involving the sentiments and habits of his constituents, he decided only within the limits that his political antennae instructed him to respect. As President he could use his office, his bully pulpit, to stretch those limits, but as commander in chief he had to conserve much of his energy for more insistent efforts. In a small measure the President affected the culture during the war. In larger measure, like other politicians, he accepted the culture as it was and as it was becoming, and the culture affected politics.

In the process the President's concessions to realities, as he saw them, reduced the chances for realizing national ideals, as he defined them. His definitions were splendid: freedom from want, freedom from fear, freedom of belief, freedom of expression, "the supremacy of human rights," the end of the beginnings of war, a "moral order" for all men everywhere. The realities were oppressive. Americans, most of whom favored the English and French, were resolved to keep out of the war that began in Europe in September 1939. The Congress revised neutrality legislation only enough to permit the French and British to buy American goods for cash and transport them on their own ships. Roosevelt, while hoping to avoid war, saw that the victims of Nazi attack would need more American supplies than they could afford. Worse, the American armed forces were unprepared for any military emergency that might arise even within the Western Hemisphere. Yet the political climate persuaded the President that he could not yet press either for more latitude in aiding the democracies or for adequate preparedness for the Army, Navy, and air forces.

Only with the German conquest of Norway, Belgium, and the Netherlands, with the fall of France and the British evacuation of Dunkirk during the nightmare spring of 1940, only then did American opinion move far enough for the President to recommend and Congress to appropriate large sums for military equipment. By then growing sentiment for conscription persuaded Congress to approve the draft, unprecedented in peacetime. Meanwhile Roosevelt had begun to devise new ways to aid the British and later, after Germany pushed eastward, the Russians. Still, the mobilization of the economy lagged, not to reach full thrust until 1942. Through 1941, moreover, the hope persisted, though Roosevelt knew how fragile it was, that American assistance would be sufficient to make direct participation in the war unnecessary. The Army and Navy still needed time to prepare for actual war, and Roosevelt was still trying to gain it for them, when the mounting antagonisms between the United States and Japan exploded in the surprise attack on Pearl Harbor.

The fact of war did not relieve the weight of the realities the President confronted daily. There were the strength of the enemy during 1942, the resulting American defeats, the continuing uncertainties within the Grand Alliance. Would Great Britain risk the invasion of the Continent? Would the Soviet Union stay the course? If not the one, then not the other, and in that event, what chance to erase the Nazi blight from Europe? What chance to roll back Japanese gains in the Pacific? Worse, what chance to keep the Germans from developing an arsenal of rockets and atomic bombs, an arsenal they might have stocked had they had three or four more years to try? At home the realities also lay heavy on the White House: the push of inflation; the impatience of the public mood; the clash of business, labor, and agricultural interests; the deepening suspicions between blacks and whites; the recalcitrance of Congress; the growing weariness of a nation fighting a war on foreign soil; withal, the persistence of politics, as partisan as ever, and in 1944, as ever, a contest Roosevelt had to win if he was to continue to lead the fight.

Those realities and others, especially the daily cost of American lives in battle, confirmed the President's commitment to winning victory as fast as possible. The need to win he had never questioned, nor had the vast majority of Americans. They saw the Nazis as evil, the

Japanese as perfidious, and both as brutal, dangerous foes, destroyers of the peace, enemies of mankind. Those enemies had to be defeated totally. The war itself, with all its agonies and all the hatred it brewed, made attractive any short cuts to the day of triumph. At one with his countrymen in that goal, the President enjoyed the unquestioning support of most of them for all the short cuts he contrived. The necessitarian view of the war, his view, whether based upon a recognition of the magnitude of the problems of fighting, or upon hatred of the enemy, or upon both, pervaded American culture and sustained the Commander in Chief and his associates in their expedient decisions. Those decisions impinged upon policies—domestic, military, and international—with far more gravity than did the ideals to which Roosevelt intermittently gave voice.

Memory also constrained Roosevelt from saying much about postwar aspirations or attempting sharply to define or vigorously to pursue them. As did others of his generation, he remembered Woodrow Wilson. He remembered the war to end wars that did not, the war to save the world for democracy that hurried part of it to Bolshevism and part of it, so soon after the armistice, to fascism. He remembered the emotions engendered by Wilson's crusading rhetoric and the collapse of mood that accompanied the broken—indeed the unattainable—promises of that rhetoric. He remembered the American rejection of the League of Nations. He also remembered, and in large measure had shared, the reinterpretations of the 1930's that found World War I needless, American involvement a mistake, Wilson a pawn of greedy bankers and munitions makers, and the public a captive of British lies. He recalled Wilson's support for an American propaganda of hatred of all things German, a propaganda that heated American spirits to a counterproductive excess. He recalled Wilson's sick rigidity when compromise might have saved the League in the Senate. In 1944 the movie *Wilson* sent Roosevelt's blood pressure briefly up close to 240 over 130, an ominous level. He remembered Wilson, and he strained to avoid what seemed to him to have been Wilson's errors.

During World War II the President tried to prevent his rhetoric from whipping up emotions he could not control. He would have preferred to employ no propaganda at all, and the little he endorsed for home consumption spoke more to the dangers of defeat than to the opportunities of victory. But men developed expectations even in the absence of a glowing public rhetoric, and private expectations during World War II became as important for American politics as had official aspirations a generation earlier.

Those private expectations and some of Roosevelt's public promises, too, emerged in part from memories that had nothing to do with Wilson. They were memories of depression, of joblessness and poverty, of hopelessness and hunger. Even the wealthy, who had not suffered, had only nightmares about the Great Depression, which, they felt, had cost them a good deal. In December 1941 young American men and women had known nothing but depression. Those of draft age, twenty years old, born in 1921, had been only eight when the market collapsed, had looked for jobs they could not find since 1936 or 1937, or had attended school and college, working at odd jobs while they studied, with little expectation of any other kind of work. Like their parents, who remembered prosperity, the young were anxious about postwar depression, a potential enemy they could not obliterate with the weapons they had at hand to blast the Japanese. The GI hoped, but scarcely dared to expect, to get home to a more affluent life than he had ever known, a better life as far as he was concerned. His parents, who were enjoying that life during the war for the first time in more than a decade, savored it the more for fear that it would vanish. That hoping and that savoring dominated the public mood in 1944 and 1945. At times the enemy seemed to be no foreign nation but the depression that the war had dispelled. The President's Four Freedoms sounded all right to his constituents, but they wanted mostly to know just what was in it for them.

The sergeant slept through the first two days of his thirty-day leave, days in which a DC-3 took him to the airport nearest home and a commuters' train carried him the remaining thirty miles. He was glad to see his father at the depot. They did not say much to each other while his father drove the six blocks he had driven back and forth every weekday for so many years. Yes, he told his father, he was really okay. Yes, he had about four weeks. No, he didn't have any plans, some good food, some sleep, some theater, some dates. His father looked older, he thought, but otherwise just the same. So did the house, the dog, and his mother, who cried though he wished she wouldn't. It was all right, he supposed, if she thought he was a hero. Maybe he was, if heroism came from surviving two years in Alaska.

The surprises began at dinner. He couldn't remember so thick a steak. Wasn't there meat rationing? Not, it seemed, if you knew the butcher and paid his price. The downtown merchant, he was told, who had had their trade so long, could find even nylons at a price. And the rationing board had given his father a classification that provided plenty of gasoline. There was plenty of wood, too, for the fireplace that kept the house much warmer than the fuel-oil ration was supposed to permit.

Controlling the excitement they obviously felt, his parents continued. They'd bought the house. He remembered seven moves during the ten years before 1941. Where had they found the money to buy the house? Business was never better, his father said. They'd always wanted to own a house.

What else had they bought, the sergeant asked. Not much, because there wasn't much they needed yet available, but they had war bonds ready to convert to a stove, a new radio, and a refrigerator. And they had had great vacations, Miami once, Atlantic City twice, the coast of Maine last summer. Salaries were controlled, but the company gave its good salesmen expense allowances that rewarded effort.

The sergeant yawned, sleepy again. He had a lot of buying of his own to do, he reflected. A new gray pin-striped suit like the one Cary Grant had worn in that movie. Maybe a Chevy convertible. Maybe a house, too, if he got married. Tomorrow he'd drive up to State to see how a T-3 looked to the girls in his kid sister's dorm.

The President and his lieutenants did use some propaganda, but propaganda to sell the war as they perceived it and pursued it, rather than to detail the nature of the peace. The President took care not to promise more than he could deliver, more than domestic and foreign politics safely permitted, more than would force him to yield his preferences to keep policy on a twenty-four-hour basis, to remain ready always for a necessary compromise. So the propaganda could not say much, and the President and his lieutenants, lacking any large or special message, used the agencies of propaganda easiest to employ, the apparatus of peacetime advertising and peacetime entertainment. Those agencies accepted their new assignments without altering their old ways. And the GI's, who had to watch indoctrination films, for General Marshall believed Americans had to know what they were fighting for, watched them without discomfort. For soldiers accustomed to Hollywood fare, the films were persuasive, just as the peacetime *March of Time* had been, just like the ads for Lucky Strikes.

The interior of the auditorium, warm as the summer day, afforded a comforting invisibility, even anonymity, for the officer trainees in the First Battalion. Going to the movies was a languid way to start another routine day. The scenes were those the newsreels had shown for years. Japanese bombers over Shanghai. Hitler jigging alongside a railway car. Pearl Harbor ablaze. Mussolini and a crowd: "Chey-do-chey, do-chey." Hitler shrieking at the uniforms in some German stadium. Counterpoint: Roosevelt and Churchill at Argentia, voice-over reciting the Atlantic Charter. More pictures, more contrasts: a town meeting, the Wehrmacht in France; a Midwestern high school, the Japanese on Bataan; a clean kitchen, Anywhere, U.S.A., with a mom removing a pie from an oven while a boy entered in khaki, limping.

Perhaps two of each twenty in the battalion were unpersuaded. "Good old apple pie," one voice whispered to the nearest discreet ear. "Why not good old bird's-nest soup?"

We who are about to die (only we hope we won't) will hate the Nip or Kraut who tries to kill us, and we'll try to kill him first. Only he's welcome to his soup or sauerkraut, just so we get home. We won't wrap the apple pie in red, white, and blue. But most people do. That's what their war's about—and to keep the blacks and the Jews out of the neighborhood when it's over, too.

It was never a lovely war. The Secretary of War particularly was annoyed by the "cherubs," Mrs. Roosevelt and her liberal friends, Vice President Henry Wallace, others here and there in Washington who thought they could reform society while the country was fighting a desperate battle. The Secretary of War approved the order that herded the West Coast Japanese, citizens or not, in American concentration camps. He had always had reservations of his own about Asians and Jews and blacks. Negroes, simple folk, lacked the brains and bravery for flying, so he believed. He was too busy to worry about their civil liberties or civil rights, though in peacetime, he knew that in principle those things had their place. The Secretary of War had plenty of support—throughout his own department, in the Navy Department, and in the White House, too.

The President preferred to talk those issues away. He bowed only when he had to, when the pressure was too strong easily to resist. On Columbus Day in 1942, a few weeks before the congressional elections, he eased the restraints he had imposed on unnaturalized Italians, enemy aliens. The Democrats needed the Italian vote. Earlier he had issued the executive order establishing the Fair Employment Practices Commission, but only because nothing short of it would stop the March on Washington that black leaders were organizing. The blacks were still discontented, but he would not desegregate the armed forces, he sent troops to put down a race riot in Detroit, he tried to keep his "missus" from bothering him with blacks' complaints. His aides already knew he did not want to read blacks' letters. He would have done more, perhaps, had he thought conditions permitted it, but he needed Southern support in Congress and he feared the effect on the war effort of agitation and counteragitation. The Negroes would simply have to wait.

Jiminez and Lopez, Mexican boys from West Texas, United States Navy Involuntary, frightened, lonesome, could not understand the rudder orders that came down to the wheelhouse from the bridge. "Left standard rudder" meant too little, too late to puzzled Mexican-American boys. So Jiminez and Lopez stood their watches near the forward gun, which they never learned to fire. Jiminez and Lopez could not read and could not write and could not speak English very well. They could not pass the tests for petty officers. So they remained simple seamen, and because they had no chance to earn a rate, no one trained them for any rate. They did not operate the radio or clean the engines or work the flag hoist. They scrubbed the deck and they peeled potatoes. They peeled them in the galley with the two Negroes on the ship, the officers' boys. Sometimes they did the officers' laundry or shined the officers' shoes.

The chief machinist's mate liked the little Mex, but he also liked to tease. "Hey, Jiminez," the chief said. "We need a bucket of steam. Take this pail and cover over to the *Omaha* and bring me back a bucket of steam." Jiminez took the pail and cover, walked over to the *Omaha*, a half a mile away, told the seaman on watch what he wanted. The seaman grinned. Down to the boiler room he directed Jiminez. The crew in the boiler room grinned. They took his bucket and gave it back to him, filled with steam, they said, and Jiminez carried it carefully back to the chief. "Christ, Jiminez," the chief shouted when he lifted the cover from the bucket, "where the hell's the steam! Don't you understand English, you stupid spick? Go over and get me that steam." Once Jiminez was out of sight again, the boys on the deck laughed until they hurt. Twice more the chief sent Jiminez on his errand. The last time, Lopez saw him return. "Jiminez," he said, after the chief had cussed him out again, "Jiminez, a bucket don't hold steam." Jiminez went away with Lopez while the chief and the boys on the deck laughed and laughed and laughed.

For those and other problems, Roosevelt had a formula. "Dr. New Deal," he had told the press, had retired, to be replaced by "Dr.

Win-the-War." On that assumption he had encouraged his subordinates to depend on available institutions to provide the guns and
ships and planes and food that victory demanded. So business
profited from the war, and labor and agriculture did, too. The President used his formula to explain why he would not fight with the
Congress about New Deal programs that his opponents rolled back.
Those programs or better ones could roll again after the war. He did
fight about taxes, which Congress refused to raise sufficiently, but he
lost. In a way that proved his point. There was not enough public
support for a liberal program for the home front, not enough to make
it worth his while to engage himself in a losing cause, not enough
even to persuade him to try to stir it up. Wendell Willkie tried that
and the Republicans rejected him for it. Roosevelt kept his own party
quiet in 1944 by dumping his vice president, who had moved well left
of Willkie. Then in 1945 Roosevelt appointed his victim to the Cabinet. Later, when necessities had diminished, the President intended
liberal voices to speak again.

To the comfort of the managers of the war, there was also little
public support for a liberal foreign policy. Mostly it was the cherubs
who protested when the President, to hasten victory, made expedient
arrangements with fascist forces in North Africa and Italy. The public wanted victory first and then security. The polls proved it. Security
was the noun that meant an absence of war and an end to conditions
abroad that bred war for the United States. Security meant safety for
Americans. In or out of public office, few Americans recommended a
postwar internationalism that might entail even a partial surrender of
national sovereignty. Some kind of an association to serve to secure
the peace as Americans made it had almost universal support, but not
a superstate, not an organization that could direct American policy.
For his part, Roosevelt expressed mainly a rhetorical concern for the
rights and roles of small nations. A postwar organization would have
to include them, of course, and bow to the *amour-propre* of China
and of France, but the President expected the big decisions to lie with
him and Churchill and Stalin, and their successors. He found the
spokesmen for a liberal peace useful as scouts to test the public's
responses, but he considered those scouts premature in their programs and unrealistic in their expectations. Again like Willkie, the
Vice President by 1944 had marched out too far ahead of the international views of the political army, and on that account also he had
had to be dumped. The President, confident that the public mood
resembled his own, was comfortable in that confidence, and tough.

The managers of the war did not use the war to convert the

governed, or as a laboratory of democracy, or as a crusade for universal peace. The war used them. They accepted the necessities that then prevailed. As a consequence, either the expectations of the governed, or the necessities of war as the governors construed them, dominated politics in the United States. At work at home or at arms abroad, Americans wanted it that way. The world would not turn brave or new that had such people in it.

1 / The Selling of the War

1. Miles from the Battlefields

The old-timer, pipe in hand, spoke in April 1943 from a column in the *Wall Street Journal* that had been bought, as a public service, by the National Distillers Products Corporation, venders of liquid fantasy. "Our town," the old-timer said, "is in the war zone . . . and fights that way! No; sir, the name of our town isn't Stalingrad and it isn't London or Chungking. . . . Our town is Middletown U.S.A. . . . On the map we're a good many thousands of miles away from the nearest Axis battlefield. But the fact is, we consider we're *in the war zone*. Yes, sir, just as surely as if we were right up front where the bullets fly . . . or the murderous little ape-men lurk in the jungle. The men of draft age have gone. . . . Yes, sir, they've said good-bye to their jobs . . . good-bye to home and loved ones—to the 'sweet land of liberty' they sang about when they were kids—and never dreamed they'd have to leave our town to defend." The old-timer asked his audience to reflect upon how well they were doing their jobs in war. Were they following regulations cheerfully, getting to work on time, saving scrap, cultivating a vegetable garden, buying bonds, and co-operating "with the F.B.I. in apprehending suspicious characters"? If they were, they were in the war zone, or so they might believe.

Elsewhere during World War II the impact of invasion or of continual attack ordinarily produced spontaneous feelings of engagement. A sense of common sacrifice, common mission, common suffering imbued besieged Leningrad and London in the time of the blitz, and spread beyond those cities through their national hinterlands. There men needed little artificial stimulation to give themselves to the anxious and arduous tasks at hand. But in the absence of a perceived and immediate danger, of a shared response to collective peril, it fell to exhortation to provide a substitute, however pale. So it was in the United States, in "our town," where the surge of outrage of December 7, 1941, so rapidly subsided. Thereafter danger remained remote, and even sacrifice had a lesser part in the general experience of war than did temporary inconvenience. The war, for Americans at home, was for four years a foreign war. From those Americans the federal government needed to elicit a larger effort than spontaneity provided, and for that purpose federal agencies and private institutions utilized techniques earlier developed by national advertising and by the media it sustained.

The importance of that exercise impressed one English visitor whose own wartime duties took him to the United States in the spring of 1943. "No one fresh from London," he later recalled, "then could fail to be overwhelmed by the contrast, by the impact of America. . . . The moment which crystallized the contrast . . . occurred in flight over the plains of Nebraska. Here, if anywhere, was normality—hundreds of miles of it and not a sight or a sound to remind one that this was a country at war. And then my . . . stewardess deposited my lunch tray in front of me and there suddenly was my Mene Mene Tekel Upharsin. As I reached avidly to attack my butter pat there, neatly inscribed on it, was the injunction REMEMBER PEARL HARBOR. It needed the butter to remind one of the guns." As one rueful American put it soon thereafter, Europe had been occupied, Russia and China invaded, Britain bombed; only the United States among the great powers was "fighting this war on imagination alone."

2. War Bonds and War Aims

Even before Pearl Harbor, others had begun to try to stimulate and to direct that imagination. They did so almost in spite of the President. His exquisite sensitivity for the rhetoric of politics notwithstanding, Franklin Roosevelt distrusted propaganda as an organized activity of government and had little interest in its systematic use. Casually rather than eagerly, he permitted a first, oblique venture initiated by his friend Henry Morgenthau, Jr., the Secretary of the Treasury. An early foe of Nazism, Morgenthau was responsible for persuading

Americans to buy government bonds to finance a part of the huge cost of military procurement. He was also convinced that Americans did not understand the stake of the United States in the war in Europe and Asia. He decided therefore, with an insight that advertising men commended, "to use *bonds* to sell the *war,* rather than *vice versa.*"

That decision owed much to Peter Odegard, a young political scientist on leave from Amherst College, whom Morgenthau recruited as a counselor to the Treasury on "mass psychology." Odegard defined the themes of the first national sales campaign for defense savings bonds. He stressed not thrift but the importance of national courage and the dangers of totalitarianism. On the radio in the spring of 1941, Morgenthau explained his purpose. There were, he admitted, faster and surer ways of raising money than by asking Americans regularly to purchase government bonds. The Treasury had chosen the more difficult course primarily "to give every one of you a chance to have a financial stake in American democracy—an opportunity to contribute toward the defense of that democracy." The bond program, Morgenthau believed, served as "the spearhead for getting people interested in the war." After Pearl Harbor, as he saw it, the program continued to contribute to national morale. "There are millions of people," he told the press in 1942, ". . . who say, 'What can we do to help?' . . . Right now, other than going in the Army and Navy or working in a munitions plant, there isn't anything to do. . . . Sixty percent of the reason that I want . . . [the bond program] is . . . to give the people an opportunity to do something." There was no better way, he argued then and later, always with Roosevelt's casual agreement, to "make the country war-minded."

Morgenthau intended the substance of the bond campaigns to be pluralistic and democratic in taste and spirit. His sales staff appealed to the taste of a national audience, often by utilizing Hollywood's most popular, and therefore usually most banal, performers. To engage particular segments of the total audience, Treasury radio programs also presented a balanced ticket of a kind, now an Irish-American tenor or comic, on one occasion Yehudi Menuhin playing Mendelssohn's Concerto in E minor—partly to say, as Morgenthau observed, "that both artist and composer are 'verboten' in dictator countries." Constantly the Secretary sought the support of a broad rather than a rich constituency. He opposed placing a bond sales booth at Rockefeller Plaza in New York because "that's the place

where they sell the most precious jewelry and . . . high price furs.
. . . Those aren't the kind of people I want to reach." He preferred
a site at Fourteenth Street, just as he preferred to decorate billboards
advertising bonds not with "a Hollywood actress swathed in mink"
but with a girl from the garment workers' union "swathed in . . .
overalls." The customers to whom the Treasury wanted especially to
sell the war were those whom the Secretary believed the New Deal
had had especially to help. He identified their interests with the
national interest and both with the whole purpose of the struggle.

That interpretation suited only Morgenthau's closest subordinates, men like Odegard and Allan Barth, the liberal journalist who
had come to the Treasury from the Washington *Post*. It seemed
quaint or irrelevant to the technicians who increasingly assumed responsibility for preparing materials for the bond campaign, particularly to the Advertising War Council of volunteer copy writers. On
the decision of the Secretary, the department relied primarily upon
publicity sponsored, as he put it, by "public spirited companies or
individuals" who bought bond advertisements costing millions of dollars a year. That system, as it was intended to, provided "a substantial amount of revenue to newspapers and magazines . . . from
banks, department stores and other institutions." Indeed, the Treasury was able during the course of the war to avoid spending "one
penny on paid advertising in newspapers and magazines or on the
radio."

Roosevelt subscribed to that policy. In 1942 he even sent a
special message of encouragement to the annual convention of the
Advertising Federation of America, where the major speaker was
Bruce Barton, sometime Republican congressman from New York
who had discovered two decades earlier that Jesus, too, was an advertising man. Roosevelt had characterized "Martin, Barton and Fish"
as symbols of Republican obstruction of the defense program. Now,
however, the President solicited the co-operation of Barton and his
associates. "For the duration," Roosevelt wrote, ". . . there are
many messages which should be given the public through the use of
advertising space. . . . If the members of your organization will
. . . assist in the war program and continue the splendid spirit
. . . which they have shown . . . advertising will have a . . .
patriotic place in the nation's total war effort." Those messages,
Roosevelt said, related to "the desire for liberty and freedom," which
could be "strengthened by reiteration of their benefits." But that
reiteration, in the words of copy writers who were trained to sell

products rather than principles, was bound to depart from the President's implicit intentions.

Roosevelt and Morgenthau had chosen a curious instrument. Advertisers had long maintained that they exposed demands rather than creating them. They traded upon basic human desires, upon appeals to sex, envy, anxiety, and they related the satisfaction of those desires to the acquisition of commercial artifacts. They exploited the dominant sentiments of the culture, they translated them, but they did not transcend them. Accordingly the savings effected by the Treasury's reliance upon advertising to promote bonds, and by promoting bonds to sell the war, carried a hidden cost in style and substance.

Morgenthau had hoped to conduct the bond campaigns without recourse to intimidation. There were to be, he said, "no quotas . . . no hysteria . . . no appeal to hate or fear." Lagging sales brought the Treasury in 1942 to the employment of quotas. Advertising associated bonds with appeals to both conventional and hysterical appetites. There was, for example, the deathless dialogue between Mr. and Mrs. Jones in a full-page advertisement of the Commercial Banks and Trust Companies of Greater New York.

> He: . . . War bonds . . . mean bullets in the bellies of Hitler's hordes!
> She: . . . War bonds mean systematic savings for us. . . .
> He: . . . You're a very practical person and I love you. But just as much as I love you, I hate Hitler and . . . Hirohito, and every War Bond means . . . a crack at those guys.

If she had doubts about how to spend her savings, Textron dispelled them: "For all women, Textron predicts a future filled with exciting promise! Today Textron is evolving the miracles of war's necessity. But from Textron's war laboratories of today will come fabulous, fantastic fabrics to adorn you and your home tomorrow." If Mr. Jones was drafted, Mrs. Jones would find solace in the message from Community, makers of the Finest Silverplate:

> Today he has a war on his hands. But the day will come . . . when your Tom or Dick or Jack comes home for keeps . . . when kisses will be real . . . ; when you may know the good feel of a tweedy shoulder . . . when crystal will gleam and silver will sparkle on a table set for two. To that day we of Community are bending every effort to speed the work of war. On that day we pledge again to make the loveliest silverware patterns.

Mrs. Jones might even share the satisfactions of military regalia: "Elizabeth Arden's newest lipstick color—Montezuma Red . . . inspired by the brave, true red of the . . . scarf and chevrons of the Women in the Marines. . . . A tribute to some of the bravest men and women in the world."

The men on the Treasury's bond staff, committed to advancing sales for 1943, utilized market research to develop improvements in their program. Their findings reflected the tenor of the bond messages financed by banks and manufacturers. Various sample polls revealed that motivations for buying bonds did not much derive from enthusiasm for the New Deal or the Four Freedoms, or even from a sense of national peril. Americans bought bonds for less lofty reasons, primarily to help a member of the family in the armed services, to invest their money safely, to preserve "the American way of life," to combat inflation, and to save for postwar purchases. Japan appeared a stronger emotional symbol of the enemy than did Germany. Americans, like the old-timer from Our Town, tended to view the Japanese as "ungodly, subhuman, beastly, sneaky, and treacherous." Future bond drives would have to emphasize the war against Japan. Consumers' preferences, in contrast to Morgenthau's, called emphatically for an appeal to hatred.

Those conclusions grew naturally out of the recourse to market research as a guide to promotional content. A decade of depression, after all, had made Americans hyperconscious about financial security and hungry for personal comforts, and the national temper had always harbored racial prejudice, not the least against the Japanese. Advertising, while perhaps exposing the demands of the culture, also reinforced them; market research confirmed a self-fulfilling prophecy. The decision to use bonds to sell the war might have had an inspirational impact if the war had been defined in terms of social justice, humanitarian striving, and an equitable peace. The implications changed when advertising sold the war as if it were fantastic gowns, flat silver, or bright-red lipstick.

Morgenthau, proud as he was of the sales record of his bond office, never admitted how completely its practices distorted his purpose. Even if he had, both he and the Congress would have rejected the advice of Colston E. Warne, for many years a guiding intelligence of Consumers' Union. Warne proposed a prohibitive tax on advertising as an efficient device to dampen demand in a time of scarcity and inflation. But such a tax would have incurred the vigorous opposition of business and of the media, a combination to which the Congress gave a sympathetic hearing in the years of war, just as in the years of peace. Roosevelt, his influence and energy inadequate for treating all the problems raised by war, gave priority to the most immediate needs for victory, and avoided battling the Congress over questions far less controversial than Warne's suggested tax. In the circumstances, the Treasury designed its bond campaigns not to attack but to employ the established system of mass communications. Though Morgenthau did not openly question that system, he sensed the failure of policy to promote his kind of idealism—"very Wilsonian," he called it. He had been talking, he told one aide in 1942, with young people, "college students . . . soldiers and sailors, and there is nothing inspirational being raised for them." He was right. That lack of idealism arose because government propaganda in general, of which the bond campaigns were only a small part, failed to overcome just the kinds of difficulties that marked the Treasury's experience.

3. Truth as a Handicap

Like many Americans of his generation, Franklin Roosevelt was determined to avoid repeating the apparent mistakes of national policy during World War I. Mindful of the calculated hysteria embedded in the propaganda of fear and hate of Woodrow Wilson's Committee on Public Information, Roosevelt shunned another public adventure in hyperbole. Though he permitted the launching of the Treasury's bond campaigns in 1941, and though he recognized that advertising by private agencies might enhance the meanings he attributed to freedom, he was initially opposed to the creation of any federal propaganda service. He changed his mind only with reluctance and only under pressure from advisers, especially Eleanor Roosevelt and Fiorello La Guardia. Their primary aim in 1941 was to provide an adequate flow of information to the American people in order to explain the growing national involvement in war-related programs, especially the build-up of the armed services and procurement for lend-lease. The President consented to a policy intended to give Americans those facts, which would presumably speak for themselves. The title of the first agency formed to execute that policy suggested as much. The Office of Facts and Figures was established in October 1941 "to disseminate . . . factual information on the defense effort and to facilitate a widespread understanding of the status and progress of that effort." That mandate remained unchanged, though circumstances did not, in the six months after Pearl Harbor.

The chief of the new agency, Archibald MacLeish, then also Librarian of Congress, brought to his desk a poet's humane sensibilities, a basic faith in sweet reason tempered by a growing alarm about the advance of fascism, and a large reputation as an advocate of American involvement in the war against Hitler. As early as 1937, arguing for immediate help for the Spanish Loyalists in their fight against Franco, MacLeish had called upon American writers and intellectuals to recognize that "the spread of fascism" was "a matter of principal concern" to them. In a controversial essay of May 1940, he attacked those who failed to accept responsibility for combating that spread which threatened "the common culture of the West." "The revolution of our age," MacLeish wrote, ". . . is a revolution of negatives, a revolution of despair . . . a revolution created out of disorder by terror of disorder . . . a revolution of gangs. . . . Caliban in the miserable and besotted swamp is the symbol of this revolution." American writers and scholars could therefore no longer remain in their closets of "objectivity and detachment," no longer cultivate only their obligations to art, for if they did not join and win the battle, they would find themselves, along with their countrymen, mired in Caliban's swamp.

Even as late as October 1941, MacLeish's position met strong opposition from those of his fellow writers who considered political advocacy a corruption of art and, more important, from that considerable minority of Americans still determined to keep the United States out of the war. Wholly apart from his personal distaste for propaganda, Roosevelt could therefore not afford to establish an agency under MacLeish that would seem committed to intervention. For his part, MacLeish had to divorce his private convictions from his public intentions. He did so by defining for the Office of Facts and Figures "the strategy of truth." "A democratic government," MacLeish announced, "is more concerned with the provision of information to the people than it is with the communication of dreams and aspirations. . . . The duty of government is to provide a basis for judgment; and when it goes beyond that, it goes beyond the prime scope of its duty." In the spirit of that doctrine, the Office of Facts and Figures during the first months of its existence concentrated on furnishing hard data, often in sparse copy, to the press and radio.

Even that information had to be controlled so as not to reveal details about production and shortages, or plans for military and naval deployment. Furthermore, unavoidable delays in recruiting the agency's staff and rivalries between that staff and its counterparts in other federal bureaus and departments impeded both the refinement and the execution of MacLeish's policy.

Those conditions frequently irritated the working press, which resented the idea of a central clearinghouse for information. That resentment peaked over two vital news stories. One pertained to the rationing of gasoline, a step vastly unpopular with a people who loved to drive and who could see no lack of gasoline except on the East Coast. Though he appealed personally to Roosevelt, MacLeish was not allowed fully to explain the order until after the release of a report by the Baruch Committee, earlier appointed by the President, a report that disclosed to friend and enemy alike that the reason for rationing gasoline was the dire American shortage of rubber. That shortage made it necessary to conserve rubber in automobile tires and to divert petroleum stocks for use in the manufacture of artificial rubber. The order remained unpopular throughout the war, and the press remained skeptical about MacLeish's enforced reticence. More important, journalists who queried MacLeish after the Japanese attack on Pearl Harbor found that he could tell them little about that disaster. He was, as the press knew, speaking under the necessary constraints of security. He was also uninformed. "I went to see J. Edgar Hoover," MacLeish later recalled, "who knew no more than I and what I knew I got from the radio. . . . It wasn't an information *agency* failure: it was an *information* failure. I have never been as frightened for the Republic as I was by midnight that night."

For its part, the press was querulous. At a conference a month later, a group of editors and publishers agreed to work through the Office of Facts and Figures but complained that the communiqués of the War and Navy departments were not sufficiently frank or full, and that they provided too little human-interest material. At that time, American, British, and Russian forces were suffering severe, continuing losses, but the editors and publishers thought that it would be enough, when full disclosures threatened to depress public morale, for the OFF to get some government official to issue a counterstatement. The press seemed to want the agency somehow to persuade the Army and Navy to tell the whole truth and to balance it with the big lie.

The OFF told as much of the truth as it could. Its policy was to make public "the maximum of information on . . . matters concerning the war, which can be revealed without giving aid to the enemy. . . . Under no circumstances does the government withhold news . . . on the ground that the news is bad or depressing." Still, in the retrospective judgment of one of MacLeish's senior subordinates, "the strategic and tactical problems involved in giving out information stymied . . . distribution." As a consequence, the agency, though blameless, "fell into disrepute because it wasn't delivering the goods."

Yet it was not a failure to deliver facts that most seriously damaged the strategy of truth. The facts were ordinarily amply delivered, especially well in the pamphlets prepared by the writers' section of OFF. But the facts were often not allowed to speak for themselves. American newspapers, long critical of the Roosevelt administration and suspicious of government news releases, interpreted the information they received according to their own preferences. Not without reason, but frequently with exaggeration, they used OFF data to report a mess in Washington—faltering war production, inefficiency and misdirection in rationing, fiscal, labor, and other policies. Newspapers and periodicals that had once preached isolationism remained hostile, in their renderings of the facts, to the Soviet Union and Great Britain and their alleged motives. In the absence of "the communication of dreams and aspirations" by the Office of Facts and Figures, the private press substituted some fantasies of its own.

The cinema was worse. From first to last during World War II, the motion-picture industry was afraid that the government might go into the business of making movies, a powerful medium of information and persuasion. It was a groundless fear. Lowell Mellett, Coordinator of Government Films on Roosevelt's White House staff and then MacLeish's deputy for films in the OFF, gave Hollywood all the assurance it should have needed. "The American motion picture," he told the industry's leaders, "is one of our most effective mediums in informing and entertaining our citizens. The motion picture must remain free. . . . I want no censorship." Rather, like the British, the American government counted upon the movies to contribute to national morale by supplying entertainment, as they always had, especially, in wartime, escapist entertainment. "We are hoping that most of you . . . will stay right here in Hollywood," Mellett said, "and keep on doing what you're doing, because your . . . pictures are a vital contribution to the total defense effort." Concurrently, Hollywood stars were to make personal appearances, and theaters to provide time, for the promotion of bond sales and the support of war-related fund drives like those for the Red Cross and the USO. Those activities made motion pictures, in the official view, an essential industry, whose brightest male names could count, if they wished, on exemption from military service. Some of them enlisted; most of them satisfied Washington by contributing to the war as entertainers and taxpayers.

About the content of wartime movies, Mellett told the producers to "use your own judgment." The Hollywood Writers Mobilization for Defense, one industry group, never subscribed to the strategy of truth. "The wartime function of the movies . . . ," the writers announced in their first policy statement, "is to build morale, and morale is . . . education . . . inspiration . . . confidence." Hollywood's duty was clear. It would "tell the American people about the war itself . . . how's it going and what for," it would "stimulate . . . initiative and responsibility," but it would also demonstrate that morale, among many things, was entertainment. That did not mean that every motion picture had to be "*about* the war, or have a war background," but it did mean that every picture, romantic or dramatic or funny, would "involve a consciousness of war." Translated, that dictum recommended business about as usual. In feature films, a few American soldiers almost always beat fifty Japanese single-handed, and the FBI invariably caught the saboteur or broke up the spy ring. The truth yielded further to art in Walt Disney's cartoons designed to persuade civilians to obtain inoculations against sundry diseases, smallpox among them. White corpuscles, of course, put up the body's resistance to disease germs. "But because most people think of blood as red," as *Fortune* reported, "and because red is a good fighting color, Walt and his men held out for red as the color of the corpuscle soldiers, and got their way."

Facts, in another special sense, became figures, for according to the Hollywood prescription, Myrna Loy could "spend her afternoon at a Red Cross class" just as easily as she could "spend it playing bridge," and the audience would be just as happy. Bob Hope, the prescription continued, could "do a routine" about the rubber shortage just as easily as he could "fall into a swimming pool," and the audience presumably would laugh just as hard. There were fantasies implicit in those formulas, fantasies that contributed to a national sense of easy security, fantasies that associated the preservation of rubber with slapstick and Red Cross nursing classes with the light comedy of the Thin Man and his Lady. Hollywood, as ever, had its persistent prescription for making "every picture . . . bear the stamp of our time."

Radio, the most influential medium of that time, was little better. To be sure, radio stations were scrupulous about following the recommendations of the Office of Facts and Figures for allocating minutes for spot announcements—brief exhortations for military recruitment, the purchase of war bonds, contributions for the USO, and volunteer assistance in sundry defense programs. But in other respects radio went its own way partly because MacLeish and his senior associates for too long regarded "radio operations as distinctly minor, not to say silly." With advertising budgets essentially unaffected by the war, radio programs continued to provide modes of entertainment commended by the advertising industry and its clients. Advertisers defined the style, for example, in some three hundred daytime series that captivated millions of women listeners. "The settings," according to one analysis of those programs during the early weeks of the war, "are middle class. . . . All problems are of an individualistic nature. It is not social forces but the virtues and vices of the central characters that move the events along. . . . No other effect than the reinforcement of already existing attitudes can be expected from such programs." Certainly no war-related effect could be expected. Rather, the escapist hypnotism of the soap operas probably dulled the impact of the factual spot announcements, and perhaps even of radio's ordinarily twisted treatment of military news. "The war," as one critic put it, "was handled as if it were a Big Ten football game, and we were hysterical spectators."

Inadvertently, the strategy of truth left the realm of dreams and aspirations to the hucksters and their associates in the media. Even if MacLeish had correctly defined the proper role for a democratic government, the habits of viewers and listeners and of Hollywood and the networks made the mere provision of information a weak instrument of communication. A separation of data from inspiration was both artificial and crippling in the opinion of most specialists on propaganda and public opinion, some of them members of MacLeish's staff. "Public relations," according to Harvard psychologist Gordon Alport, "advertising and public opinion work are war industries and ought to be mobilized." Max Lerner, political scientist and historian of ideas, called for "a bold and affirmative attempt to use democratic persuasion . . . as a form of political warfare." The foremost American student of propaganda, Harold Lasswell, who was also a major influence within the Office of Facts and Figures, believed that propaganda had to have "a large element of fake in it. . . . That only truthful statements should be used . . . seems . . . an impractical maxim." Of the same mind, Ralph Ingersoll, of

PM, urged a policy of "propaganda *for* the truth." Sherman H. Dryer, an astute critic of radio, summarized the case. "The strategy of truth," he wrote, ". . . is a handicap. . . . Truth . . . will enhance the integrity of our officialdom, but it is a moot question whether it will enhance either the efficiency or the effectiveness of our efforts to elicit concerned action from the public."

By early 1942 the Office of Facts and Figures had moved close to that conclusion. The entry of the United States into the war had in itself removed some of the constraints of October 1941 and given new urgency to the mobilization of public sentiment. Lasswell charted the new course. "The *Strategy* of Truth in communication," he proposed, called "for the *tactics* of clarity and vividness." Literalness was not the truth, especially in drama. Attempting now vividly to sell the war, the OFF authorized a series of radio programs which all the major networks broadcast beginning in February 1942. But the new tactic brought the agency, in the absence of any quick and easy alternative, to employ precisely those talents which had previously served the networks and their advertisers. Each of the dramas in the series, "This Is War," was written by or under the supervision of Norman Corwin, a veteran writer-director of the Columbia Broadcasting System, and they were all produced by H. L. McClinton, an executive of N. W. Ayer and Sons, a New York advertising agency.

Corwin and McClinton doubtless had less time and budget than they wanted, but their series reflected those liabilities less than it did the constant style of American radio entertainment. The first program, "America at War," struck *Variety* as a "hyperdermic of vitamins," a broadcast "written, acted and directed with angry intensity." That quality raised exactly the problems MacLeish had hoped his original strategy of truth would avoid. As one critic said, "America at War" made too much of American virtues. "We Americans," one actor had read, "are affable enough. We've never made killing a career, although we happen to be pretty good with a gun. . . . A sentimental people; a sympathetic people . . . We show it and we act it." The critic was not so sure about that. Raymond Moley, that tireless detractor of all things associated with Franklin Roosevelt, had even larger reservations about "America at War." It was, he wrote, "a highly imaginative simulation of the sounds and emotions of war preparation. It was as if Mr. MacLeish . . . had handed a man who wanted 'accurate' information about Denmark a copy of 'Hamlet.' "

The program that followed provoked similar criticisms. There was much of exhortation, less of interpretation of war aims, and, in spite of more banality, little success in reaching beyond the 10 per cent of Americans who were the best educated in the nation. "To the Young," a drama which won some plaudits, began on a note of young love. Then the voice of a boy: ". . . That's one of the things this war's about." Girl: "About us?" Boy: "About *all* young people like us. About love and gettin' hitched, and havin' a home and some kids, and breathin' fresh air out in the suburbs . . . about livin' an' workin' *decent, like free people.*"

That mindless definition of war aims proved to be less controversial than the stronger message in "Your Army," a broadcast that Leo C. Rosten, one of MacLeish's associates, considered a model for future ventures. They should stress, he wrote, "more . . . along the lines of this kind of simple and forthright description and . . . less . . . 'lofty principles.' " A friend of the President disagreed. The forthright facts in "Your Army," he complained, "would depress people to the point of suicide."

That friend had doubtless preferred the second show in the series, "The White House," which antagonized hardened anti-Rooseveltians. It was, Raymond Moley wrote, "an impressionistic account of parts of Mr. Roosevelt's life, with a dash of Abraham Lincoln thrown in to suggest a parallel . . . that Mr. Roosevelt was lonely and working hard. . . . What has Mr. Roosevelt's career as a state senator to do with the making of tanks? . . . When are we going to . . . get down to the hard facts about this particular war?" For Moley, as for many like him, vividness was the exclusive province of the administration's opposition.

It remained also at times an embarrassment for MacLeish, who could never bring himself to agree with Lasswell. He had seen Lasswell only once, MacLeish later wrote: "He talked to me for an hour in my office . . . and left me glassy-eyed." So, too, MacLeish could not endure the contrived emotionalism of Madison Avenue dramatists or feel comfortable with the prospect of suburban love and life as the objective of the war. In May 1942, Arch Oboler, one of Norman Corwin's writers, speaking before the Institute for Education by Radio, called for "hate on the air." MacLeish, upset, explained in a public letter that the Office of Facts and Figures, in accordance with the Christian doctrine of hating sin but loving the sinner, stood not for hatred of Germans or Japanese, but for hatred of evil. But that admirable distinction, as MacLeish must have realized, was beyond the grasp of many of those engaged in selling the war and most of those whom they were trying to reach.

It was also beyond the reach of the strategy of truth, which

MacLeish had had in any event to modify. Though he was not prepared personally to advocate fakery, he had concluded that the
American people needed assistance in associating their best aspirations with his best reading of the facts. Moley would not have liked
the letter MacLeish wrote Roosevelt on May 16, 1942:

> . . . One of our recent Intelligence Surveys . . . contains in-
> formation I thought you would find particularly interesting. The more
> I think about it, the more I think that it is information which ought
> to be made available to the American people. It is a heartening thing
> for any man to know that the best thoughts he thinks are shared by
> many others. . . .
>   Less than one person in ten now wants to follow an isolationist
> policy in the post-war world. Eight persons in ten are prepared to take
> a full and active share in some kind of world organization. The Four
> Freedoms . . . have a powerful and genuine appeal to seven persons
> in ten. . . .
>   Thus the American people show themselves idealistically in favor
> of guaranteeing freedom of speech and religion all over the world, of
> maintaining a world police force to guarantee against future wars, of
> helping to secure better working and living conditions for people all
> over the world, and guaranteeing that all nations get a fair share of
> raw materials. Furthermore, four out of five people believe that this
> country should and will help to feed the hungry peoples of the world
> after the war is ended.
>   The same survey, however, reported another situation which is
> closely related. . . . The same people who express these ideals . . .
> have their private misgivings. Seven out of ten expect to be personally
> worse off after the war. Six out of ten expect lower wages. Three-
> quarters expect there will be fewer jobs. This feeling that even the
> winning of the war may bring economic disaster presents a storm
> warning. It has been, and can increasingly be, fostered consciously
> by the divisionists; unconsciously by many others. But the fact that
> those who hold the doubts hold also to the ideals seems to me to
> present a golden opportunity. No divisionist can divide the American
> people once they become convinced that their ideals are practical,
> that they can be attained . . . when we win. . . .
>   A full knowledge of what we are fighting for, coupled with as-
> surance that we can win our goals, can be a positive measure in win-
> ning the war. And this, I submit, is the key to unity in fighting this
> war—the keynote to our ideological offensive.

For MacLeish, then, as for Lasswell, literalness was not the
truth, but for MacLeish in May of 1942 time had run out. Dissatis-

fied though he was with the original strategy of truth and with the first ventures on the radio of the tactic of vividness, he knew that he could not effect the change he commended to the President. Restrictions on the dissemination of information, particularly about Pearl Harbor, though no fault of OFF, had cost the agency much of its credibility. The dramatization of "This Is War" subtracted further from its stature. Both the principle of truth and MacLeish's own sense of purpose got lost in the sentimentalism and commercialism of the media and their masters.

As the head of the Office of Facts and Figures, MacLeish had tried to carry out the kind of information policy the President seemed to prefer. When the execution of that policy drew fire, MacLeish, like others in different agencies before him, took the brunt of the attack directed at the White House. He was vulnerable because of his personal identification with artists and intellectuals and with controversial issues so many of them advanced, issues like vigorous world organization and civil rights. He had also been indiscreet. Angered by the biased reporting of the Patterson-McCormick newspapers, particularly their "defeatist" and their anti-British and anti-Russian policies, MacLeish had punched back in two strong public addresses. "I was certainly ill advised," he later reflected, "but it was a pleasure."

MacLeish was caught, too, as others had been before him, in the clash of competing bureaucracies in Washington. The Office of Facts and Figures had had the authority only to co-ordinate, not to control, government information programs. The executive order creating the office, MacLeish felt, "read like a pass to a ball game." Accordingly the War, Navy, and State departments, the Co-ordinator of Inter-American Affairs, the Division of Information within the Office of Emergency Management—all those and other agencies had continually circumvented the decisions of the unofficial interdepartmental committee that MacLeish had appointed and through which he had had to operate. The result, as he put it to Roosevelt, was a Tower of Babel. Samuel Rosenman, one of Roosevelt's closest counselors, Budget Director Harold Smith, and Attorney General Francis Biddle agreed with MacLeish about both the need to reorganize the information programs and the desirability of creating a new agency with clear, central authority.

As Roosevelt moved toward the establishment of that agency, the Office of War Information, in the spring of 1942, MacLeish asked him to make it "the occasion of my elimination." The new policy-making post was to go to Elmer Davis, a veteran newspaperman and radio newscaster whose dry voice and sparse style symbolized reportorial accuracy and integrity. The operating position below Davis for the domestic field, MacLeish believed, should fall to "a bang-up executive with a technical knowledge of radio, movies, press, etc. . . . Obviously, I am not the man."

The problem was less obvious than MacLeish suggested. He probably was not the man to run the new agency. But a "bang-up executive" with a technical proficiency in the media was more vulnerable than MacLeish had been to precisely those distortions of information policy which had most diverted propaganda from the kinds of objectives that he held precious. The Office of Facts and Figures had suffered, to be sure, from organizational inefficiency common to federal agencies in the early months of the war, and from the hostility of the press to any departures from the open practices of peacetime. The mission of the OFF had suffered more from those facets of American culture that Madison Avenue had long cultivated, and from the banalities and absurdities that advertisers and their hired writers had long made the stock of their trade, the stuff of the dreams they nourished.

4. Prescriptions for War Information

The executive order of June 1942 establishing the Office of War Information defined a seemingly broad mandate: "to coordinate the dissemination of war information by all federal agencies and to formulate and carry out, by means of the press, radio and motion pictures, programs designed to facilitate an understanding in the United States and abroad of the progress of the war effort and of the policies, activities, and aims of the Government." The facilitation of understanding implied much more than the mere broadcast of facts and figures, and implied, too, as it affected the policies of the government, an obligation to defend as well as to explain them. That obligation in turn, however appropriate for wartime propaganda, naturally whetted the suspicions both of the press and of political partisans who feared that the interpretation of government policies would spill over into advocacy of the Roosevelt administration. Particularly in addressing domestic rather than foreign audiences, Elmer Davis had to cut a narrow path between a neutrality toward policy that would impede understanding and an advocacy that would provoke Republican tempers. He had also to satisfy the reasonable demands of the press for war news without alienating powerful federal agencies eager to bend information to their own advantage. He had further to work through the media, which had already, as he knew, frequently perceived the war as melodrama. Yet Davis had to proceed with the authority only of a co-ordinator.

Within a few months, Davis organized his agency along clear, functional lines. He divided domestic operations among seven desks, each responsible for providing information about a particular phase of the war effort. The deputy at each desk represented OWI in dealing with the bureaus or departments making policy in his area of concern. Each was to maintain a program "to give the public a clear and accurate picture of . . . the war," and those programs were "to be carried out" by another set of officers, "the chiefs of the media bureaus." Overseas operations, charged with reaching both friendly and enemy peoples, were organized by region, with major desks responsible for Europe and for Asia. Many domestic information programs, Davis presumed, would also serve to educate the world about American wartime policies and goals.

Every program, domestic or foreign, was subject to the explicit approval of the Board of War Information, a kind of agency cabinet that met daily. Initially it included Davis, MacLeish, Robert Sherwood, Milton Eisenhower, and Gardner Cowles, Jr. Of that able and prestigious group, MacLeish and Eisenhower, the younger brother of the commanding general of the European Theater of Operations, did not long remain with the agency. Sherwood, the renowned playwright, an intimate of Harry Hopkins and a speech writer for Roosevelt, was responsible for overseas operations. Cowles, a close friend of Wendell Willkie, was a prominent publisher whose properties included two Iowa newspapers, a radio station, and *Look* magazine. He served under Davis as the director of the media desks of the domestic bureau, a position from which he came to exercise a large influence within OWI. His cabinet, Davis hoped, would generate clear, continuous and penetrating policies, and would also have sufficient authority to permit OWI to control information programs throughout the government.

Those hopes were circumscribed by the same conditions that had affected OFF. Almost from the outset, Davis's chief advisers disagreed about the appropriate direction of information policy. Depressed by the flaccidity of the liberal spirit in the United States, MacLeish in November 1942 urged Davis that "the American people should make up their minds now—meaning as soon as possible—meaning before the war ends—as to the things they are fighting for." Further, MacLeish believed that OWI, even at the risk of attacks upon it, should "accept responsibility for the job of putting before the people . . . the principal issues which must be decided, in a form which will excite and encourage discussion." Eisenhower and Cowles were dubious. "We know," Eisenhower wrote Davis on December 1, 1942, ". . . that in the end the people as a whole will decide . . . and we may also be certain that we will not be completely satisfied with the decision. For that decision will be a compromise between widely divergent economic and political views." OWI should therefore develop no blueprint of its own but concentrate on disseminating the relevant facts that would encourage discussion and "all shades of opinion. . . . We must maintain our policy of objectivity. . . . O.W.I. . . . should continue to be thought of primarily as an *information* agency." As Eisenhower put it directly to MacLeish, "Our job is to promote an understanding of policy, not to make policy." MacLeish replied that "the basic question is whether O.W.I. is to be a mere issuing mechanism for the government departments. I gather it is your position that it should be. On that I disagree . . . emphatically." But Davis and his colleague forced MacLeish to retreat to an unsuccessful effort to link OWI to a State Department planning group. OWI itself, as Eisenhower had understood, had no mandate from Congress or the President to undertake to define the purpose of the war or to try to sell its definition.

Indeed, its mandate as an "issuing mechanism" was far from clear. Officials in the War and Navy departments always contended that Davis could "advise" but not "direct." He was unable to bring the Army or the Federal Bureau of Investigation to release more than skeletal information about the Nazi saboteurs landed in the United States from German submarines, captured and then tried by a military tribunal. The Navy ordinarily issued communiqués that gave the impression of calculated reticence about American losses. Admiral Ernest J. King, the Navy's chief, so it was said within OWI, would like to have released only one statement during the entire war—the one announcing victory. Davis defended the Navy's communiqués, such as they were, rather lamely as being "much closer to the complete story than those of the enemy." It would be "most unfortunate," he continued, if "resentment of any failings" of the Navy in the subsidiary field of information were to "undermine public confidence in their capacity" to fight and to win the war. Yet Davis obviously felt some resentment himself. In contrast to the Navy, he explained, the Office of War Information followed "the guiding principle . . . that the American people have a right to know everything that is known to the enemy, or that would not give him aid if he found out. . . . We believe that the better the American people understand what the war is about, the harder they will work and fight to win it. We are not press agents for the government. We expect to set forth . . . the difficulties with which both the military and civilian branches of the government are faced, and their shortcomings as well as their successes."

That expectation clashed with the professional habits of the Army and Navy, while Davis's concern for the integrity of information contrasted with the tactical purpose of Colonel William Donovan, the head of the Office of Strategic Services. Donovan conceived of his activities abroad as an "adjunct to military strategy," a vehicle for planting stories and rumors that would cripple the enemy and enhance the reputation of the United States. His branch of psychological warfare by its very nature conflicted with the comparatively benign propaganda mission of the OWI, for psychological warfare had often to bury rather than expose the truth about American prowess and policy.

Prodded by Donovan, the War Department moved to obtain control over both psychological warfare and foreign propaganda. "Both of these functions," Secretary of War Stimson wrote Roosevelt in February 1943, "are definitely weapons of war." Davis and Donovan, Stimson complained, had "differed vigorously . . . as to the scope and jurisdiction of their separate duties. As the head of the War Department, I am in the position of the innocent bystander in the case of an attempt by a procession of the Ancient and Honorable Order of Hibernians and a procession of Orangemen to pass each other on the same street. I only know that every Army commander . . . in a foreign theater, if the present difficulties persist, will be subjected to great embarrassment and danger to his operations." As a solution, Stimson proposed placing representatives of the OWI and the OSS on a committee which the Joint Chiefs of Staff would dominate. Donovan had agreed to that plan, but Davis had not, partly because Stimson, with magisterial disdain, had insisted upon selecting the OWI representative whom Davis was titularly to appoint. The issue was never really resolved. Roosevelt did not approve Stimson's proposal, but Davis could not establish operational jurisdiction over the information practices either of the OSS or of the armed services. Stimson was more imperious but no more zealous than others in asserting his independence of the Office of War Information. William M. Jeffers, a brusque, tough business executive whom Roosevelt had made Rubber Director, attempted unsuccessfully in April 1943 to kill a story about the continuing shortage of automobile tires, an account based on Jeffers's own report to the President of the previous February. Perhaps in part because MacLeish had been wounded by government disingenuousness about rubber, Davis was furious with Jeffers. "Mr. William Jeffers," Davis announced on April 19, 1943, "tried to stop me from telling the American people facts about rubber which had been certified as correct by his own office. So long as I am here I propose to tell the people the truth as accurately as I can ascertain it, whether Mr. Jeffers likes it or not." Davis had done so, and Jeffers did not like it. "I can get along very well on the rubber program," he replied through the newspapers, "without Mr. Davis."

Tougher by far than Jeffers, Secretary of the Interior Harold Ickes defied OWI late in 1943 by publishing over its objections an article in *Collier's* dealing with the United Mine Workers' strike in the bituminous coal fields. Ickes criticized the miners' demand for higher wages as inflationary, which it was, but he blamed the strike also on the "unbending posture" of the War Labor Board. The walkouts, he wrote, constituted "a black—and stupid—chapter in the history of the home front," a phrase Davis had wanted deleted. The article, Davis complained to the White House, was "one to which this office has serious objections, on the grounds that some of the statements are at variance with the Government's policy in the coal matter." Because Davis was supposed to "facilitate an understanding" of that policy, he argued that Ickes had violated appropriate jurisdictional lines. For his part, Ickes, whose pungent phrases had already made his point, explained innocently that he had not intended to ignore the opinions of the OWI, but an "unconscionable" delay in receiving them had permitted the article to go into print without the recommended deletions.

The episodes with Jeffers and Ickes revealed some of the difficulties that plagued Davis. He was seeking at once to get federal agencies to tell the truth and to persuade them to do so in a manner that would not undercut official policy, with which they at times disagreed. Yet Roosevelt had given him the authority only of a coordinator. Furthermore, the problem was not just jurisdictional.

There were questions of substance at issue. Davis felt obligated to admit the persistence of shortages, which harassed administrators were endeavoring to overcome in the face of partisan but often accurate charges of ineffectuality. Davis felt equally obligated to explain, and however inadvertently thereby to support, government policies on inflation, which others besides Harold Ickes considered inequitable in various of their applications. One OWI publication on inflation advocated the Treasury's position on taxation, which struck conservative congressmen as an attack on legitimate business profits and personal income. The middle road was hard to find and harder to follow, especially where the military were involved. Confident that the American people could take bad news, Davis wanted to tell them more about casualties and losses than did the armed forces, who prevailed. Their publicity, as the Director of the Budget pointed out, continually overplayed victories and thus generated undue public optimism and complacency.

Conflicts over the content of information also arose, during the first year of Davis's tenure, both within the OWI and between the agency and the media. The questions of the comics and the movies exemplified the latter problem. The Bureau of Intelligence of the OWI found the content of American motion pictures continually distorting, continually frivolous in addressing the issues of the war. Living in luxury, Andy Hardy, absorbed by adolescence, was oblivious to the war. When ideological conversation occurred between good and bad Americans billeted in one of Hollywood's fictional cantonments, it tended to terminate with an "Oh, yeah" and a sock on the jaw. Hollywood's policemen and federal agents constantly followed false leads while amateur sleuths apprehended enemy spies and saboteurs. Even in *Casablanca*, the message of the Czech freedom leader was buried beneath the cynicism of Rick, so memorably conveyed by Humphrey Bogart. It was still the same old story: with few exceptions, *Wilson* and *Mission to Moscow* for two, films designed for the box office carried no message of purpose or idealism. On the contrary, producers during the war relied upon the kind of entertainment—love stories, situation comedies, adventures—that had filled theaters in peacetime. The movies, Archibald MacLeish concluded, were "escapist and delusive," a contributing factor to the failure of Americans to understand either the origins or the objectives of the war.

After surveying the comics, two psychologists reported to the OWI that most of the heroes had donned uniforms (Superman and

Lil' Abner were exceptions), but did not understand global war. In most comic strips, Americans won their battles alone, allies were nonexistent or subordinate, the enemy was a pushover. Only two popular comics satisfied the analysts. In "Terry and the Pirates" the Chinese were admirable fellows. Even better, Joe Palooka had buddies in the Army who made fluent statements against race prejudice and Anglophobia, though Joe himself was becoming rather too bloodthirsty. Comic strips conveyed especially unfortunate messages about the home front: "The effect of the war on the civilian is only a subject of humor. The home front's actions and thoughts and the demands on the civilian, have yet to receive serious and sympathetic treatment." Aside from "Gasoline Alley," the comics had neglected civilian defense, various shortages of civilian goods, and the recruitment needs of the Army. Those matters were ordinarily either the object of derision or the occasion for near hysteria, as in "Orphan Annie." Yet the comics reached millions of Americans who read almost nothing else. The OWI therefore published some comic books of its own, primarily for distribution to civilians overseas, though many copies reached American soldiers. One such, "The Life of Franklin D. Roosevelt," a Republican congressman condemned as unfair political propaganda, a waste of the taxpayers' money, and—intending no irony—an unhappy imitation of "Tarzan of the Apes." Earlier, the *Wall Street Journal* had defended the comics on the ground that the artists would lose their readers if their heroes did not triumph single-handedly. The OWI, not the comics, according to the *Journal,* was unrealistic. By that standard, Davis, like MacLeish before him, had to suffer the treatment of the war according to habits of the media that he was powerless to reform.

There were those within the OWI who accepted just about that standard with regard to the role of advertising. Gardner Cowles, Jr., in February 1943 told the annual advertising awards dinner that advertising was needed to win the war. If advertising was not yet fully organized for that purpose, he said, the fault lay with the OWI and "would be remedied very soon." Within two weeks, the agency had taken a long step in that direction. The OWI planned to publish a new magazine, *Victory,* for circulation in Europe and among liberated peoples there. Copies also reached American troops in England. *Victory,* OWI announced, would contain paid advertising copy, partly to defray the cost of publication, partly for other, illuminating reasons. Advertising, the agency said, "lends an appearance of authenticity to a magazine," and German, Italian, and Japanese propaganda journals therefore carried advertising. (By implication, the United States had to be as authentic commercially as were its enemies, an odd point for an American propaganda agency to make.) "Institutional advertising," moreover, told "an exceedingly powerful story of the war effort of American business." (Perhaps sometimes it did, but one senior advertising executive had complained publicly about "schoolboy pictures of zooming American bombers winning the war thanks to Filch's Bolts and Nuts . . . stomach-turning copy.") Finally, "the Office of War Information felt that American business should have the opportunity to keep alive trade names and brands throughout the entire world in anticipation of a peaceful international trade." Doubtless that noble war aim would indeed organize the energies of American advertising, as Cowles had promised.

The influence of advertising men within OWI ultimately demoralized the writers in the publications bureau. They had been laboring, many of them, since the early weeks of OFF, to prepare pamphlets and other copy about the war that would transcend, on occasion contradict, the typical output of the media and their sponsors. They admired Elmer Davis and respected Gardner Cowles, but they found their position intolerable after Cowles brought in as his senior assistant William B. Lewis, a former vice president of the Columbia Broadcasting System. With support from Cowles and ultimately Davis, Lewis curtailed the production of pamphlets and turned instead to relying upon the initiative of the press in reporting about domestic programs. That decision reflected adversely particularly on Henry F. Pringle, the head of the writers' division, a liberal and iconoclastic journalist and historian. Pringle and his associates felt they had used the press effectively. They cited the wide reporting derived from their pamphlets about American aircraft, drinking in the armed forces, the shortage of doctors, and the importance of employing Negroes in war industry. Even "creative pamphlets," such as that on the Nazi occupation of Warsaw, had received extensive publication. The real issue, they felt, arose because Lewis intended to abandon the careful provision of information about issues and substitute "advertising techniques." Cowles admitted as much. The writers, he said, "resented the fact that I brought in experienced advertising men and sought to persuade industry to use its advertising program to help war information."

That resentment, along with an accumulation of other grievances, in April 1943 provoked a spate of resignations. Among others,

Henry Pringle, Harold Guinzberg, president of Viking Press, Henry Brennan, former art editor of *Fortune,* Edward Dodd, a vice president of Dodd, Mead, and Arthur Schlesinger, Jr., then a precocious but already an incisive historian, left the OWI. As Lewis had intended they should, they also left the field to the advertisers, the men and the ideas that had earlier distorted the publications and broadcasts of the Treasury's bond program and of the OFF. One development that precipitated the crisis prompted Pringle to a splendid summary of the issue at stake. Lewis had grouped a number of desks under a new Bureau of Graphics and placed in charge Lieutenant Commander Price Gilbert, formerly advertising manager for Coca-Cola. Pringle produced a mock poster to characterize the change. It displayed a Coca-Cola bottle, wrapped in the American flag, with a legend below: "Step right up and get your four delicious freedoms. It's a refreshing war."

5. The Politics of Propaganda

The departure of Pringle and his friends from the domestic branch of the OWI preceded by only a few months the decision by Congress substantially to eliminate that branch itself. Republican gains in the fall elections of 1942 gave a coalition of Republicans and Southern Democrats control over both houses of Congress in 1943. Both partners to that coalition had long been suspicious of the OWI, which they viewed as a propaganda agency not for the United States but for Roosevelt and the New Deal. Some of the most conservative Republicans sat on the Appropriations committees of the Senate and the House, the committees that had the first chance to attack Roosevelt by denying funds for operations he commended but they detested, among them the National Youth Administration, the National Resources Planning Board, and, not the least, the OWI.

The attack began in February 1943 when Republican Senator Rufus Holman, of Oregon, condemned an OWI film, "A Prelude to War," as "personal, political propaganda," window dressing for Roosevelt's unannounced but pending fourth-term bid. The magazine *Victory,* Holman continued, wasted money by publishing articles about the President's warm-hearted personality and Mrs. Roosevelt's social graces—subjects, Holman suggested, scarcely "terrifying . . .

to the Japs." Worse still, *Victory* had contrasted the New Deal favorably with reactionaries, including Herbert Hoover, and had praised Roosevelt's foreign policy. Representative John Taber, of New York, ranking Republican on the House Appropriations Committee, contended in March that *Victory* was created largely to "provide jobs for the faithful," who were now "contributing generously to the paper shortage that is threatening the press." In general, Taber said, OWI films, radio scripts, and publications were "partly drivel, partly insidious propaganda against Congress and for a fourth term," and on occasion "along communistic lines."

Davis ventured a rather lame defense. The article in which the New Deal's opponents were called "reactionaries," he said, he had only just read, "culpable negligence" on his part, for "what was said could have been better selected." In any event, the issue of *Victory* containing that article had been sent only to England, Australia, Egypt, "some to Hawaii," and 5,000 to American troops in Great Britain, the last at the Army's request. "The magazine," Davis explained, "is designed for foreign countries, and if you discuss the war effort of this country, you have to say something about the President. . . . It is utterly misleading to read into it propaganda other than propaganda for the foreign front." More important, OWI leaflets, building on some of the themes in *Victory* and dropped on enemy troops from American airplanes, had helped to persuade some Germans, Italians, and Japanese to surrender. Taber remained dubious, and Davis, obviously worried, warned all of his staff members, as he should have in any case, scrupulously to avoid anything that could be construed as partisan political activity.

The attacks and rejoinders continued intermittently through the spring, while Davis, harassed, managed inadvertently on several occasions to offend the press, whose support he needed. In that time Representative John Ditter, a Pennsylvania Republican on the House Appropriations Committee, described the OWI staff as "thousands of starry-eyed zealots out to sell their particular pot of gold to a bewildered people. . . . I shall propose . . . that O.W.I. . . . go out of the field of producing radio programs, movies, and magazines, and leave that to those who know how." Davis called Ditter uninformed, but the congressman's attitude at the least fit the mood within the OWI that led to the resignations of Pringle and his associates. Senate criticism of the OWI pamphlet on inflation as partisan, pro-Treasury, and anti-Congress elicited an apology from Cowles, who said the publication "had a number of arguments . . . which I think were the views of certain political factions" and concluded that the pamphlet should indeed not have been issued. Southern Senators also objected to the pamphlet *Negroes and the War,* which praised the Works Progress Administration and the National Youth Administration for assisting blacks during the Depression, commended blacks for their contributions to war industry, and described the New Deal as ringing "like a pleasant bell in the ears of the American Negro." Republican Senators, Henry Cabot Lodge, of Massachusetts, for one, considered the pamphlet to be partisan pleading, subsidized at the taxpayers' expense. Davis, after some argument, conceded that he would have written the text differently had it been his to do, and the OWI promised also to "eliminate publications on currently controversial subjects and . . . confine them to an authoritative statement by the appropriate agency."

That promise offered the Congress too little, too late. On June 18, 1943, with Southerners and Republicans standing together, the House of Representatives voted 218 to 114 to abolish the appropriation for the domestic branch of the OWI. Davis threatened to resign unless the Senate restored the funds, but the Senate Appropriations Committee agreed to provide only $3 million for the fiscal year 1944, and earmarked that money so as to deny any of it for publications, films, or radio scripts. The House then lopped off $250,000. Davis had enough only for a small operation, inadequate even for persuading the media to co-ordinate policies that they generated themselves. The OWI had to close twelve regional branches, roll up its motion-picture bureau, and abandon publications altogether. As Davis put it, "We shall be . . . passing back a good deal of the work that we have been doing to . . . the motion picture industry, radio and various others." "We call for more help from advertising," one of his subordinates later added. ". . . Advertising informs and persuades and can be repeated again and again until action is forthcoming." The action of the Congress had returned to the media and to those who bought advertising space the whole field of domestic propaganda, a field they had monopolized in peacetime and the government had entered, when war began, only partially, temporarily, and superficially.

Congress spared the foreign branch of the OWI, for which it appropriated a handsome $24 million, almost as much as Davis had requested. That generosity from critics of Roosevelt to an operation directed by Robert Sherwood revealed a hesitancy to challenge a propaganda effort that seemed potentially effective for persuading enemy troops to surrender. "Nobody is satisfied," said one Democrat on the Senate Appropriations Committee, "but this time we had to decide on the basis of ignorance, assuming that the Sherwood office . . . may be half as good as it says it is."

Within three weeks the doubters abandoned that tentative assumption. Following the Allied invasion of Sicily, Benito Mussolini, the increasingly unpopular Italian dictator, was forced to resign. The British and American governments then began prolonged secret negotiations for Italian surrender, which from the start pointed toward an accommodation with King Victor Emmanuel and his effective deputy, Marshal Pietro Badoglio, both fascists. The architects of that policy believed it would speed Italian capitulation, embarrass German forces in Italy, and enhance the probability of co-operation of Italian citizens as well as soldiers with Allied troops as they later moved on to an invasion of the Italian peninsula. But that argument, which never much persuaded Italian and other European liberals, was apparently not communicated at all to the OWI. Through its shortwave facilities, the Voice of America, OWI on July 25, 1943, the day of Mussolini's resignation, beamed to Italy and the rest of Europe statements attacking both the King and the Marshal. The broadcast quoted an imaginary American political commentator, "John Durfee," to the effect that the United States would continue the war against Italy no matter who ruled that country, whether Mussolini, the Marshal, "a high-ranking Fascist," or "the Fascist King himself." The broadcast also quoted Samuel Grafton, a columnist for the New York *Post,* who described conditions in Italy with admirable but untimely precision:

> The moronic little King, who has stood behind Mussolini's shoulder for twenty-one years, has moved forward one pace. This is a political minuet and not the revolution we have been waiting for. It changes nothing; for nothing can change in Italy until democracy is restored.

For the OWI the broadcast was a disaster. The State Department, always jealous of its role in foreign policy, had been waiting for a chance to demonstrate the clumsiness of OWI activities abroad. Sympathetic as the department was to a collaboration with Badoglio, the chance seemed at hand. The War Department, for its part, could interpret the broadcast as another case of embarrassment for a field commander—Eisenhower was guiding negotiations in Italy—growing out of the independent adventures of OWI. Worse, Roosevelt had a curious affection for royalty and a strong predisposition, revealed earlier in North Africa, to co-operate with presumably powerful but undemocratic leaders whose assistance promised to hasten the end of the war. In that spirit the OWI itself had without reflection entitled its magazine not "Freedom" but "Victory."

Acting as fast as he conveniently could, the President at his press conference of July 27 denounced the OWI broadcast of the previous day. The broadcasting division of the agency, he said, had failed to consult him, Secretary of State Cordell Hull, or Robert Sherwood. The subordinate officials in the OWI who originated the broadcast, Roosevelt said, had unfortunately complicated delicate diplomatic negotiations and imperiled the lives of American fighting men. In the House of Commons that day Prime Minister Winston Churchill told his colleagues, "We would be foolish . . . so to act as to break down the whole structure and expression of the Italian State . . . and find ourselves without an authority with which to deal." Those statements delighted the press. The *Wall Street Journal*, speaking in the voice of American conservatives, concluded that the OWI had been trying to throw Italy "into the hands of extreme left-wing radicals"; the agency was therefore patently inept or dangerously subversive.

Unhesitatingly, James P. Warburg, deputy director of the overseas branch of OWI in New York City, accepted full responsibility, with his associate Joseph Barnes, for the controversial broadcast which expressed their views. Sherwood would not repudiate his two aides. At a conference with Hull on July 28, Sherwood promised that thereafter "anything the least bit controversial will be referred to the State Department and the Joint Chiefs of Staff." But, Sherwood also said, with regard to future OWI statements about Victor Emmanuel, "no epithets, but no blinking of the fact that he supported the Fascists and . . . Mussolini," and about Badoglio, "no secret of his Fascist identification." Elmer Davis remained uncharacteristically silent.

There the matter seemed to rest. On the suggestion of Milton Eisenhower, Roosevelt in September went through the motions of directing the military to recognize that Davis had the authority to determine what information was to be released to the public, but the same directive allowed the Army and Navy before any release to disagree with Davis and take their case to the White House. As a consequence, no real change in policy occurred, nor had the President succeeded, as Eisenhower had wished, in restoring "the prestige and strength of O.W.I." On the contrary, hostile congressmen continued to charge the agency with incompetence, partisanship, and radicalism. With the cross-channel invasion of France impending early in June 1944, the War and State departments worried more than ever about propaganda for European consumption. Bowing to their influence, Davis undertook what he gently called a "realignment" of the New York office of the overseas branch. That reorganization involved the discharge of three officers, among them Joseph Barnes and James Warburg, the originators of the celebrated broadcast about Badoglio and the King. Incensed, Sherwood would not agree to the firing of his subordinates. With Davis's permission, Sherwood took his case personally to the President. For his part, in a long memorandum, Davis told Roosevelt that the time had come to decide "who is director of O.W.I."; unless the issue was resolved in his favor, he would resign. Roosevelt supported Davis, Barnes and Warburg left the OWI, and in February Sherwood went to London, later—in September—to resign, ostensibly to assist the President during his campaign for re-election.

Friends and admirers of both Sherwood and Davis on occasion mused about the personal pique, out of character for both men, that led to their falling-out. Yet it is hard to avoid the conclusion that that falling-out was not primarily personal. Rather, the events surrounding it identified it as the deferred culmination of the crisis over Victor Emmanuel. Sherwood in the end felt obliged to protect his subordinates, men who had spoken, however indiscreetly, the vivid truth. Davis in the end felt obliged to protect his agency from the more powerful State and War departments, and to maintain his own authority over it. Roosevelt, as MacLeish could have attested at the time, never much believed in propaganda anyhow, and always "got his Dutch up" when members of his official family fell into public controversy. But after the crisis, overseas as at home, information policy was reduced primarily to efforts to undermine enemy morale, and the OWI, in the observation of one British expert, remained "that most unhappy of Washington's wartime agencies."

Neither an independent information policy nor an agency to generate it had ever had a chance to succeed or a mandate to try. Roosevelt considered the OWI at best unimportant and at worst impertinent. As Milton Eisenhower had told MacLeish, the OWI, whatever its merit, had no right to an independent view. The direction of foreign policy was the President's responsibility. The broadcast of statements that belied his intentions constituted insubordination and misled Europeans about American priorities, which put victory first. Yet OWI support of other of the President's policies, as his opponents in Congress argued and as Elmer Davis grudgingly conceded to them, became in some instances indistinguishable from partisan pleading. In a democracy in wartime no public agency could seem to promote the President's candidacy. As Davis came also to realize, the authority to co-ordinate the dissemination of information was no authority at all. In wartime Washington, as Vice President Henry Wallace once wryly remarked, a co-ordinator was only a man trying to keep all the balls in the air without losing his own. Against the greater power of the State, War, and Navy departments, Davis could not prevail. And whatever the content of OWI releases for home consumption, the independent media could operate according to their conventional commercial criteria.

Against those odds the OWI accomplished about as much as its director should have expected. It squeezed more information out of the armed services than they would otherwise have provided. Its leaflets and booklets heartened resistance groups in Europe and, in the judgment of the Army, speeded the surrender of enemy soldiers. But the propaganda of words said less than did the impact of deeds, and no democrat, at home or abroad, could find in a leaflet much solace for the American decision, however exigent for military purposes, to embrace that moronic little king. *Fortune* was right when it complained that overseas American propaganda lacked guts.

So, too, within the United States, especially after the apotheosis of Coca-Cola, the OWI contributed little to a civilian people who in most cases and at most times were fighting the war on imagination alone. Yet that imagination did not lack fuel. It was provided partly by copy writers, to be sure, but also by authors and playwrights who drew pictures of the enemy that inspired terror and hatred of increasing fervor as the war went on.

6. Pictures of the Enemy

During World War II Americans at home had little trouble hating their enemies. Their abiding prejudices toward Asians, their irrational dread of a "yellow peril," focused after Pearl Harbor on the Japanese. The Nazis, for their part, had been consistently loathsome for many years before December 1941. To be sure, Americans were reluctant to enter the war in Europe or to provide asylum for Jewish victims of Nazi terror, but from 1939 forward their sympathies lay preponderantly and increasingly with Germany's foes. The private propaganda of the war years derived from those existing attitudes, as well as from perceptions of actual circumstances, while it simultaneously strengthened the resulting biases. In the climate that developed, loftier purposes of the war ordinarily got lost. In 1942 nine of ten Americans could name no provisions of the much publicized Atlantic Charter. A majority admitted they had no "clear idea of what the war is all about," a confession that still affected two Americans in five as late as 1944. But few questioned the need to fight. The enemy in Europe and Asia had embarked upon a "rampage of destruction"; Nazism was "an aggressive force that had to be stopped."

Hostility toward the Japanese received continual reinforcement from the reports of Japanese cruelty during the exhausting battles for Bataan and Corregidor in 1942 and later from other journalistic accounts about island warfare and the suicide bombings by Kamikaze pilots, who seemed to Americans to express an insane martial spirit. Between 1937, when the war in China began, and 1942, the conviction grew "that Japan was a dangerous foe, not only of China, but of the United States itself." The "ominous shadow" of the Oriental schemer, devious and slant-eyed, now fell only across the Japanese, who assumed in the public mind the characteristics once attributed to that dreaded fictional Chinese villain, Fu Manchu. A sampling of public opinion in 1942 disclosed that Americans considered the Chinese diligent, honest, brave, and religious, in contrast to the Japanese, who were deemed treacherous, sly, cruel, and warlike.

The national media confirmed that stereotype. *Time*, for one example, instructed its readers about "How to Tell Your Friends from the Japs." The Japanese were hairier than the Chinese; "the Chinese expression is likely to be more placid, kindly, open; the Japanese more positive, dogmatic, arrogant. . . . The Japanese are hesitant, nervous in conversation, laugh loudly at the wrong time. Japanese walk stiffly erect . . . Chinese more relaxed . . . sometimes shuffle." Even quasi-scholarly treatments of Japanese culture stressed the warlike ethic inherent in it. Officially Washington did not consider the United States a party to a racial war, but the War Production Board approved an advertisement published in 1943 that called for the extermination of the Japanese rats. The comic strips depicted the Japanese as teeth and spectacles, a subhuman species, and yet (except in "Terry and the Pirates," where they won half the time) no match for the gallant Americans. Similarly, in Hollywood's battles the Japanese, if they prevailed, heavily outnumbered the

Americans and usually tortured and mutilated their captives. Consequently any schoolboy in "our town" could dream of enlisting someday to fight alongside John Wayne for the annihilation of "the murderous little ape-men."

The image of the Germans in wartime America had more complicated contours. Though Americans almost universally despised Hitler and his associates, the Germans as a people at first evoked little of the animosity so commonly expressed against the Japanese. Hostility even toward the Nazis had developed too slowly to satisfy the best-informed American observers of events in Europe during the late 1930's. Vincent Sheean in his best seller of 1939, *Not Peace but a Sword,* strained to awaken a "sickly conscience" in the United States about Nazi pogroms. The book reminded its readers of the looting, violence, murder, and wholesale arrests in Germany. It described the concentration camps and their horrors, a long step, as Sheean noted, toward Hitler's announced goal of exterminating the Jews. The Nazi treatment of the Jews, he warned, had a "counterpart which is bad for the balance of mankind. . . . For . . . in . . . a mass crime with a mass criminal and a mass victim, all must suffer: the victim immediately in blood and spirit, the criminal by his . . . irretrievable degradation to savagery, and the remainder of the species by the necessity of either subduing the savage or submitting to him." Appeasement, Sheean argued, while it seemed in 1938 to protect the personal comfort of the English and French, "served as the wing for the ostrich." The time might come "when even the New World cannot redress the balance of the Old."

The playwright Lillian Hellman rang a similar alarm in her popular *Watch on the Rhine,* first produced in April 1941. Her hero, a German fighter in the anti-Nazi underground, soliloquized against the material selfishness that held men back alike from the inconvenience and from the risk of resisting the Nazis. "Does it make us all uncomfortable," he asked, "to remember that they came in on the shoulders of the most powerful men in the world? Of course. And so we would prefer to believe they are men from the planets. They are not. . . . They are smart, and they are cruel. But given men who know what they fight for . . . ," given courageous democrats, the world could be saved.

Though Sheean believed that message reached most of his contemporaries, Hellman felt the need to make her point again in 1944. European and American financiers and diplomats who had temporized with the Nazis provided her targets in *The Searching Wind.* As the play indicated, she was equally disgusted with those who had sought the co-operation of fascists during the campaigns in Sicily and Italy. She contrasted Sam, a wounded hero of the Italian front, with his decadent father, a professional American diplomat. The older man, like his fashionable friends, had failed even rhetorically to resist the fascist take-over of Rome and the Nazi annexation of the Sudetenland. While he reflected about his ambiguous past, Sam, his son, visibly uncomfortable physically and emotionally, cut off his ties to his family just as surgery was about to amputate his festering leg. His first home, Sam said, was not in any embassy or fancy school, but in the Army; his first friends were his Army "buddies." In the mud in Italy they had revealed to him the shallowness of his parents and their companions—"all old tripe who just . . . pretend they are on the right side. . . . They sold out their people and beat it. . . . My God, Sam . . . if you come from that you better get away from it fast, because they made the shit we're sitting in." "I am ashamed of both of you . . . ," Sam told his parents as he left them. "I don't want to be ashamed that way again."

The courage to resist evil against long odds characterized the Norwegians in John Steinbeck's novel *The Moon Is Down*, a best seller in 1942 and 1943. The men and women of Steinbeck's Norwegian town, bitter about the local Quisling who had helped the Germans take over the harbor and the coal mine, "moved sullenly through the streets." The rage they subdued broke out now and then in small acts of sabotage, in escapes by sea to England, and, as the novel began, in the murder of a German captain by a miner. During his trial, a mockery of justice, the miner was asked whether he was sorry: "Sorry? . . . I'm not sorry. He told me to go to work—me, a free man! I used to be an alderman. He said I had to work." As the mayor said, the murder "was the first clear act." The miner's private anger provoked a public anger. The "free man's" widow permitted a German officer to court her until she had a chance to kill him. "This is no honorable war," one of her friends told her. "This is a war of treachery and murder. Let us use the methods that they have used on us." Armed with small sticks of dynamite dropped by parachute from British airplanes, the free men and women and children of the little town blew up the coal mine. "The flies," the local doctor said, "have conquered the flypaper."

Steinbeck wrote in praise of courage but also, as Hellman did, about the evil of the Nazis. The Nazis in his Norwegian town were capable of love, fright, and nostalgia; some of them were wise, though invariably also loyal, Germans; most of them were arrogant and brutal. They also deserved the treachery and murder visited upon them. Over time that mandate for revenge prevailed in American culture. In 1941 American emotions were running counter to the President's decision to concentrate on the war in Europe first. Germany was then the more dangerous but Japan the more despised antagonist. Gradually, with the experience of the actual fighting and the impact of a supporting imagery, the Nazis and ultimately all Germans acquired in the public mind the attributes of unequivocal evil.

The Moon Is Down picked up the theme of the evil of Nazism that the movies and other novels had already struck more starkly. "The lines between right and wrong, between good and evil," as one film critic noted, "were clearly drawn—and we were on the right, the angelic side . . . [in] a fight between a free world and a slave world." In Above Suspicion (1941) the novelist Helen MacInnes introduced a sophisticated Nazi. "I haven't changed so very much," he told two English acquaintances. "I'm still interested in literature and music. . . . I haven't become a barbarian. . . . Politically, well, I've progressed. . . . I've seen the stupidities committed in the name of idealism. . . . People are made to be led. . . . With strong leadership, they can achieve anything. At first they must take the bad with the good; in the end they will forget the bad, because the ultimate good will be so great for them." As the novel progressed, those British acquaintances, traveling in Germany, observed less cultivated Nazis. "Did it really prove greater efficiency," the Englishman asked himself, "to walk with a resounding tread, to open doors by practically throwing them off their hinges . . . ? Here was the value of it—it made you look and therefore feel efficient. The appearance of efficiency could terrify others."

MacInnes, depicting the Nazis, leaders and followers, intellectuals and policemen, only as cruel bullies, also suggested that those were essentially German traits. In Nürnberg her heroine reflected about the "cold-blooded beasts" of German history. Later she connected past and present: "If only the methods of hate and force had been resisted at the very beginning: not by other countries . . . but by the people of Germany, themselves. But, of course, it had been more comfortable to concentrate on their private lives. . . . It was easier to turn a deaf ear to the cries from concentration camps, to harden their hearts to the despair of the exiles. . . . And now it had come to the point where other peoples would have to do the dying." There were few good Germans in the novel. "You and I," the heroine's husband said, "don't hate the Nazis because they are German. We hate the Germans because they are Nazis." In 1942, in Assignment in

Brittany, a memorable adventure story, MacInnes depicted only bad Germans, now in the stereotype that obtained in other light fiction and in Hollywood films throughout the rest of the war. The Gestapo agent who interrogated her hero was "shark-faced," a "razor puss." The Bretons distrusted foreigners but they despised not just the Nazis but all the "filthy Boches."

So it was also in Glenway Wescott's *Apartment in Athens,* a popular novel of 1945. The Prussian captain, "merciless . . . efficient," who moved into the flat of a Greek family, gorged himself at dinner, fed crumbs to the birds, and sent his scraps to a fellow officer's dog rather than leaving them for his hungry, reluctant hosts. "The unnerving thing," in the view of the Greek father of the harassed family, "was the great Prussian manner, serene and abstract . . . with insincerity in it somehow, but combined with absolute conviction." The Greek spoke not of Nazis but of all Germans: "I have come to the conclusion that the Germans are cruel." They were also violent; the captain struck and whipped the Greek's "high-strung little boy." And they saw themselves as supermen. "We grow more and more capable," the captain told the Greek. ". . . We will have the future!" At the end of the novel, the Greek, in prison awaiting execution, wrote a last letter to his wife. She should beware of "the good moods of Germans," their occasional efforts to please: "The likeable and virtuous ones are far worse than the others . . . because they mislead us. They bait the trap. . . . The Germans . . . will let us come up again for a season; then . . . mow our lives down again in a disgusting . . . harvest." She was to instruct the Americans in that lesson, to stop "foolish political men" from "babbling about permanent peace." There would be no peace as long as there were Germans, "hard-working, self-denying Germans, possessed of the devil as they are."

American readers found the same teachings in the novels of the English author Nevil Shute. In *Pied Piper* (1942), a book with a large American circulation, Shute described the perils of an aging Englishman attempting to help a group of children escape from France during the German blitzkrieg there. The conquerors, as in so many novels, were deadly serious, efficient, arrogant, and "one thing was very noticeable; they never seemed to laugh." Except when they killed. As dive bombers attacked a bus filled with French civilians, "a few bullets flickered straight over Howard and his children on the grass. . . . For a moment Howard saw the gunner . . . as he fired at them. He was a young man. . . . He wore a yellow students' corp cap, and he was laughing as he fired." Later a Gestapo officer threatened the old man: "Before evening you will be talking freely, Mister Englishman, but by then you will be blind, and in horrible pain. It will be quite amusing for my men." The Germans in Shute's account were sadists, as Steinbeck and MacInnes also implied. Their depravity, those authors also insinuated, included homosexuality, too. In *Most Secret* (1945), Shute expanded the indictment. The novel was one of hatred, vengeance, and horror. The Germans executed a shopkeeper in a small French town: "But why did they do it? . . . It was, of course, because they were Germans." They also executed all those in the surrounding area who were too old or feeble to work. The Germans were "murderous . . . uncivilized . . . without decent codes of conduct." A French Jesuit explained that the Germans did not worship Jesus. Instead "they make sacrifices of living goats to Satan . . . they bow to false gods of war, to Mithras and to Moloch. . . . These are the works of Satan." In France, the priest went on, the Germans had sown "lies and deceit in every form . . . sexual immorality . . . bribery, false witness . . . corruption. First they destroy the souls of men and then they occupy the country." There was but one solution—fire: "Before that power of fire all powers of heresy, idolatry, and witchcraft must recoil. . . . No other weapon purges evil from the earth. . . . In past centuries the Church wielded one great cleansing weapon against heresy and infamy and all idolatry . . . the weapon of temporal fire." That was the "most secret" weapon that the English prepared in special chemicals placed upon war craft disguised as fishing boats and sprayed mercilessly over terrified Germans, who were deservedly cremated in the flames.

American authors were also warning their audience against the diabolical German enemy. So it was with that perennial best-selling series, Upton Sinclair's saga of Lanny Budd, a superhero who served Franklin Roosevelt in *Presidential Agent*, a novel of 1944 set in the international politics of 1936–37. Lanny's incredible adventures took him into the highest councils of the German state. He met Hitler, Göring, Goebbels, Hess. Theirs was a scheming, dangerous, evil, and, not the least, prurient existence. "My husband," Frau Goebbels told Lanny, "cheats in love as he does in everything else in the world. He is one of those lewd men who want every young woman they see; he wants a new one every night." Another high-ranking Nazi, so Lanny learned, was "one of those whose *Liebenleben* is somewhat different . . . unusual." As for Hitler, according to Frau Goebbels: "The great man is impotent; he, the most powerful in the world, cannot accomplish what his commonest soldier . . . can do. He is frightfully humiliated . . . he becomes . . . hysterical . . . he foams at the mouth; he blames the woman. He makes her do unspeakable things. . . . Such is our *neue Ordung.*" Such it was and more, so Lanny concluded. Hitler had converted Germany into the instrument of "ancient forces of tyranny and oppression which had ruled the world before the concept of freedom had been born." Hitler, Lanny warned FDR, "had managed to infect with his mental sickness a whole generation of German youth. . . . I made a remark in a woman friend's hearing: 'There will be nothing to do but kill them.' . . . It is literally true; they are a set of blind fanatics, marching, singing, screaming about their desire to conquer other peoples." "Monsters and madmen," the Germans had devised the means for the mass production of torture, "a wholesale procedure . . . intended to terrorize . . . the whole continent of Europe. *Heute gehört uns Deutschland, morgen die ganze Welt!"*

Roosevelt stood, Lanny believed, in contrast, "the champion of democracy. . . . Roosevelt versus Hitler! These two had not created the forces, but led them and embodied them." Americans versus Germans, and as for the Germans, there was nothing to do but kill them.

Hellman, Steinbeck, MacInnes, Shute, and Sinclair doubtless reached a far larger audience than did the Office of War Information. They purveyed a different propaganda. They were not entirely wrong; Nazism was evil. But Sinclair's point was simplistic. His war was waged between Americans and Germans, Roosevelt and Hitler, absolute good and absolute bad. That was the kind of war in which many Americans came to believe. Missing from the picture of the enemy that the novelists painted was the gentle conviction of Archibald MacLeish, his reminder to his countrymen that Christian doctrine called upon man to hate sin but to forgive the sinner.

The picture of the enemy that Americans were shown left Japanese and Germans alike stripped of their humanity. It invited support for unconditional surrender. It recommended recourse to expedients that would hasten total victory. *Fortune* could not have complained that the hard sell of wartime literature lacked guts. There was in it, to be sure, little of the inspirational, but it fed the imagination of the American people in their haven far from the front. Roosevelt did not plan it that way. But the comics, the films, the adventure stories provided a powerful stimulant for the attitudes the President forbade federal agencies to cultivate, attitudes that nevertheless made his constituents receptive to other policies he decided he had to pursue.

2 / Homely Heroes

1. The GI's: American Boys

On September 21, 1943, War Bond Day for the Columbia Broadcasting System, Miss Kate Smith spoke over the radio at repeated intervals, in all, sixty-five times, from eight o'clock in the morning until two in the morning the next day. Her pleas to her listeners, some 20 million Americans, resulted in the sale of about $39 million worth of bonds. The content of her messages, according to a convincing analysis of her marathon, was less important than her person. Her listeners responded as they did in large part because for them she symbolized, in heroic proportions, values they honored: patriotism, sincerity, generosity. In that, of course, she was not alone. Edward L. Bernays, the premier public-relations counselor in the United States, accepted a commission during the war from the Franklin Institute "to give Benjamin Franklin greater fame and prestige in the hierarchy of American godhead symbols." As Bernays went about his business of persuading local communities to name streets, buildings, even firehouses after his subject, he found his task easy, for, as he put it, "our society craves heroes."

War accentuated that craving, especially for those at home who sought symbols on which to focus the sentiments they felt or were, they knew, supposed to feel—symbols that would assist the imagination in converting daily drabness into a sense of vicarious participation in danger. The battlefield provided a plenitude of such symbols, of genuine heroes who were then ordinarily clothed, whether justly or not, with characteristics long identified with national virtue. The profiles of the heroes of the war followed reassuring lines, some of them perhaps more precious than ever before because they had become less relevant, less attainable than they had been in a simpler, more bucolic past. Some others, less sentimental, were no less reassuring, for they displayed the hero as a man like other men, not least the man who wanted to admire someone whose place and ways might have been his own, had chance so ruled.

No leap of a reader's imagination, however, could easily find believable heroes in the Army's official communiqués. Though they sometimes mentioned names, those accounts supplied only summaries of action that generally obliterated both the brutality and the agony of warfare. Robert Sherrod, who landed with the marines at Tarawa and wrote a piercing description of that ghastly operation, deplored the inadequacy of American information services. "Early in the war," he commented, "one communiqué gave the impression that we were bowling over the enemy every time our handful of bombers dropped a few pitiful tons from 3,000 feet. The stories . . . gave the impression that any American could lick any twenty Japs. . . . The communiqués . . . were rewritten by press association reporters who waited for them back at rear headquarters. The stories almost invariably came out liberally sprinkled with 'mash' and 'pound' and other 'vivid' verbs. . . . It was not the correspondents' fault. . . . The stories which . . . deceived . . . people back home were . . . rewritten . . . by reporters who were nowhere near the battle." Bill Mauldin, the incomparable biographer of the GI, made a similar complaint about reporting from Italy. Newspapers, he recommended, should "clamp down . . . on their rewrite men who love to describe 'smashing armored columns,' the 'ground forces sweeping ahead,' 'victorious cheering armies,' and 'sullen supermen.' " W. L. White, who interviewed the five survivors of Motor Torpedo Boat Squadron 3, the group that evacuated MacArthur from Corregidor, quoted Lieutenant Robert B. Kelly to the same point: "The news commentators . . . had us all winning the war. . . . It made me very sore. We were out here where we could see these victories. There were plenty of them. They were all Japanese. . . . Yet if even at one point we are able to check . . . an attack, the silly headlines chatter of a victory."

The resulting deception was not inadvertent. While the Japanese early in 1942 were overpowering the small, ill-equipped American garrisons on Pacific islands, the armed services invented heroic situations, presumably to encourage the American people, who might better have been allowed to face depressing facts. So it was with the mythic request of the embattled survivors on Wake Island: "Send us more Japs." That phrase, which the motion pictures tried later to immortalize, had originated merely as padding to protect the cryptographic integrity of a message from Wake to Pearl Harbor describing the severity of the American plight. So, too, in the case of Colin Kelly, a brave pilot stationed in the Philippines, who died in action when the Japanese attacked. The Army exploited his valor by exaggerating his exploits, a ruse soon exposed to the desecration of Kelly's memory. His heroism, like that of the marines on Wake Island, deserved better treatment than it received. It deserved the truth.

The truth about American soldiers, heroic or not, centered in their experience in the Army, in training, in the field, under fire. In contrast to the official communiqués, the best independent reporting revealed that truth, which was often comic or poignant when it was not triumphant or glorious. It was harder to find out much about the men themselves, their lives before they had become soldiers, their homes and parents, rearing and calling, character and hopes. About those matters even the best reporters had ordinarily to work from partial evidence and had to write, given the wartime limits of time and space, selectively.

In the first instance, from among all the men in arms, the heroes selected themselves. Their bravery, self-sacrifice, and sheer physical endurance earned them a martial apotheosis. Usually that was the end of the story, except for a parenthesis identifying the hero's home town. But on occasion, moving to a next stage, correspondents at the front used what data they had to endow the soldiers they knew with recognizable qualities of person and purpose. In the process, truth became selective. Whether consciously or inadvertently, the reporters tended to find in the young men they described the traits that Americans generally esteemed. Those in uniform shared with their countrymen a common exposure to values dominant in the United States and to the special circumstances of the Great Depression, just ended. They had a sameness that in some degree set them apart from servicemen of other countries. But the necessarily selective reporting about them, governed as it was by the comfortable conventions of American culture, made the GI's and their officers more than merely representative Americans. It freed them from the sterile anonymity of official communiqués, but it also made them exemplars of national life, heroic symbols that satisfied the normal social preferences and the wartime psychological needs of American civilians.

Early in the war John Hersey, a man of scrupulous professional integrity, undertook to write an account of General Douglas MacArthur and his beleaguered troops. Hersey wrote from the Time-Life building, a station he eagerly abandoned in his later reporting for positions at the front. Even from New York he did his best to be accurate, to go beyond the wire-service reports and learn about his subjects from their parents and their friends. He did not invent material, but he did arrange it in conventional modes typical of much war reporting, modes he discarded as soon as he had seen the war at first hand. *Men on Bataan,* a widely read book that Hersey quickly found embarrassing, frankly encouraged American hero worship. The American people, Hersey began, "adored their MacArthur as if he were a young genius who had just flown across the dim Atlantic . . . or as if he were a big and perfect slugger . . . or . . . some new sheik of the silver screen. . . . In wonderful healthy American ways they worshipped MacArthur." Hersey wanted to keep the MacArthur legend from distortion, to prevent the symbol from becoming what the man was not. Yet the data he selected supported American heroic stereotypes. So it was that MacArthur and his mother had been very close; so it was that MacArthur had played for West Point in the first Army-Navy baseball game; so it was that MacArthur's formula for offensive warfare derived from baseball: "Hit 'em where they ain't."

As for MacArthur's men, Hersey wrote: "They are average people; they are a cross-section; they are you. They are wonderfully brave and . . . they encompass the highest human values." They also had done homely American things before they were killed. One of them swam "strongly . . . and . . . would box with anyone just for fun." Another "clowned on the track at State College, and clowned in the jalopy of a plane that he . . . bought . . . and clowned about wanting to be a good doctor. . . . Mostly his life was making people laugh." Another, "a gentle boy . . . was a friend of beauty. He . . . read avidly . . . wrote in secret . . . loved natural things. . . . Music was his passion." One "played football with heathen ferocity" before he became a preacher. Still another was "an average American boy from an average modest American home where there was more than average love and peace. He liked to run and to swim and to hear the good clear crack of a golf ball untopped. He was a born collector. . . . He played the saxophone. He liked his United States . . ." and he died bravely while bombing and strafing a Japanese airstrip.

Again and again in American war reporting, as Hersey's vignettes suggested, soldiers entered combat prepared by the playing fields of their high schools and colleges. That convention appeared alike in spontaneous dispatches from the beachheads and in the contrivances of public relations. The National Physical Fitness Committee naturally adopted the theme. The committee's plan for a series of radio broadcasts of football and basketball games "called for between-halves interviews . . . with a local hero returned from the fighting front who would tell how his playing for Biffle Manor had helped him fight the Germans and how the memory of what his old coach had told him helped him in combat." *Time* ignored the coaches but during 1942 alone observed that the victor in the first American dogfight over Europe "had been a high-school athlete"; one of the pilots in the first major American air raid in France "had played tackle on Alabama's Rose Bowl team"; a marine private whose tank the Japanese overran in New Guinea was a "one time guard in the Huron (South Dakota) high school football team, who liked to bake chocolate cakes when he was at home."

"These Are the Generals," a series of articles published during 1943 in *The Saturday Evening Post,* linked athletic prowess, academic mediocrity, and success in command. There were exceptions. For one, Lieutenant General Joseph T. McNarney, Deputy Chief of Staff, "the shrewdest poker player and the most ruthless executive in the United States Army," according to the recollection of a childhood friend "was reading Shakespeare when the rest of us were still struggling with the McGuffey's Third Reader." But McNarney was atypical even for generals not at the front. Major General Walter R. Weaver, chief of Army Air Forces Technical Training, made "physical education . . . almost a fetish," had under his command "perhaps more All-American football players" than did any other branch, had "flunked French at West Point . . . and just skimmed through at the bottom of the class." Lieutenant General Mark W. Clark, Eisenhower's chief deputy, negotiator of the controversial American deal with French Admiral Darlan, and later commander of the Fifth Army, had not, as the *Post* described him, excelled in athletics at West Point, or in academic subjects either—he graduated 110 out of 139—but he was "solidly built, with strong shoulders," he avoided cigarettes and coffee, and fishing was his "favorite recreation." As for Eisenhower himself, "at West Point he played halfback . . . until he broke a leg in a game with the Carlisle Indians and Jim Thorpe. His scholarship . . . was not conspicuous." Better still for the folklore, "Old Blood and Guts," Major General George S. Patton, by reputation the toughest and most dashing of them all, "did not exactly set V.M.I. intellectual standards. He majored in polo, football and horsemanship." Later, at West Point, Patton "took five years to cover four" and "continued to specialize in football and horsemanship, dismounting long enough to set an intercollegiate 220-yard-dash record."

The reportorial emphasis on sports as a significant part of the heroic image contributed to a process of mythologizing that had begun long before the war. A culture that had made heroes of its athletes could hardly avoid making athletes of its heroes. Men who had found their boyhood models in Babe Ruth or Red Grange or Bobby Jones now served under Weaver or Clark or Patton, who did play important if not decisive roles in early American victories in the war. Journalism did not create them, but it helped to make them appealing symbols for Americans in the service or at home.

No less appealing and far more numerous were the GI Joes who slogged through the mud and filth of infantry warfare in the tropical Pacific, in Africa, in Italy, and in France. They were, as a group, unwilling, nonmilitaristic heroes, harassed by bureaucracy and red tape as well as by the enemy. Rawness, brutality, and fatigue characterized their lives. "Forget about the glorified picture of fighting you have seen in the movies," the New York *Times* observed. ". . . The picture you want to get into your mind is that of plugging, filthy, hungry, utterly weary young men straggling half dazed and punch-drunk, and still somehow getting up and beating the Germans." The battles themselves, said *The Saturday Evening Post*, were "mixed-up, confused . . . bewildering." But the foot soldier, however much an anonymous number on a dog tag, was indispensable for victory in every military theater: "Only Bill and his buddies could get in there, on foot, the way God made them."

God had made them Americans, "just plain Americans . . . from plain American families," and their Americanism moved with them across both oceans. Ernie Pyle, the GI's most beloved reporter, interpreted even the defeats that the troops suffered as natural functions of American virtue. Military discipline alienated the creativity and individualism of American youth. "We . . . aren't a rigid minded people," Pyle wrote from North Africa. "Our army didn't have the strict and snappy discipline of the Germans. . . . It takes a country like America about two years to become wholly at war." Even war did not dampen the American soldier's impulse for kindness, generosity, and affection. "Our boys," Pyle wrote, "couldn't resist the sad and emaciated little faces of the children, and that was when they started giving away their rations"—and going hungry themselves.

The GI hero, a reluctant hero, a folk hero, was typically "a coffee merchant by profession, a radio actor by avocation, and a soldier by the trend of events." Even wounded veterans remained "just ordinary American boys . . . friendly and enthusiastic and sensible . . . as normal as if nothing had happened." So also with the marines with whom John Hersey moved into a hostile valley on Guadalcanal: "When you looked into the eyes of those boys, you did not feel sorry for the Japs: you felt sorry for the boys. The uniforms, the bravado . . . were just camouflage. . . . They were just American boys. They did not want that valley or any part of its jungle. They were ex-grocery boys, ex-highway laborers, ex-bank clerks, ex-schoolboys, boys with a clean record . . . not killers."

Nevertheless, as every observer reported in one phrase or another, "no group of men . . . [could] ever top in spirit and courage the kids from the streets of American cities, from American farms . . . Sunday-school rooms, pool-rooms, and business offices." American valor, as it was supposed to, responded to long odds. The GI "fought well when the going was good," the New York *Times* wrote. "When the going was bad he fought savagely." Victory in the Solomons, *Life* reported, "was heroic. . . . In the best historical tradition, a small American force pitted itself against a bigger enemy and came off the winner."

While courageous, the GI's were humble and modest, effective fighting men not out of false pride but out of duty accepted in the face of danger. In Italy Bill Mauldin's Willie moved forward. "Do retreatin' blisters," he asked Joe, "hurt as much as advancin' blisters?" So it was also with another Bill at the Battle of Buna: "Bill was a Wisconsin boy. He loved the blue lakes of Wisconsin, the cool winds, the bright farmlands. He never considered himself a warrior." Still, he "charged through . . . chasing a bunch of Japs, hot and tired and blazing mad," and he was killed: "It was the first rest he had had in a long time. But he would never go back to his girl in the farmlands of Wisconsin."

Perhaps because they were by the 1940's so dominantly an industrial and urban people, Americans selected their heroes disproportionately from the ranks of country boys. MacArthur's men on Bataan, according to Hersey, were "average people . . . a cross-section," but his descriptions referred continually to their rural roots. Hersey did not, he later observed, choose those heroes or invent their backgrounds: "They chose themselves. . . . Writing in a time of defeat and jitters, I was trying . . . to fill the need we had for heroes to bolster our morale in a war against palpable evil." As it worked out, they were rural heroes. One captain "was the product of the land . . . born of a long line of pioneers"; a chaplain was also of pioneer stock; a lieutenant "never could get away for long from the farm on which he was born"; a soldier "was reared on the old Slagle farm, where generations of Slagles, country boys and proud of it, had been reared"; other soldiers "grew up muscular" in New Mexico. "The range of their background was as broad as America," Robert Sherrod wrote of the marines at Tarawa, but his "hard-boiled colonel," he noted, was "born on a farm" and his bravest captain came from a small town. Ira Wolfert, in *Battle for the Solomons,* provided background information about only two of the dozens of men he mentioned. One was an accountant who loved the blues; the other "a farm boy out in Wisconsin." Of the relatively few heroes whom *Time* chose for special attention in 1944, one was a share-cropper, another "a big, silent farm hand."

The strains in American culture that related the virtuous to the rural or the outdoors or the gridiron recalled the images of the early twentieth century, of the Rough Riders and Theodore Roosevelt's "strenuous life." Similarly, *Life* and the New York *Times,* comment-ing upon the long odds against victorious GI's, evoked the cult of the underdog, the sentiments that in times of peace had often given an allure to the antitrust laws or, for the apolitical, to the Brooklyn Dodgers. The victory of character over hard work, over the long odds of the society or the economy, had provided, too, the stuff of the folklore of success, the scenario for the poor boy whose struggle to overcome the handicaps of his background won him fortune and fame. That kind of struggle, though rarely successful, had particularly marked American experience and consciousness during the 1930's. It was a part of the civilian past of most soldiers, and, naturally enough, a part frequently remarked by war correspondents.

The habit of joyful hard work, one ingredient of the cult of success, had always beguiled *The Saturday Evening Post,* which built its circulation not the least upon continual publication of updated Alger stories. The *Post* found an illustrious example of its favorite theme in Dwight David Eisenhower. As a boy in a household of modest means, he had "always had plenty to do. They had an or-chard, a large garden, a cow, a horse, and always a dog. The boys did all the outdoor work, milked the cow and . . . helped with the housework. . . . They also all found additional jobs. . . ." Dwight pulled ice in the local ice plant, or helped near-by farmers. "It taught them a lot," their mother said. By implication, Sherrod and Hersey said as much about their young heroes on the Pacific islands who had faced the vicissitudes of the Depression as they faced the ordeals of the jungle. There was, for one, "Hawk," a marine captain, promoted from the ranks, killed at Tarawa. Before the war, "he . . . was

awarded a scholarship to the Texas College of Mines. . . . Like
most sons of the poor, he worked. . . . He sold magazines and
delivered newspapers. . . . He was a ranchhand, a railroadhand, and
a bellhop." A sergeant on Bataan had "had to drop out of school in
his first year at high because there were eleven mouths to feed at
the . . . house. He took a job delivering milk." His lieutenant had
"worked his way through . . . college," his chaplain had "worked
as a bricklayer . . . water boy on a road gang . . . farmhand
. . . newsboy . . . longshoreman . . . meatcutter." Another officer
"had a paper route . . . would stop and chat with old ladies and
run down to the store for them. . . . He was stroke . . . of the
crew at U.C.L.A. . . . He drank only a little beer and smoked not at
all." Still another was "a perfect example of a man who always knew
what he wanted to do and finally did it. What he wanted to do was fly
superbly. His desire started when he was a cub scout. . . . He began
to make crude airplane models and gradually the models got better.
. . . He won prizes. . . . He had some fun on the side . . . swimming, hunting, fishing, tennis and golf. . . . He used to work nights,
as a store clerk or theater usher or night clerk. . . . By day he
studied aeronautical engineering."

Aviators, when they won attention as heroes, shared many attributes of the foot soldiers but also represented uncommon qualities,
those of a glamorous elite. The pilots and navigators, bombardiers
and gunners were special men. They had to pass rigorous physical
and mental tests. They received rapid promotion and high hazardous-
duty pay. Instead of mud or jungle heat or desert cold, they enjoyed,
at least part of the time, the amenities of an air base and always the
romantic environment of the sky. There, exploring a vertical frontier,
operating complex, powerful machinery, they flew into sudden and
explosive danger. As Ernie Pyle observed: "A man approached death
rather decently in the Air Force. He died well-fed and clean-shaven."

The Saturday Review of Literature, in an editorial entitled "A
Hero for America," praised the elite qualities of its subject: "We are
sending most of the best of our young brains into the air, into a
combat where they retain to an unusual degree their individual responsibility, into intense danger which they face with the developed
imagination of the highly educated man." Nevertheless, "under the
heroes' uniforms," *The Saturday Evening Post* believed, were "the
lineaments of kids who were in filling stations and stores, at soda
fountains and on farms, kids from high school and maybe college,
who went off to war in the flush of youth." Don Gentile, whom
Eisenhower called a "one man air force," was "soft-spoken and self-

effacing, a rather naive and quite unworldly boy . . . a new type of American . . . who comes into his magnificent own when he becomes air-borne." Withal, "they have the look of veterans," *Life* wrote, "these youngsters in their late teens and early twenties. . . . They have looked death in the face repeatedly. . . . They . . . well know the odds for or against their survival." But even the heroes of the air were "like people," the *Post* decided. ". . . They get scared and excited and desperate, like other people. That's what makes their achievements remarkable."

They fit the pattern of American values remarkably well. To a flier of any nationality, John Hersey knew, his machine was his life: "It falls down, he falls down." "The first side of any airman," he wrote in *Life,* referring only to Americans, "is his love of his machine. He gets to know his plane. He learns exactly what it can do, where it will take him, how fast, how sweetly in the sky." The real Jimmy Doolittle, Hersey went on, was not merely the leader of the first, daring raid over Tokyo; he was an aeronautical engineer with a doctorate from MIT, he was "the scientist of flying, the man who . . . dedicated his life to his machine." He was also "a regular guy" with a great "capacity for having fun when it's time to play." He had a "great heart. His smile . . . goes all the way from his forehead to his cleft chin." Above all, he knew his plane. More than the cleft chin or big heart, the aviator's facility with and affection for machinery stamped him for Americans as a hero, an amalgam of Dan'l Boone and Henry Ford.

Of all soldiers, the aviators especially functioned as part of a group of equals co-operating in a perilous mission. Joined in the social compact of combat, they were at ease in that union because they had learned as Americans—on playing fields, in sales organizations, along assembly lines—what war also taught. Major Jim Howard, "a maverick one-man air force" who had knocked down six Nazi interceptors in one extraordinary day, was "an apostle of teamwork. It sounds bromidic to say it's teamwork, not individualism, that counts with us. I guess only a pilot can understand." So could the crew of a bomber. "This," John Steinbeck wrote during a venture in reporting the war, "is the kind of organization that Americans above all others are best capable of maintaining. The bomber team is a truly democratic organization. . . . The best . . . team is one where each individual plays for the success of the mission." Ernie Pyle reached the same conclusion about bombers and fighters alike: "Basically it can be said that everything depended on teamwork . . . Sticking with the team and playing it all together was the only

guarantee of safety for everybody." Yet the aviators were not different from the doggies. Everywhere the GI fought for the buddies he
served with. Sherrod put it very well: "The Marines . . . didn't
know what to believe in . . . except the Marine Corps. The Marines
fought solely on *esprit de corps.*"

The make-up of both ground and air units, as they were ordinarily described, preserved the American convention of the melting
pot. In war films, as one critic pointed out, "the composition of these
teams . . . was quite consciously . . . American, with each group
invariably composed of one Negro, one Jew, a Southern boy, and a
sprinkling of second-generation Italians, Irish, Scandinavians and
Poles." *Time* reported that the list of those who raided Rouen
"sounded like the roster of an All-American eleven. . . . There were
Edward Czeklauski of Brooklyn, George Pucilowski of Detroit,
Theodore Hakenstod of Providence, Zane Gemmill of St. Clair, Pa.,
Frank Christensen of Racine, Wisconsin, Abraham Dreiscus of Kansas City. There were the older, but not better, American names like
Ray and Thacker, Walsh and Eaton and Tyler. The war . . . was
getting Americanized."

Time liked to think so, as did most Americans, but the selectivity of war reporting continually discriminated against those Americans who had grown up outside of the dominant culture. That
discrimination hurt those whom it slighted and spurred their protests
to the federal government, often to the White House. Under pressure
from Negro leadership and at the direction of President Roosevelt,
the Office of War Information tried to compensate. It prompted the
Army and Navy public-relations desks to issue "feature material
about Negro combat troops." Little came of that suggestion, little
even from independent reporters. The brave blacks who fought for
the United States on every front received recognition almost exclusively from Negro correspondents writing for Negro audiences. In a
parallel attempt to compensate, one OWI officer asked Henry Luce to
publish stories about Jewish heroes with surnames "which would
. . . identify them as Jews" and "a tactful reference to their faith."
The OWI also called for magazine articles about Japanese-American
casualties and asked "Wally's Wagon," a syndicated column, to
"carry a piece about that fellow with a Japanese face who . . . lost
a son fighting in Italy." But such stories remained exceptional. The
conventions that guided the ordinary descriptions of American military heroes reflected cultural preferences that overlooked those ordinarily proscribed at home.

In that as in many other ways, those conventions provided reassurance. The GI as athlete, farm hand, mechanic, soda jerk, college student, and organization man—the GI as an ordinary American boy—gave the American people exactly the symbol with which they could easily identify, the brave soldier in whom they could see themselves and in whom they could take personal pride. The selectivity of war reporting sharpened lines already drawn in life. GI's were for the most part ordinary Americans, products of the culture they represented and seemed to reinforce.

Like their countrymen at home, who suffered much less, American soldiers overseas derived from the hard experience of war, of fear and deprivation and nostalgia, many of the hopes they harbored for peace. That, too, the correspondents revealed in their dispatches from the lines.

2. In Foreign Foxholes

Of all the war correspondents, Pyle, Hersey, and Mauldin wrote most intimately and extensively about the men they knew, about their hopes and dreams in the context of their fright and hardship. "In the magazines," Pyle wrote, "war seemed romantic and exciting, full of heroics and vitality . . . yet I didn't seem capable of feeling it. . . . Certainly there were great tragedies, unbelievable heroism, even a constant undertone of comedy. But when I sat down to write, I saw instead men . . . suffering and wishing they were somewhere else . . . all of them desperately hungry for somebody to talk to besides themselves, no women to be heroes in front of, damned little wine to drink, precious little song, cold and fairly dirty, just toiling from day to day in a world full of insecurity, discomfort, homesickness and a dulled sense of danger. The drama and romance were . . . like the famous falling tree in the forest—they were no good unless there was somebody around to hear. I knew of only twice that the war would be romantic to the men: once when they could see the Statue of Liberty and again on their first day back in the home town with the folks."

The GI's shared, in Pyle's words, "the one really profound goal that obsessed every . . . American." That goal was home. Before the landing in Sicily they talked to Pyle about their plans: "These gravely yearned-for futures of men going into battle include so many things—things such as seeing the 'old lady' again, of going to college . . . of holding on your knee just once your own kid . . . of again becoming champion salesman of your territory, of driving a coal truck around the streets of Kansas City once more and, yes, of just sitting in the sun once more on the south side of a house in New Mexico. . . . It was these little hopes . . . that made up the sum total of our worry . . . rather than any visualization of physical agony to come." After Sicily, after Anzio, in the mud of Italy, those hopes persisted, sometimes more as reveries about the past than ambitions for the future, sometimes hardened by a gnawing cynicism about American civilians. A black soldier had had a "dear John" letter from his wife, who had found another man at home. "Any guy overseas," that soldier told Alfred Duckett, a perceptive black reporter, "who says he's in love with his wife tells a damn lie. . . . He's in love with a memory—the memory of a moonlit night, a lovely gown, the scent of a perfume or the lilt of a song." Yet "even a memory," Duckett added, "is a straw to clutch—even a memory."

Faked memories did not serve. "We all used to get sore," Bill Mauldin wrote from Italy, "at some of the ads we saw in the magazines from America. . . . I remember one lulu of a refrigerator ad showing a lovely, dreamy-eyed wife gazing above the blue seas and reflecting on how much she misses Jack. . . . BUT she knows he'll never be content to come home to his cozy nest (equipped with a frosty refrigerator; sorry, we're engaged in vital war production now) until the . . . world is clear for Jack's little son to grow up in. . . . Like hell Jack doesn't want to come home now. And when he does come home you can bet he'll buy some other brand of refrigerator."

Insofar as possible, the doggies brought a semblance of home to their bivouacs. "The American soldier," Pyle wrote, "is a born housewife. . . . I'll bet there's not another army in the world that makes itself a 'home away from home' as quickly as ours does." Mauldin agreed: "It's strange how memories of a peacetime life influence these makeshift houses. If a soldier has fixed himself a dugout . . . and has cleaned it up and made it look presentable, his visitors instinctively feel that this is a man's house, and he is its head." "Now that ya mention it . . . ," Mauldin's Willie said in one cartoon depicting him and his buddy, Joe, in a foxhole with a storm battering their helmets, "it does sound like the patter of rain on a tin roof." Joe and Willie, dignified in Mauldin's ironic portrayals of their lot, spoke the language of persisting discomfort but the language also of men displaced from home. "Joe," Willie said, "yesterday ya saved my life an' I swore I'd pay ya back. Here's my last pair of clean socks."

Soldiers in the armies of all nations in all wars have yearned to go home, but the GI's sense of home was especially an American sense. "Our men," Pyle wrote, ". . . are impatient with the strange peoples and customs of the countries they now inhabit. They say that if they ever get home they never want to see another foreign country." Home for the soldier, according to the New York *Times,* was "where the thermometer goes below 110° at night . . . where there are chocolate milk shakes, cokes, iced beer, and girls." The GI had had enough of crumpets and croissants: "Tea from the British and vin rouge from the French . . . have only confirmed his original convictions: that America is home, that home is better than Europe." Even the sophisticated missed homely American fare. Richard L. Tobin, a correspondent for the New York *Herald Tribune,* had arrived in London only a few days before he complained, like the GI's, about English food: "What wouldn't I give right now for a piece of bread spread with soft butter, heaped with American peanut butter, and accompanied by a big glass of ice-cold milk!"

Food, of course, was metaphor. Its full meaning was best expressed when John Hersey went into that Guadalcanal valley with a company of marines. "Many of them," Hersey wrote, "probably had brief thoughts, as I did, of home. But what I really wondered was whether any of them gave a single thought to what the hell this was all about. Did these men, who might be about to die, have any war aims? What were they fighting for, anyway?" Far along the trail into the jungle, "these men . . . not especially malcontents" gave Hersey his answer. "What would you say you were fighting for?" he asked. "Today, here in this valley, what are you fighting for?"

. . . Their faces became pale. Their eyes wandered. They looked like men bothered by a memory. They did not answer for what seemed a very long time.

Then one of them spoke, but not to me. He spoke to the others, and for a second I thought he was changing the subject or making fun of me, but of course he was not. He was answering my question very specifically.

He whispered: "Jesus, what I'd give for a piece of blueberry pie."

. . . Fighting for pie. Of course that is not exactly what they meant . . . here pie was their symbol of home.

In other places there are other symbols. For some men, in places where there is plenty of good food but no liquor, it is a good bottle of Scotch whiskey. In other places, where there's drink but no dames, they say they'd give their left arm for a blonde. For certain men, books are the things; for others, music; for others, movies. But for all of them, these things are just badges of home. When they say they are fighting for these things, they mean they are fighting for home—"to get the goddam thing over and get home."

Perhaps this sounds selfish. . . . But home seems to most marines a pretty good thing to be fighting for. Home is where the good things are—the generosity, the good pay, the comforts, the democracy, the pie.

Hersey, a decent man, listed democracy, but soldiers usually talked about creature comforts, secure routines, even affluence. There were three sailors Ernie Pyle knew. One wanted to build a cabin on five acres of his own in Oregon. Another wanted to return to earning bonuses as a salesman for Pillsbury flour. As for a third, a photographer before the war: "His one great postwar ambition . . . was to buy a cabin cruiser big enough for four, get another couple, and cruise down the Chattahoochee River to the Gulf of Mexico, then up the Suwannee, making color photos of the whole trip." A marine lieutenant colonel in the South Pacific had simpler fancies: "I'm going to start wearing pajamas again. . . . I'm going to polish off a few eggs and several quarts of milk. . . . A few hot baths are also in order. . . . But I'm saving the best for last—I'm going to spend a whole day flushing a toilet, just to hear the water run."

Home spurred the troops to fight. Even the self-consciously reflective soldiers, who linked the real and the ideal as Hersey did, stressed the palpable. *The Saturday Evening Post* ran a series by GI's on "What I Am Fighting For." One characteristic article began: "I am fighting for that big house with the bright green roof and the big front lawn." The sergeant-author went on to include his "little sister," his gray-haired parents, his "big stone church" and "big brick schoolhouse," his "fine old college" and "nice little roadster," his piano, tennis court, black cocker spaniel, the two houses of Congress, the "magnificent Supreme Court," "that President who has led us," "everything America stands for." It was a jumble: he mentioned "freedom" one sentence after he wrote about "that girl with the large brown eyes and the reddish tinge in her hair, that girl who is away at college right now, preparing herself for her part in the future of America and Christianity." The jumble satisfied the *Post* and its readers, who would have liked less the findings of the Army Air Corps Redistribution Center at Atlantic City. Returnees there in 1944, a representative group of men, "surprisingly normal physically and psychologically," in the opinion of the physicians who examined them, felt contempt for civilians, distrusted "politicians," and resented labor unions. According to the Assistant Secretary of War for Air, "there is very little idealism. Most regard the war as a job to be done and there is not much willingness to discuss what we are fighting for."

The Assistant Secretary thought indoctrination lectures would help. On the basis of his own experience, Ernie Pyle would probably have disagreed:

Awhile back a friend of mine . . . wrote me an enthusiastic letter telling of the . . . Resolution in the Senate calling for the formation of a United Nations organization to coordinate the prosecution of the war, administer reoccupied countries, feed and economically re-establish liberated nations, and to assemble a . . . military force to suppress any future military aggression.

My friend . . . ordered me . . . to send back a report on what the men at the front thought of the bill.

I didn't send my report, because the men at the front thought very little about it one way or the other. . . . It sounded too much like another Atlantic Charter. . . .

The run-of-the-mass soldiers didn't think twice about this bill if they heard of it at all. . . .

We see from the worm's eye view, and our segment of the picture consists only of tired and dirty soldiers who are alive and don't want to die . . . of shocked men wandering back down the hill from battle . . . of . . . smelly bed rolls and C rations . . . and blown bridges and dead mules . . . and of graves and graves and graves. . . .

The mood of the soldiers conformed in large measure to the mood of Washington. There was, as Henry Morgenthau had said, "little inspirational" for young men and women. The President, deliberately avoiding talk about grand postwar plans, concentrated on victory first and almost exclusively. So did the GI, for he knew that he had to win the war before he could get home, his ultimate objective. He felt, the New York *Times* judged, "that the war must be finished quickly so that he can return to take up his life where he left it." There was not "any theoretical proclamation that the enemy must be destroyed in the name of freedom," Pyle wrote after the Tunisian campaign; "it's just a vague but growing individual acceptance of the bitter fact that we must win the war or else. . . . The immediate goal used to be the Statue of Liberty; more and more it is becoming Unter den Linden."

Winning the war, his intermediate goal, turned the soldier to his direct task, combat. There impulses for friendship and generosity had to surrender to instincts for killing and hate. "It would be nice . . . to get home," one pilot told Bob Hope, ". . . and stretch my legs

under a table full of Mother's cooking. . . . But all I want to do is
beat these Nazi sons-of-bitches so we can get at those little Jap bastards." The hardening process of training and danger, in Marion
Hargrove's experience, made "a civilian into a soldier, a boy into a
man." "Our men," Pyle concluded, "can't . . . change from normal
civilians into warriors and remain the same people. . . . If they
didn't toughen up inside, they simply wouldn't be able to take it."
The billboard overlooking Tulagi harbor carried the message: "Kill
Japs; kill more Japs; you will be doing your part if you help to kill
those yellow bastards."

Bill Mauldin was more reflective: "I read someplace that the
American boy is not capable of hate . . . but you can't have friends
killed without hating the men who did it. It makes the dogfaces sick
to read articles by people who say, 'It isn't the Germans, it's the
Nazis.' . . . When our guys cringe under an 88 barrage, you don't
hear them say 'Those dirty Nazis.' You hear them say, 'Those goddam Krauts." Mauldin understood hate and hated war:

> Some say the American soldier is the same clean-cut young man
> who left his home; others say morale is sky-high at the front because
> everybody's face is shining for the great Cause.
>
> They are wrong. The combat man isn't the same clean-cut lad
> because you don't fight a Kraut by Marquis of Queensberry rules.
> You shoot him in the back, you blow him apart with mines, you kill
> or maim him . . . with the least danger to yourself. He does the same
> to you . . . and if you don't beat him at his own game you don't
> live to appreciate your own nobleness.
>
> But you don't become a killer. No normal man who has smelled
> and associated with death ever wants to see any more of it. . . . The
> surest way to become a pacifist is to join the infantry.

War, Bob Hope thought, made "a lot of guys appreciate things
they used to take for granted," and Pyle believed that "when you've
lived with the unnatural mass cruelty that man is capable of . . .
you find yourself dispossessed of the faculty for blaming one poor
man for the triviality of his faults. I don't see how any survivor of
war can ever again be cruel." Mauldin put it more bluntly: "The vast
majority of combat men are going to be no problem at all. They are
so damned sick and tired of having their noses rubbed in a stinking
war that their only ambition will be to forget it." Consequently Mauldin was not much worried about the adaptability of the veteran:

> I've been asked if I have a postwar plan for Joe and Willie. I
> do. . . . Joe and Willie are very tired of war. . . . While their bud-

dies are . . . trying to learn to be civilians again, Joe and Willie are going to do the same. . . . If their buddies find their girls have married somebody else, and if they have a hard time getting jobs back, and if they run into difficulties in the new, strange life of a free citizen, then Joe and Willie are going to do the same. And if they finally get settled and drop slowly into the happy obscurity of a humdrum job and a little wife and a household of kids, Joe and Willie will be happy to settle down too. They might even shave and become respectable. . . .

Indeed they might. The GI, a homely hero, naturally decent and generous, inured slowly to battle and danger, would be in the end still generous, still trusting, wiser but still young, dirtier but still more content in his office or factory or on his sun-swept farm. He was as plain, as recognizable, as American as the militiamen of the past, he was the conscript citizen—competent enough but fundamentally an amateur, a transient, and an unhappy warrior. He was the essential republican, the common good man. He was the people's hero.

Like them, he had little visible purpose but winning the war so that he could return to a familiar, comfortable America, to what an earlier generation meant, more or less, by "normalcy." The priorities of the soldier resembled official policy in its serried emphasis on operative hatred rather than ultimate charity, on unconditional victory, and on an unconditional homecoming. The Germans and the Japanese, the targets of heroic striving, were dragons to be slain, after which the hero could return to his fair lady in her fair land. The content of heroic imagery said little about freeing the oppressed, an objective to which some Union soldiers had given high priority eighty years earlier, or about making the world safe for democracy, the stirring purpose of so many doughboys in 1918. The hero of World War II stood for blueberry pie and blond sweethearts, for the family farm and for Main Street, for perseverance and decency—for Americanism as a people's way of being. Brave men, those heroes, and nice guys, too, but it was in the folk culture of the national past, ebullient still in their own day, that they fought for their brave new world.

3. Cossacks and Comrades

The bravery of the individual GI had its equivalent in the valor of individual Englishmen, Germans, and Russians. The manufacture of heroes by the press and radio of other nations followed special contours. Each national pattern reflected different cultures and traditions, different social structures, different wartime anxieties. The differences in turn afforded a rough measure of the particularly American qualities mirrored in the image of the GI.

The British, according to the findings of one American research group, defined their heroes with characteristic understatement. Ordinarily the hero was a unit or command—a ship of the fleet or squadron of the Royal Air Force. When an individual was mentioned, he was usually singled out because he was commanding officer. Others were cited by name for conspicuous bravery, but without details and much less frequently than in American practice. Even after Dunkirk, that most gallant hour of retreat, the British press mentioned heroic exploits primarily in order to generalize about "the spirit of every ship and man engaged."

The Germans, in contrast to their usual procedure, reported the surnames of soldiers within victorious units while those units sped through France in 1940, but neither then nor ever said much about their battlefield heroes except to note their martial bravery. Like British reporting, German reporting, in its case governed of course by official Nazi orders, ordinarily stressed the bravery of military groups rather than of individual soldiers. When the Wehrmacht met first trouble and then reverses in Russia, the units were made to supply in valor what they lost in irresistibility. More and more, Hitler alone was allowed heroic qualities other than intrepidity, though Rommel enjoyed a short season of personal triumph. So did Reinhard Heydrich, the second-in-command of the Gestapo, whose death led to fierce retaliation against the Czechs. Partly perhaps to condone that terrible purge, Hitler and Gestapo chief Himmler described Heydrich after he had died in terms usually reserved for *der Führer* himself: "He was a character of rare purity with an intelligence of penetrating greatness and clarity. . . . He was filled with an incorruptible sense of justice. Truthful and decent people could always rely on his chivalrous sentiment and humane understanding." But the pattern of German propaganda, as American analysts noted, rarely permitted competition at the top: "Neither Goering nor Goebbels, neither Himmler nor Ribbentrop, nor, least of all, any of the generals ever reaches the exalted stage of a hero whom the nation might worship." Only Hitler was a "demi-god."

The Russians, at least until the closing days of the war, gave Stalin a less singular status. He even shared his glory gladly with his ablest generals. Furthermore, Russian reporting, while no less subject to official control than was German, most resembled American. The

Soviet press and radio, in the eyes of American analysts, built up particular workmen or soldiers as heroes, as they had during the peacetime five-year plans, when they systematically praised prodigies of productivity. Official Russian communiqués regularly devoted half their space to the details of individual exploits. Apparently aware that even in a Communist state mass achievement depended on personal enterprise, the Russians stressed the importance of the individual in the war. Especially from the home front, they provided heroic figures against whom ordinary workers might measure themselves.

Whether in the factory or in combat, the Russian folk hero of World War II, as he emerged in the pages of *Pravda* and *Izvestiia,* could not be mistaken for an American GI, but the two shared various characteristics, particularly their youth, their commitment to hard work, their victory over superior forces, their self-consciousness about belonging to a team, their identification with ordinary people. "Ruthlessly bold in battle," Flight Lieutenant Nikolai Terekhin was "modest and quiet," a twenty-five-year-old commander of a squadron in which "the air crew are all youths." The product of a "simple and ordinary" background, Terekhin had dreamed about flying since his boyhood, as had his men, who regularly defeated the Germans in spite of "their numerical superiority." Pilot-Officer Peter Mel'nikov, thrice wounded in action, also twenty-five, looked "significantly younger," but in talking about warfare quietly became "business-like and gloomy," and eager to rejoin his "unit . . . my comrades who are flying." Estrafuid Mamedov, an artillery lieutenant, twenty-two years old, was "the embodiment of the new generation of our homeland." Born into a plowman's family, he attended his local school and agricultural institute, "studied diligently and persistently," joined the Red Army, and on one occasion alone killed seventy Germans "against great odds." As he put it, "that's my whole biography, the same as everyone else's."

"Companions in battle," the crew of one tank knew their jobs well. "The youngest member of the team" was responsible for "checking the apparatus." Alert even during rare moments of relaxation, the men—each mentioned by name—moved forward in spite of some casualties "in an unequal battle" with German air and armored forces. In another tank, "a friendly team," the four-man crew had had their "friendship . . . strengthened in constant combat." They worked together with accuracy and familiarity, "they understood each other with half words."

That kind of teamwork also characterized an ammunitions factory in the Urals where the dedicated selflessness of the workers enhanced the Soviet cause. One foreman, young Tokarev, a "cheerful, convivial man," had worked in a porcelain factory as a painter. "He knew the secrets of color, loved his trade, and if there had not been a war would never have abandoned it. With the factory he travelled . . . to the Urals and there changed his profession. It was necessary to prepare arms for war with the enemy. . . . In a short while Tokarev became a good revolver maker and then an adjuster. Because of his furious attention to his work, he was marked as a man who would succeed." Others who had changed their trade had experienced difficulties but soon overcame them "with their will to conquer and to be worthy of the front." In another gun factory, one group of workers had set a "heroic" example of productivity. Among the men and women within that group, whom *Izvestiia* described at length, one supervisor explained the source of their extraordinary morale. It was "the knowledge of the responsibility for the fate of Russia, for the fate of our children."

The accommodation of the worker to his wartime task paralleled the accommodation of the soldier to combat. Like the GI's, the Russian troops were depicted as naturally peaceable men. They developed the qualities of warriors partly out of the hardening of combat, partly from a sense of personal identity with the very soil of Russia, which the Germans had desecrated, partly out of hatred for the barbaric invaders and a consequent zeal for revenge. War made one of two brothers mature and "coarse," lined his face, imposed a gravity of soul. His still younger sibling, acting head of the family, for their father had moved beyond the Urals to work in a factory, watched the Germans come to their village and hang his old teacher. He, too, joined the Army; "he must wreak vengeance . . . with his young hands."

Outstanding dealers of death to "the fascist barbarians," Stalin's First Cavalry during one evening of rest let their reveries flow: "In the memories of the Cossacks there was born as in a warm vision the spaces of the native Kuban Steppe, the alluring chains of mountains of the Caucasus, the exciting blue smoke of the foothills, the blindingly white huts . . . the sweet faces of near ones. Like a heavy golden wave the wheat flows under the sun. The caps of the sunflowers burn. The maize rustles and rises up like a mighty forest. In the pastures are innumerable flocks and herds of horses. Such are the strengths of the people. . . ."

Young Toknev, a common foot soldier, represented that

strength and its potential. "Swarthy, small in height, black of hair,"
Toknev came from the faraway Caucasus:

> At first he could in no way adjust to the new circumstances . . .
> to the strict absolute demands of military discipline. In the art of war
> he showed no particular interest. . . . He could have remained . . .
> in modest service if there had not occurred a tremendous transforma-
> tion. . . . Evil enemies thrust themselves onto our land, and Toknev
> was reborn. . . . In the boiling of battle suddenly he revealed him-
> self. . . . This happened rather unexpectedly. . . . He was laugh-
> ing. . . . He shouted: "Comrade Commander . . . look, look! I
> have killed two. . . ." He rejoiced in his strength, that he was a war-
> rior, that two out of a horde of violators and plunderers lay dead. . . .
> This small proud man, who in times of peace had peeled potatoes
> . . . learned the military art not theoretically but in the very process
> of fighting. . . . Was he thinking . . . about the faraway native
> skies, the mountains of home? . . . He was thinking of the . . .
> battle. . . . In that small body lived the indefatigable energy . . .
> and beat the fiery, magnificent heart of a Soviet patriot.

That patriotism flowered not only in the heartland of the Russian past, in Muscovy and in its besieged cities, but also and vigorously in the further reaches of the Soviet Union, areas remote from
Moscow and Leningrad and from the cultural traditions of the czarist
era. It was a sentiment that embraced the land and its people, but the
Soviet government and Communist party, too. "Armenian female
workers for the Army," *Pravda* reported, were "at one and the same
time Armenian and Soviet patriots, willing to sacrifice their lives out
of hatred for the Germans." A young flier, falling to sleep, *Pravda*
related, recalled his youth in the Ukraine, then the destruction there,
"the gray ash of burned villages in the white snow, the singed earth
covered with wounds. . . . He squeezed his fists . . . and began to
think about the Germans whom he had killed . . . in the air battles
of this year. There were already twelve." The next day he downed six
more. Another ace, weary after a difficult mission, remembered the
"narrow streets . . . the old hut in the shape of a fir tree, the gray
head of his old mother. . . . Dear Siberia! Beloved native town."
Still another, "a young Communist," derived his ferocity in the air
from his "real Bolshevik passion." A natural passion, *Izvestiia* suggested in an article about an Azerbaijanian hero, for in his case, and
by implication in all cases, the "homeland has given him everything
. . . provided   for   his   family . . . educated   him . . . taken

[him] as a candidate member of the Communist party." One thought
therefore obsessed him: "to destroy the enemy, not to surrender one
. . . clump of Soviet land."

As time passed, as Russian resistance first delayed, then
stopped, then turned the German advances, *Pravda* and *Izvestiia* published less about individual heroes, military or industrial, though
some received attention as long as the war lasted. The two newspapers, sensitive to the shifting priorities of the war in the Soviet
Union, tipped the weight of reporting about heroic figures from
the battlefields to the home front increasingly in 1942 and thereafter.
With that gradual change, an early emphasis on the youthfulness of
gallant soldiers gave way in part to recognition of the perseverance of
older factory hands. So, too, as victory approached, the theme of
membership in the Communist party recurred more often. But those
modifications did not significantly alter the general pattern, the persisting image of the Russian hero, especially the continuing motif of
his cold hatred for the Germans.

During the summer of 1941, *Pravda* commented on the "special
stamp" of the citizens of Leningrad, the objective of one rapid German thrust. Alexei Kabanov, a Leningrad factory worker turned
soldier, carried that stamp. He was one of those "steadfast, manly,
courageous political" men, a Bolshevik, a political worker who provided "the cement of the army, its backbone" in the "cruel war with a
villainous, treacherous, strong enemy. The German fascists! Alexei
Kabanov experiences a feeling of deep hatred for them." Fifteen
months later the Russians were counterattacking at Stalingrad during
the epic struggle for that city. *Izvestiia,* commenting on one young
officer, wrote that "hatred for the enemy is the basis of life of our
people." In an interview with an Army surgeon, the newspaper asked
what he had learned about the temperament of the heroic wounded.
The doctor had discovered, he said, the human will, as with the scout
Ivan Karmalok, shot through the shoulder, bleeding profusely:

> His brain worked with a furious heat. I will run—he thinks—
> while there is still blood left, only I can't fall, no I won't fall. The
> damned ones are moving on. . . .
> Karmalok was . . . trembling from frenzied anger. It seemed
> that if he should meet a German . . . he would tear him apart with
> his teeth. . . .
> Ivan Karmalok was a regular rank-and-file soldier, he wasn't
> counted on for any particular heroism. . . . He was the native son
> of . . . the Great October Revolution. . . . Not to die, to avenge

the enemy was what Karmalok wanted. To live! . . . He is a hero of the human will.

Another hero, a sergeant in the artillery, explained why he personally hated the fascists. He had been a gardener in a Ukrainian village. "On one of those days when the white apple blossoms had already fallen," he said, "when the cherries had ripened into color, the war began. . . . I cried when I parted with my beloved garden, and then for the first time I hated the damned Germans." More important, they later destroyed that garden and he saw elsewhere "burned towns and villages, the corpses of women in the ditches, a child with his legs torn off. . . . There my anger . . . grew." Still later he received a letter from a brother, a tankman, "about the fascist plundering committed against our family. Under the German occupation of our birthplace, the fascist monsters . . . tortured our twelve-year-old brother . . . but our parents, the bandits shot because they had . . . sons in the Red Army. I believe that you will wreak vengeance on the rabid enemy for his crimes. . . . The monsters of Hitler will pay for everything with their blood."

Marching westward into Germany in October 1944, the Red Army, according to *Pravda*, nourished the bitterness born of the German invasion and occupation of Russia:

The flame of hatred burns in the heart of Soviet fighters. . . . On the approaches to the Prussian border . . . I saw a gray-haired youth —the lance corporal Afancsiz Chuzhinov. As a sixteen-year-old adolescent he saw . . . how they took his mother from prison and shot her in the woods. . . . Chuzhinov heard the cries of his mother before death and saw her tortured body. The adolescent became gray and became a warrior.

Victor Zakclsluk, the first to reach the Prussian border, said that he remembered at that moment his beloved sister Maria, shot by the Germans in Kiev. . . .

Many letters were written by Soviet fighters on the joyful day of crossing the Prussian border. They are written . . . in Russian, in Ukrainian, in white Russian, in Lithuanian, Tator, Armenian, Bashkir. They are all addressed to Comrade Stalin; they are all alike . . . and all different. On this glorious day of victory people did not forget about the past and they think of the future.

Mikhey Minchenkov writes: "There was a time when you saw the German viper walk in our land, then you would gnash your teeth, and your spirit became anguished: the damned Prussian gorges on our bread, our fat, and yet he imagines he is a special breed, that he is better than everyone else in the world. I hate the Germans. They burned my native village, now . . . I shoot into the Germans who walk on the Prussian soil. I am reporting to you, Comrade Stalin, that I have already killed fifteen Germans on their soil."

Those Russian soldiers, writing Stalin in many languages, those folk heroes for all the Russian peoples, craved revenge. Home, for them, was not potato soup, but a cottage burned, a mother dead, a town pillaged. That was home, too, for the workers moved beyond the Urals. It was a far different home from Our Town, U.S.A., where activity at the front meant planting victory gardens or buying government bonds. The Russian peoples, more than any other, suffered the pain of Nazi ravages and bore the burden and the cost of the fight against Hitler. For them, unconditional victory was less a doctrine than an impulse; they needed little imagination to fight the war.

Nevertheless, *Pravda* and *Izvestiia,* selecting qualities that identified their heroes, emphasized those that fed the red flames of revenge. Those papers served as a kind of Soviet OWI and a kind of Russian Ernie Pyle, too. They depicted a soldier hero whom the Soviet state could approve and the Russian people recognize. Like the GI, the Russian soldier remembered the countryside he knew best and missed the trade he had abandoned (and doubtless the sweetheart, too, though *Pravda* never suggested that he was forlorn away from women or a natural housewife in his bivouac). Like the GI, the Russian did not muse about the brotherhood of man, the rights of nations or free peoples, or the fabric of a just and lasting peace. Like the GI, he had little clear purpose but winning the war so that he could go home, but first he intended also to punish the enemies who had defiled his past. The past he yearned to rebuild. He was less likely than was the GI to try quickly to forget the war.

It was the war and what it did to men rather than some profound variation in national character that differentiated Russians and Americans. The people of the two countries, whether soldiers or civilians, responded to the experience they had, and the Russians had much the worst of it. On a sunny July Sunday in 1944 George F. Kennan, that most observant of Americans in Russia, boarded a crowded train leaving Moscow for a short journey into the surrounding countryside. He could overhear snatches of the conversations on the platform of his car: "Someone had read in the morning paper the new decree about marriage and divorce, and the idea of premiums for large families was giving rise to a series of hilarious comments among the women." A similar decree, had it been necessary in the United

States, would surely have provoked a similar response among Brooklyn women sunning that same day on Jones Beach. But on another
Sunday Kennan watched some 50,000 German prisoners "being
marched several miles across town. . . . The object of the operation
was plainly to make a spectacle of the men before the population of
the city." Caesar had paraded captives in ancient Rome, but Kennan
shrank from the scene of exhausted men, pushed along in the hot sun
at a fast clip, now and then forced to stumble or to faint. "It was not
a very great brutality, as brutalities of war go," he recalled. "The
Germans . . . had done many times worse, and on a scale far
greater, with the Russian prisoners. . . . Still, I came away . . .
shaken, saddened, and unsatisfied. These prisoners were young men.
. . . Surely they had never been consulted about the great issues of
this war, still less about the abominations of the Nazi system. . . .
Was it right . . . to punish them? . . . Was brutality either sanctioned or sanctionable as a measure of revenge? . . . I recognized,
at that moment, that I stood temperamentally outside the passions of
war. . . . Wherever I lived . . . it was primarily against people's
methods rather than against their objectives that [my] indignation
mounted. . . . I found myself concerned less with what people
thought they were striving for than with the manner in which they
strove for it."

There spoke a private spirit, lonely whether he was in Moscow
or in Washington. In the Soviet Union the elite in high office in
government and the Communist party alone decided issues of national and world policy, the theoretical considerations molding them,
and the manner in which they were to be pursued and achieved.
Others who may have thought about those questions had no safe way,
if any way at all, of broadcasting their conclusions, even in metaphors
attached to brave men in arms. In the United States there was a
chance for literate and persuasive men who so desired to suggest to
the American people more than was offered them in OWI releases or
in journalistic profiles of American fighting men, profiles that could
not have satisfied and may have saddened Kennan. Only the homely
qualities attached to common soldiers and by extension to the common man could provide metaphors for understanding the significance
of both the butter and the guns, metaphors pointing to ways of living
in peace or war. Writers of fiction observed fear and fatigue, hatred
and revenge. They selected from all they felt and saw in order to
create what seemed more genuine to them than any actuality they had
known. In so doing, they told a special truth.

4. The War in Literature

Seen close at hand or from afar, war grated on the sensibilities of the artist. It always had. World War II, probably no more or less brutal or futile than most wars, marked man, as all wars did, as capable of inflicting horror as well as of abiding it. Like some other men in arms, some Americans rejoiced in the process of hurting and killing, others were indifferent to the viciousness of their compatriots. It was not easy for the sensitive to find in the war redemption for any war, for the United States, or for mankind.

The artist, as even *Fortune* realized, saw the war not "as an intellectual problem of production and politics, battles and strategy, but as a vivid emotional reality. The battlefield of this war is the whole earth, and all the people who live on it are soldiers." Modern art depicted the death of cities both in and as "the agonies of innocent people." A terrified mother and child fled Marc Chagall's blazing Russian village. Max Ernst painted Europe "dissolving in slow ruin." Rufino Tamayo conveyed the savagery of war in his renderings of beasts. Distorted images fit a distorted world, as Picasso had demonstrated after he saw in Spain the preliminary stages of the war. His *Guernica* told the whole story—the painting's tortured forms conveyed the grief, despair, and pain the artist felt.

So, too, in literature, an Army psychoanalyst, the protagonist in Edward Newhouse's story "Time Out," expressed what the author felt, as did so many returning soldiers: "I didn't go to the party in the neuropsychiatric ward, but I understand it was a pronounced success. The tobacco was handed out and the popcorn eaten. The only awkward occurrence was Sergeant Lanigan's deliberate smashing of the goldfish bowl. I'm afraid he'll tell his mother all about it and I'll have to explain that her boy still has bad moments but that he'll be perfectly all right in time. In six months or a year I may even have him ready for another war. All he needs is a breather. . . ." Lanigan, shot down over France, had seen his crewmates die, made it back to England, flown six more missions, and then cracked, scared literally almost to death, shocked, sickened, crippled—to what end? Not to fight again—in any event, General Patton would have written him off as wanting in discipline. Not for blueberry pie, as John Hersey would have been the first to say. Loathing the war, the American artist might try, but never find it simple, to define a purpose in it large enough to satisfy, even to appease, a Lanigan, his doctor, and his mother.

There was a girl in William L. Worden's "Mist from Attu," a girl who had had an affair with a young man killed there "on a mission made necessary by madmen . . . for a land no man had ever wanted badly." Before she had learned that he died, and at just the time of his death, she met another young man, a sailor on leave, who loved her and married her. The story, told by a parson, suggested that God's will was done, that there had occurred between the dying soldier and the living sailor a transmigration of the soul. Before the battle, the dead man's buddies had been discussing religion:

. . . There was the boy from Brooklyn. "Me," he said, "I don't know much about this religious stuff; but I know what I'll do if I get knocked off."

One of the others egged him on, "What you planning that's so hot, Frankie?"

"I'm going right back to Brooklyn," Frankie said, "and have a look around my girl's house. There's any guy there, he's going to get haunted like crazy."

The second man said Frankie was too fat to be a good haunt, but a third boy took it up, saying flatly he believed he could come back to earth if killed. . . .

. . . One of the oldest . . . spoke quickly at last. "I don't think any of these things," he said. "I figure heaven has to be some kind of reincarnation. Every guy's done something that, ever since, he's been wishing he had a chance to fix right again. . . ."

For a soldier, for a time, the cleanliness and quiet of home as it might become seemed more inviting. J. D. Salinger in "A Boy in France" had his boy muse in a foxhole:

"When I take my hand out of this blanket," he thought, "my nail will be grown back, my hands will be clean. My body will be clean. I'll have on clean shorts, clean undershirt, a white shirt. . . . A gray suit with a stripe, and I'll be home, and I'll bolt the door. I'll put some coffee on the stove, some records on the phonograph, and I'll bolt the door. . . . I'll open the window, I'll let in a nice, quiet girl . . . not anyone I've ever known—and I'll bolt the door. . . . I'll look at her American ankles, and I'll bolt the door. I'll ask her to read some Emily Dickinson to me . . . and . . . some William Blake . . . and I'll bolt the door. She'll have an American voice, and she won't ask me if I have any chewing gum or bonbons, and I'll bolt the door."

"He stared up into the sky again, the French sky, the unmistakably French, not American sky." He read an old letter from his younger sister with a postscript: "Please come home soon." He "sank back into the hole and said aloud to nobody, 'Please come home soon.' Then he fell . . . asleep." That was a believable boy, as real as Hersey's marines in the valley, as homesick, as American. He was a clean, decent, sensitive, middle-class boy. As they appeared in fiction, though no less real than he, some Americans were not.

In novels about the war, as in so much wartime journalism, any company in any regiment might display the ethnic mix of the American people. The novels, however, were frank about prejudice. "They shouldn't teach . . . immigrants today all about democracy," said Angelo in James Jones's *From Here to Eternity*, "unless they mean to let them have a little of it. . . ." Even James Gould Cozzens, whose *Guard of Honor* was atypical in its affection for the Army, hung his novel on an ethnic theme—white persecution of blacks. For no valid reason, the swashbuckling hot pilot at Cozzens's air base in Florida punched a Negro trainee; a Negro journalist was expelled from the base; Negro officers had to attend a segregated club. Cozzens, no reformer, made the blacks in the novel vulgar in their speech, drew their outstanding white champion as a boor, and had his hero work out a token accommodation that preserved *esprit de corps*. But Cozzens's reporting was honest: "A Negro," one captain said, "happens to be a member of a relatively inferior race; physically, mentally, every way." Obviously Cozzens, while prepared to do little to contain or to alter that attitude, lamented it. John Hersey condemned it. In his *Bell for Adano*, which was published during the war, the GI's were often condescending or hostile to the Italians, and General Marvin, a villainous but American type, shouted about the book's decent protagonist, Major Joppolo: "Joppolo . . . Remember that name. That goddam Major's a wop, too. . . . He's a goddam wop himself." For Hersey, Joppolo was also a hero.

Perhaps predictably during a struggle against Nazism, prejudice in the novels of World War II ordinarily took the form of anti-Semitism in its American guises. Irwin Shaw created a heroic victim, Noah Ackerman, in *The Young Lions*. Ackerman, though sensitive and tubercular, managed to join the Army. At a Florida base his lisping Southern sergeant, a Nazi type in American khaki, told him where he stood: "All raht, Ikie . . . Ah'll tell you heah an' now. Ah ain't got no use for Niggerth, Jewth, Mexicans or Chinamen. . . . Move, Ikie. Ah'm tahd of lookin' at your ugly face." Later, his savings stolen,

Ackerman offered to fight the thief. A small man, Ackerman had to
take on, one by one, the ten largest in the barracks. "I want," he said,
"every Jew . . . to be treated as though he weighed two hundred
pounds." So did Shaw, who had Ackerman lose his fight, desert for a
time, return, and prove himself under fire.

In Norman Mailer's *The Naked and the Dead*, Jews fared less
well. So did mankind. The platoon in that novel symbolized an
America that Mailer despised—a country sour with prejudice, locked
in injustice, revealed at its basest in his "time machines"—excursions
into the past of his soldiers. There was Julio Martinez: "Little Mexican boys also breathe the American fables, also want to be heroes,
aviators, lovers, financiers. . . . What can a Mexican boy do in San
Antone? He can be a counterman in hash-house; he can be bellhop;
he can pick cotton in season; he can start store; but he cannot be a
doctor, a lawyer, big merchant, chief. He can make love." There was
Polack Czienwicz, born where "it smells like a urinal in the hallway,"
fatherless at ten, at eighteen prosperous in the numbers racket. There
was Sergeant Sam Croft from Texas—"whelped mean," his father
said: "The first time Croft ever killed a man he was in a National
Guard uniform. There was a strike on . . . in the oil fields, and
some scabs had been hired. They called the Guard. (The sons of
bitches started this strike come from up north, New York. They's
some good boys in the oil fields but they got they heads turned by
Reds, an' next thing they'll have ya kissin' niggers' asses.) The
guardsmen made a line against the gate to the plant and stood sweating. . . . The pickets yelled and jeered. . . . A stone lofts through
the air. . . . All right men, the lt. pipes, fire over them. Croft sights
down his barrel. He has pointed his gun at the chest of the nearest
man. . . . The shot is lost in the volley, but the striker drops. Croft
feels a hollow excitement." There is Roy Gallagher from South Boston, who worked in a warehouse and joined the Christians United:
"We gotta stick together, or we'll be havin' our women raped, and the
Red Hammer of Red Jew Fascist Russia WILL BE SMASHING YOUR
DOOR DOWN. . . . Who takes away your jobs, who tries to sneak up
on your wives and your daughters and even your mothers . . .
who's out to get you . . . 'cause you ain't a Red and a Jew . . .
who don't respect the Lord's name. . . . Let's kill them! Gallagher
shrieks." There is Joey Goldstein from Brooklyn whose immigrant
grandfather laughed to himself because "this America is not so different." "I think a Jew is a Jew," he told Joey, "because he suffers. *Olla
Juden* suffers. . . . Suffer. It is the only world Joey Goldstein ab-

sorbs." There were others—Mailer's humankind—among them Roth, another Jew, restless in his Jewishness, like Goldstein a butt for Croft and Gallagher. While three men attempted to move a gun, one, not a Jew, let it go; the gun was lost, Croft blamed Goldstein:

> "Listen, Izzy."
>
> "My name isn't Izzy," Goldstein said angrily. . . .
>
> "Listen, you've done nothing but have ideas about how we could do something better. But when it comes down to doing a little goddam work, you're always dicking off."

There was a sense of futility about Goldstein's efforts to serve his country as a soldier, a sense of futility to which Croft contributed but which he also shared. That sense, in life as in the Army, sent men searching for someone to make suffer; that sense, a function of American society as Mailer saw it, made men cruel.

Like American society in the proletarian literature of the 1930's, the armed services in the protest novels of the war gave to the cruel and incompetent power over the lives of the helpless and the sensitive. Captain Colcough in Shaw's *The Young Lions,* a paranoid authoritarian, grabbed Ackerman's copy of *Ulysses,* which he called a "filthy, dirty book." In Jones's *From Here to Eternity,* in the words of the critic Maxwell Geismar, "there is not a single decent or efficient army man above the rank of sergeant . . . except perhaps the . . . Jewish Lieutenant Ross." At once comic and malign, Captain Morton in Thomas Heggen's *Mister Roberts* exercised an absurd and intolerable authority. "A Captain," Heggen wrote, "is not a person. . . . He is an embodiment. He is given stature, substance, and sometimes new dimension by the massive, cumulative authority of the Navy Department which looms behind him like a shadow."

The shadow behind that shadow, as John Horne Burns suggested in *The Gallery,* was man himself, man as war exposed him. In war the ridiculous ruled. Major Motes, a stiff Virginian, the vain commander of an office to censor the outgoing mail of soldiers in North Africa, was able to make it seem that his men were so busy that none could be spared to fight in Italy. When the major and his staff at last reached Naples, the city was secure and their censors, now checking the mail of prisoners of war, were not GI's but Italians: "dottori and professori and geometri and ragionieri and studenti," so hungry they fainted at their work. But Motes only resented the turnover, the replacement of Americans by Italians, whom he despised. "Those Ginsoes expect us to serve em a lunch," one officer complained. "They're like nigras and must be kept in their place," Motes replied. "A year ago the Italians were at war with us," another junior officer remarked. "Now they're reading the mail for their own prisoners of war. . . . What an obscene comedy." "The whole war is obscene, goddam it," Major Motes said. "All Europe and its parasitic population are obscene . . . like the nigras. . . ."

War rather than authority released the brutality as well as the vanity to which Americans in every rank were susceptible. Mailer's Croft, by nature a sadist, found in war an outlet for his temperament. At one point, he and Gallagher captured a Japanese. Croft gave him a cigarette, examined pictures of his family, offered him water. The captive thanked him. Then casually Croft killed him, to the horror of Gallagher. At another point, the platoon hunted for souvenirs on the rotting bodies of dead Japanese. In the process, Martinez, a man at once brave and soft, a naturally gentle man, extracted the gold teeth from the jaw of a corpse by smashing it with the butt of his rifle. To Mailer's horror, war brutalized even the basically decent.

For Mailer, moreover, the system—the Army and society alike—seemed omnipotent, the war pointless. His General Cummings embodied the system, a structure based on power and lacking humaneness. Traditional liberal values, personified in Lieutenant Hearn, had no place in the system; no chance for survival. Cummings in the privacy of his quarters treated Hearn as an equal "and then at the proper moment . . . established the fundamental relationship of general to lieutenant with an abrupt . . . shock like the slap of a wet towel." The Army, he told Hearn, functioned best "when you're frightened of the man above you, and contemptuous of your subordinates." Individuality was "just a hindrance." Moved to defiance but reduced to a juvenile gesture, Hearn surreptitiously snuffed out a cigarette on the general's inviolable floor. Cummings, aware of who the culprit was, gave Hearn the choice of picking up the butt or standing a court-martial. Hearn, humiliated, obeyed. Later, Cummings ordered Hearn to take a group of men on a perilous mission over a rugged mountain. Hearn was killed on that operation, which proved to be as needless as it was murderous. "The only morality of the future," Cummings had said, "is a power morality. . . . The only way you can generate the proper attitude of awe and obedience is through immense and disproportionate power." That was, Mailer indicated in his vignette of Cummings's past, "a peculiarly American statement." The son of the richest man in a Midwestern town, Cummings learned a false virility from his father, his military school, West

Point, and learned, too, to follow the line of conventional ambition, to command, to succeed. For his wife he developed a cold hatred. Freud he found "rather stimulating. The idea is that man is a worthless bastard, and the only problem is how best to control him." Hitler had "the germ of an idea. . . . He plays the people with consummate skill." Cummings's fancies late in the 1930's saw the ranks ahead of him—"colonel . . . brigadier . . . major general . . . lieutenant general . . . general? *If there's a war soon it'll help.*" It killed Hearn.

Though other novelists expressed less despair than did Mailer, they, too, offered little solace to Lieutenant Hearn and men like him. James Jones and Thomas Heggen, each in his own way, either had the virtuous destroyed or divorced them from the service and the state. Theirs at the best was a sense of futility. Prewitt, the underdog, individualistic hero of *From Here to Eternity,* rebelled against Army conventions—American conventions, but knowing his Army, he still considered those conventions inescapable "in the Armies that are coming," and knowing his country, he said he "still loved it." In the end, in spite of the probable costs, Prewitt asserted his patriotism. In that process he, too, was killed by military policemen who did not understand his motives. Mister Roberts, driven by the abominable captain of the *Reluctant* to symbolic acts of rebellion, alone of the officers on the ship stood up for the crew as well as against their tormenter. Alone of the officers, a courageous and a democratic man, he recognized an enemy other than the captain, applied for transfer to a ship engaged in the forward area of the Pacific, and then was killed.

Despair, futility, absurdity, death—Randall Jarrell put them together in a wartime poem:

In bombers named for girls, we burned
The cities we had learned about in school—
Till our lives wore out; our bodies lay among
The people we had killed and never seen.
When we lasted long enough they gave us medals.
When we died they said, "Our casualties were low."

They said "Here are the maps"; we burned the cities.

It was not dying—no, not ever dying;
But the night I died, I dreamed that I was dead,
And the cities said to me: "Why are you dying?
We are satisfied, if you are; but why did I die?"

The easy answers were unpersuasive. Colonel Ross, speaking for James Gould Cozzens, sounded like any one of Mauldin's weary and disenchanted GI's: "His war aim was to get out as soon as possible and go home. . . . He did not need to know about their [the Nazis'] bad acts and wicked principles. Compared to the offense they now committed . . . by shooting at him and so keeping him here, any alleged atrocities of theirs . . . were trifles." Herman Wouk, in the postwar best seller *The Caine Mutiny*, disliked the authoritarian structure of the Navy as much as Cozzens respected the orderliness of the Army, but Wouk was prepared to tolerate even the despicable Captain Queeg because he and those like him made civilians into effective fighting men. Only by fighting could the virtuous but disciplined turn back the enemy, stop the horrors of Nazism, keep "Mama out of the soap dish."

But if Mailer was right, the United States had already succumbed to the sickness that infected Germany, and if Jones and Heggen were correct, the Army and the Navy were spreading that affliction. Victory and homecoming, necessitarian answers to a questioning of the purpose of the war, promised less than enough to compensate for the terrors that Americans in arms both suffered and inflicted upon others. The soldiers and the officers in the most pungent fiction of the war, unlike the GI's in war reporting, were not nice guys, not emblems of a benign folk culture, not capable of perceiving any brave new world. The writers, by and large, perceived none for them, or for their readers—not even, as with the Russian soldiers, the brutal satisfaction of revenge.

The most prolific of the writers, author of six books about the war, John Hersey, knew the anguish and shared the doubts of so many of his fellows. He also observed the courage and found the faith to see some glimmering redemption arising from brutality—and horror. His journey through the war, "upstream," as he recalled it, took him beyond the pain and yearnings of the soldiers, through the shock of recognizing so much that was ugly in the social actions of Americans, and out beyond despair and futility and death to new affirmations.

The innocence of *Men on Bataan*, Hersey's first book on the war, was his as much as his subjects'. Far away from their experience he corresponded with their families and took from the letters of devoted mothers the data with which he tried to find their sons' characters. Predictably, the sons emerged not only as ordinary but also as exemplary Americans. Still, as Hersey pointed out twenty-five years later, the men on Bataan were brave men, whose mothers had a right

to be proud. He withdrew the book from print not because his soldiers were undeserving, but because he discovered soon after writing
it that it presented an inadequate picture of war and an inaccurate
portrait of MacArthur. In contrast, Hersey remained satisfied, as he
should have, with his reporting in *Into the Valley*. On Guadalcanal he
had been rather surprised, he recalled, by the reflections he provoked
about blueberry pie, but in the circumstances in which those reflections occurred, they were, after all, not so surprising. The marines,
however nostalgic, had both at the time and in the context of recollection faced the brutality of war with fortitude. Their nostalgia characterized them no more than did their repulsion—and his—at that
brutality.

In Sicily, Hersey also recalled, the GI's seemed to recognize the
Germans as the real, the brutal enemy. But in Sicily he could not
escape observing that Americans were brutal, too. Several soldiers
had manned an antiaircraft gun during a devastating, night-long
bombardment near Salerno. Shortly after dawn, General George
Patton strode by, inspecting his troops, and "ate them out because
they did not have leggings on." He was, Hersey said, "a vile man."
He was also, of course, the model for General Marvin in *Bell for
Adano*, General Marvin who hated "wops," who played mumblety-
peg on an antique mahogany table, a selfish, loud, domineering man,
in the literature of the war the first ugly American:

> Probably you think of him as one of the heroes of the invasion:
> the genial, pipe-smoking, history-quoting, snazzy-looking, map-carry-
> ing, adjective-defying divisional commander; the man who still wears
> spurs even though he rides everywhere in an armored car; the man
> who fires twelve rounds from his captured Luger pistol every morning
> before breakfast. . . .
>   You can't get the truth except from the boys who come home
> and finally limp out of the hospitals and even then the truth is bent
> by anger. . . .
>   But . . . General Marvin showed himself during the invasion
> to be a bad man, something worse than what our troops were trying to
> throw out. . .

He was not the only bad man. Three GI's, their generous
motives drowned in wine, smashed furniture, sculpture, Venetian
glassware, and a painting—"a lot of Eyetalian junk." Others cheated
the Italians in the black market for American cigarettes and chocolate, and with the profits bought Italian girls. Hersey was not exaggerating. "I remember that my heart finally broke in Naples," wrote
John Horne Burns in *The Gallery*. ". . . Our propaganda did every-

thing but tell us Americans the truth: that we had most of the riches of the modern world, but very little of its soul. . . . We couldn't resist the temptation to turn a dollar or two at the expense of people who were already down." Or resist the urge to be vicious in the frustration of war, as Alfred Hayes suggested in *All Thy Conquests:* Schulte, one of his characters, about to attempt to rape an Italian girl, addressed her escort: "When I was getting shot at what were you doing? You and your goddam white teeth. You and your goddam hair oil . . . You and the goddam vino we drink. You and the goddam broads we get doses from. You and your goddam country."

Barbarian prejudice and greed, as Hersey witnessed them, infected the genteel, too, like Lieutenant Livingston, U.S.N.R., a product of Kent School and Yale, "a fellow who would do anything for you if he likes you, but he was rather choosy in his friends," and excluded "meatballs" from a lesser social background. The soldiers and their officers represented the American people, including civilians whose preoccupation with business-as-usual the GI's so resented. So did Hersey, who also understood that part of the fault lay in American society. *Bell,* he later wrote, "was the first novel, or book of any kind, during the Second World War, to suggest that the American hero . . . might be a dangerous shit. . . . The novel was written the way it was written because of when it was written—when the tide of war was just beginning to turn, when our good cause (we had one in that war) was still on a shaky footing."

At Hiroshima, as Hersey reported so vividly in 1946, the most dangerous men were Americans. It was not a matter of national character. Rather, a concern for the unit—for the American armada that might have had to attack Japan—provided the rationalization by which President Truman, Secretary Stimson, and their military advisers excused the resort to the awful weapon. The brutalizing effect of prolonged war dulled their sensitivities, and the sensitivities of most other Americans, to the inhuman implications of what they had done, to the terror of conventional war with its napalm weapons as well as the terror of the ultimate instruments.

The ultimate brutality, whatever the weapon, was man's, as also was the ultimate good. So Hersey said in *The Wall,* whose Warsaw Jews and their German tormentors could have been any men. "By 1948," Hersey recalled, "when we were getting plenty of bulletins about our heroes being evil, I was writing *The Wall,* because . . . the time had come to remind people that in the face of evil men could be both venal and ruthless, as some Jews in the Warsaw ghetto were, and steadfast and courageous, as some had been." "I have decided," Hersey had one brave Jew say, as the novel moved toward its end, "that nationalism is not enough for a man to live by. . . . Extreme nationalism can be as frightful in a Jew as in a German." Or an American, or any other man who permitted his concern for the unit—the platoon, the country—to eclipse his concern for mankind. So, too, succinctly, wrote Marianne Moore:

Hate—hardened heart. O heart of iron,
  iron is iron till it is rust.
There never was a war that was
  not inward. I must
fight till I have conquered in myself what
  causes war. . . .

American boys, and their ordinary American wives and parents and friends, usually received a different message. It was not their fault. Within the United States, Americans never saw the enemy. The nation did not share or want to share in the disasters that visited Europe and Asia. The war was neither a threat nor a crusade. It seemed, as *Fortune* put it, "only a painful necessity." For most Americans fighting it, the war, although it was an awful personal catharsis, had no point. It was a dangerous but inescapable interruption, unreal in terms of life as it had been and should again become. "The American soldier," *Fortune* commented, "is depression-conscious and . . . worried sick about postwar joblessness. Just as he lives and fights out of 'pride of outfit' . . . so would he like to get home as soon as possible. He hates the Japanese; he may detest Hitler; he is not particularly conscious of what fascism is. . . . The American soldier is in the war because his country has been attacked. . . . But the soldier himself is most deeply interested in the quest for personal security." Orientation programs little affected that attitude, for "no army can do what its civilization has not done."

American civilization, or parts of it, had lent credence to Mailer's Croft and Cummings, not to his Hearn. Parts of that civilization had also instructed the GI's and shaped their manner and their dreams. Brave men, they were often cruel and violent, too. Nice guys, they cared less for Christian friendship than for staying alive, helping their buddies, and getting home to a job, a girl, and a cold Coca-Cola. They were, after all, just like Americans at home. There, too, the war could signify only what the culture ordinarily endorsed.

3 / Getting and Spending

1. The Return of Prosperity

"The hand that signs the war contract is the hand that shapes the future." So reported a Senate committee studying wartime changes. There was much truth in the exaggeration. The assignment of war contracts determined which business firms would thrive, which would languish, and determined, too, which areas of the country would feel the impact of industrial expansion and which would not. On a broader scale, the sheer volume of war contracts created such an economic demand for goods and services, and assured such a concurrent demand for labor, that investment in new productive facilities and employment of extra workers rapidly liquidated the conditions of depression that had characterized American life during the twelve lean years before Pearl Harbor. Full employment and prosperity, for their part, permitted Americans, in spite of wartime restrictions, to begin to buy many of the necessities and some of the comforts they had been unable to afford for so long, and to dream about buying others once an end of war made those again available. Indeed, the ebullient prosperity of the war years encouraged a mood of buoyancy, with attendant political implications, as significant as was the more remarked "miracle of production" that furnished the sinews of battle. Within the arsenal of democracy, government expenditures made business vastly better than usual and restored the circumstances for a carnival of consumption, which newly prosperous businessmen and their friends ordinarily attributed to the blessings of free private enterprise.

Essential though that enterprise was, the agency of recovery, as John Maynard Keynes had demonstrated theoretically, was the huge and growing deficit resulting from federal spending for military and other programs. Taxes defrayed only 50 per cent of federal costs. The balance was raised by borrowing, with precisely the kind of effect that Keynesian economics had predicted. The deficits, though arising not from a deliberate pursuit of economic theory but from the exigencies of war, spurred the investments that produced the goods and provided the jobs that lifted the economy to full capacity at a level higher than ever before.

Statistics told part of the story. By the time of Pearl Harbor, military spending had already reached a monthly rate of $2 billion. In the first six months of 1942, federal procurement officers placed orders for $100 billion of equipment—more than the American economy had ever produced in a single year. In 1942 alone the proportion of economic activity devoted to war production grew from 15 to 33 per cent, and by the end of the following year federal spending for goods and services exceeded the total product of the economy in 1933. The federal budget, about $9 billion in 1939, rose to $100 billion in 1945. Under that impact, the gross national product, $91 billion in 1939, reached $166 billion in 1945. During the war, 17 million new jobs were created, the index of industrial production rose 96 per cent, of transportation services, 109 per cent. Production for civilian use, while diminishing, remained so high that Americans knew no serious deprivations. The industrial sector, while supplying the war-related needs not only of the United States, but also, in considerable measure, of the nation's allies, in effect superimposed its new obligations upon its prewar output for the American market. At the peak of the war effort in 1944, the total of all goods and services available to civilians was actually larger than it had been in 1940.

The prodigies of production involved not only new investment in industrial plant and machinery, even the creation of new industries like synthetic rubber, but also increased mechanization to enlist technological energies where manpower reserves had disappeared. Even with mechanization, the mounting need for labor, after absorbing the unemployed, called on the civilian working force for extra hours, at extra pay of course, and for the recruitment of unprecedented numbers of women as well as adolescents and those previously engaged in domestic chores and other marginally useful tasks. Full employment under wartime conditions assured better, more dignified, more remunerative employment. The civilian spending that accompanied that development further heated the economy, which moved before the end of 1942 to a worrisome inflation that did not wholly subside, in spite of vigorous federal efforts to control it, through the rest of the war years.

2. The Wartime Consumer

By stimulating the economy, the war did wonderful things for the American people. After the drab years of depression, Americans in 1941 and thereafter found themselves enjoying conditions many of them could scarcely remember. There were plenty of jobs. Business and farm profits were rising, as were wages, salaries, and other elements of personal income. At every level of society, men and women, even children, had money to spend, for luxuries if they were rich, for amenities long denied them if they were of moderate means, for small conveniences, decent food, and some recreation if they were workers. In spite of taxes, rationing, and price and wage controls, the wartime surge of buying was exciting in part because for so long most Americans had had to stint. It was also frustrating because wartime shortages denied Americans much of what they wanted. The contemplation of the end of deprivation after the war fostered dreams that achieved a partial fulfillment in the immediate consumption of such goods as there were. While they awaited, sometimes petulantly, a postwar consumers' nirvana, while they feared a postwar depression but fussed about wartime inconvenience, Americans indulged their appetites as broadly as they could afford to.

Less than a year after the Japanese attack on Pearl Harbor, government spending had worked its miracle. The average income per family in American cities (an average based upon salaries, wages, rents, dividends, and interest) had soared above the depression levels of 1938. In Hartford, Connecticut, it had moved from $2,207 to $5,208; in Boston, from $2,455 to $3,618; in New York, from $2,760 to $4,044; in Chicago, from $3,233 to $3,776; in San Francisco, from $2,201 to $3,716; and in Los Angeles, from $2,031 to $3,469. The largest reported relative jump occurred in Washington, D.C.— from $2,227 to $5,316.

In the national capital the influx of new recruits to government work and of transients with government business was generating a boom no less visible than those that airplanes brought to Hartford and Los Angeles and ships to New Orleans and Seattle. One transient, the novelist John Dos Passos, told his Washington cab driver he was in a hurry. "Hurry," the cabbie said, scornfully. "Nobody ever used to be in a hurry in Washington." No one could hurry in its queues. There were queues for housing, for buses, outside of restaurants and theaters. Sometimes even a queue was a privilege, as Dos Passos suggested in reporting that at one laundry the proprietor abruptly informed a young man who had entered: "We don't take no new customers." Another young man, overheard at a lunch counter, told his neighbor: "I've been in this town four days and I haven't found a room yet." Washington, in another view, was "a bivouac." It was filled and filling with strangers, emptying young men into the services, receiving young women to do all the typing and mimeographing and filing that the Army and Navy and their supporting agencies required. Though it had no industry, it was not much different from any other city. "I haven't been to a dance in six months," Dos Passos heard one girl tell another; "Washington's the loneliest city."

Looking for better jobs, following war industries, or following servicemen to locations near their camps and training stations, Americans moved faster and in greater numbers than ever before. On the highways, buses carried peak loads of passengers—in 1942, for the first time ever, more even than the railroads. The railroads for their part filled their coaches and had demands far beyond the supply for Pullman reservations. They made money on passenger traffic in 1942 for the first time in fifteen years. River cargoes went way up; air transport went into the black in the first quarter of 1942 and stayed there. The boom cities of the South, the Southwest, and the Pacific Coast flourished. New Orleans, typical of the others, saw its population rise 20 per cent in 1942 and then keep rising. Bank deposits reached a record high. War contracts that year were huge, with Higgins Industries, builders of boats and ships, alone receiving over $700 million. And boom begot more boom, as war workers had to have $100 million worth of new housing.

Baton Rouge, Seattle, Los Angeles, Philadelphia, Detroit, and San Diego, like New Orleans and Washington, staggered under the new prosperity. Schools, short of teachers, could not absorb all the children needing schooling. Public transportation had difficulty handling all the new workers in all the new neighborhoods. There were not enough houses or apartments, too few parks and playgrounds, insufficient places at night clubs and bars. Life was hurried and strange and sometimes harassed, and for those wistful for home or old friends or a boy in uniform, Washington was only one of dozens of loneliest cities.

Shortages, rationing, the hustle and strain and anxiety of daily urban living during the war grated upon Americans. Naturally they complained, though their lot was better than that of any other people in the world, and better, too, than it had been only a few years ago. They ate bountifully. Industry and government co-operated in arranging programs that packed vitamins into workers' lunches, and provided free milk, snacks between meals, and other inducements to spur output and combat absenteeism. But appetite often overruled hunger, and well-fed workers, blue collar and white, tended to care less for the snacks they were given than for the steak they could not buy. Yet even the scarcity of steak and bacon, or, later, of cigarettes and shoes, carried its benefits. Wage earners who could not purchase what they missed instead paid off what they owed. "The pawnbroking business," the *Wall Street Journal* reported in October 1942, "has fallen upon dark days."

The race for consumer goods began right after Pearl Harbor. With incomes and prices both rising, along with rumors of impending shortages, rationing, and controls, Americans began to stock up at a record rate. A study of fourteen cities showed hoarding under way in many items: food (especially sugar, canned meats and vegetables, coffee, tea, spices, and olive oil), rubber goods (used automobile tires, gaskets for jars, garden hoses, golf balls, galoshes, girdles), household supplies (soap, linen, furniture, blankets), clothing (particularly men's suits and shoes), and a miscellany of rifles and shotgun shells, typewriters, and paper clips.

The War Production Board, anxious about wool supplies early in March 1942, issued an order forbidding men's suits to include an extra pair of trousers, a vest, patch pockets, or cuffs, and calling for the manufacture only of single-breasted and somewhat shorter jackets with narrower lapels. Those prescriptions were designed to save 40 to 50 million pounds of wool a year. To the casual observer, according to the WPB, only the absence of cuffs would be noticeable, a deliberate tactic, since "an abrupt style change," the agency feared, "would start a buying rush for 'victory suits' which would defeat the conservation aims of the order." The tactic was not wholly successful. By the beginning of April, sales of men's suits, which merchants had not hesitated to advertise, were running at three times the normal volume.

Women were also spending freely, especially women war workers, of whom some 6.5 million, most of them middle-aged and married, entered the labor force between 1941 and 1945. They had yet to question traditional attitudes toward women's roles. Except for a few feminists, they did not protest the continuing disparity between men's and women's wages or the lack of day-care centers for the children of working mothers. Instead, they accepted the conclusion of the Children's Bureau that in war, as in peace, "a mother's primary duty is to her home and children." On several accounts that conventional definition suited almost all newly employed women. They did not, for the most part, intend to remain in the working force, although after the war many of them did not leave and others soon returned. They had, in the main, a special satisfaction, a continuing sense of the importance of their wartime tasks, of the nation's need for their labor, a point that the federal government continually stressed. Perhaps most important, they were beguiled by their new income. Their earnings, supplementing the military pay or the wages of their husbands, gave their families, usually for the first time, the chance to buy, or to save in order later to buy, the conveniences, the comforts, the small luxuries that had become a part of American middle-class expectations and a mark of middle-class status. After the disheartening years of depression, the prospect of achieving that status, and the immediate excitement of available cash to spend at will, seemed to many women, as to many men, not just satisfying but almost miraculous.

By 1943, 5 million more women were employed than in 1941, and the wages of women factory operatives were up over 50 per cent for the same period. They had some $8 billion more in pocket than they had had. Retailers had doubled their sales of women's clothing and found price no barrier as women sought quality as well as quantity. As cotton, wool, and nylon supplies dropped, the government ordered a 10-per-cent reduction in the amount of cloth in a woman's bathing suit, an objective reached by banishing the billowing bathing skirts of the 1930's and concentrating on two- rather than one-piece outfits. "The difficulties and dangers of the situation are obvious," the *Wall Street Journal* observed. "But the saving has been effected—in the region of the midriff. The two-piece bathing suit now is tied in with the war as closely as the zipperless dress and the pleatless skirt." No complaints were reported about the sacrifices thus entailed.

When clothes and cars were unavailable, Americans spent their money on entertainment and recreation. The motion-picture industry by the summer of 1942 had become one of the best "war babies" on the stock market. Theater earnings climbed as war workers flocked to films. The shortages of gasoline and tire rubber added to the attraction of neighborhood theaters, where air-cooled interiors provided a substitute of a kind for a drive to the shore or mountains or a trip to a resort. Also, American producers were relieved for the while of the competition from abroad. Their only worry was a possible loss to the services of celebrated male stars. Resorts, too, did not much suffer, for trains and buses carried eager clients to them. Night clubs and cafés, in spite of rising prices, were crowded. With work weeks prolonged to accelerate war production, Sunday became a rich day for entertainment. In spite of wartime restrictions, as one report noted, "Americans are finding fun—and lots of it." Indeed, during the war New York was really "fun city." There the boom in demand for hotel rooms and seats in restaurants and theaters reflected a rising influx of servicemen, executives, and war workers seeking relaxation and amusement.

By the summer of 1943 the quest for pleasure had not abated. Though thousands of men had joined the service, others took up the slack in memberships in country and golf clubs. To be sure, a lack of labor and equipment forced managers to narrow fairways and let the grass grow longer in the roughs; to look for younger caddies; to trim luncheon and dinner menus. But there were generally few vacancies and large profits. New England resort hotels, faced with a quota of food 30 per cent below normal and with a dearth of waitresses, nevertheless cheered up when the Office of Price Administration allowed residents of the Northeast enough gasoline for one vacation trip. Those who stayed home and those who returned energized the markets for music and books. Prices of musical instruments rose rapidly; a robust black market developed for grand pianos; sheet-music sales climbed to a ten-year high; phonograph records were scarce. So, too, book clubs set sales records in 1943, as did bookstores. Publishers also began vigorously to market paperbound editions in drugstores and on newsstands. Riding the wave, Wendell Willkie's *One World* sold faster than any book in history. Two and a half months after publication, the cloth edition had sold 250,000 copies, the paper edition 1 million. From newsstands and chain stores, civilians and servicemen alike bought a growing number of the twenty-five-cent titles of Pocket Books, Avon, and Dell, with murder mysteries leading the way and self-help and health books close behind. Largely because of the new mass market, publishers, in spite of the paper shortage, did very well, some 40 per cent better in 1943 than in the previous year.

So did race tracks. Though many tracks had to close because gas rationing made them inaccessible, wagering in 1943 and 1944, and the profits to the tracks, climbed to their highest levels. Even *Fortune* took notice. At any track on any day, it reported, the crowds were large and characteristically varied and colorful: "There are the people, the low life and the swank, the sweating, shoving mob at the rail and in the stands. There are the touts and tarts and the priests and the housewives and the hundreds of servicemen (admitted free)." There was one face of wartime America, where "the winners smugly walk off to the well-stocked bars, the soda-fountains, the hot-dog stands, and the restaurants." Wagering at the mutuel machines at the tracks exceeded $1 billion in 1944 (up two and a half times the 1940 figure), with resulting gains in the share of the handle that went to the states that legalized betting. The amount bet illegally through bookmakers could only be estimated as between twice and ten times the legal figure. In California, where the legal handle at the only track running in 1944 averaged about $420,000 a day, "the illegal handle through bartenders, barbers, elevator-men and poolrooms" was perhaps as much as $10 million a day.

There were other ways, too, of enjoying the new affluence. It was no problem to obtain scarce and rationed goods at a price in the black markets. Boneless ham in Washington, D.C. sold on the black market for $1.25 a pound, almost twice its legal ceiling price. With patience, a buyer could find nylon hose in most cities for five dollars a pair. In Philadelphia, without recourse to rationing coupons, a determined customer could buy five-dollar shoes for about seven dollars. Those "willful violations," as an official of the Office of Price Administration said, were hard to prevent, since they occurred largely within regular retail channels on the basis of quiet collusion between the merchant and his patron.

In wholly legal markets, Americans during 1942 bought $95 million worth of pharmaceuticals, $20 million more than during the previous year. That increase did not arise from an abundance of colds, indigestion, or headaches. Rather, as the druggists knew, people simply had "more money to spend." For the same reason, jewelry sales mounted, depending on local circumstances, between 20 and 100 per cent. Though the volume of business from the wealthy declined, war workers avidly purchased diminishing stocks of cigarette cases, lighters, rings, silverware, and watches, particularly jeweled models. "People are crazy with money," one Philadelphia jeweler said. "They don't care what they buy. They purchase things . . . just for the fun of spending."

Though shortages became more severe in 1944, the rush to buy continued. Over-all retail sales in the first six months were up 8 per cent compared to 1943. The average sale in department stores, two dollars before the war, became ten dollars. American men shaved more often and with fresher blades in spite of the shortage of steel. Just before the new luxury tax became effective in the spring of 1944, R. H. Macy & Company enjoyed a buying wave for furs, jewelry, cosmetics, and handbags. During 1944 the food industry made significant gains in its sales of breakfast cereals, baked goods, mixes, spices, and better grades of coffee, a rationed commodity. Supermarkets—of which there had been only 4,900 in 1939—were constructed so rapidly in the face of low stocks of building materials that by 1944 there were more than 16,000. Even nature smiled: there was a record crop of cherries, quickly sold, a guarantee of "more pie all around." Well before Christmas 1944, holiday purchasing had stripped retailers' shelves. On December 7, the third anniversary of Pearl Harbor, Macy's had its biggest selling day ever. The Federal Reserve index of department-store sales for November had exceeded the high of any month in any previous year. Textiles, jewelry, and clothing moved especially fast. "People want to spend money," one store manager said, "and if they can't spend it on textiles they'll spend it on furniture; or . . . we'll find something else for them." Then as in earlier years it was "delayed consumer demand," a function of deprivation during the Depression, in the opinion of the *Wall Street Journal*, that accounted for the wartime baby boom.

There was no change in 1945. In spite of OPA curtailments of apparel production, in spite of high prices, the Easter dash for clothing again set records, up 7 to 50 per cent, depending on the city, over 1944. So, too, high prices and shortages failed to deter diners-out. One hotel menu had "ox tongues, tails, nothing between," but restaurants, like department stores, thrived.

For the well-to-do, those who were accustomed to spending, those whose possessions permitted them to spurn more jewelry or furs, the conditions that hurt the most were usually those that removed housemaids from mansions to production lines and prevented annual excursions to the Riviera or Switzerland. Both the wealthy and the comfortable who enjoyed food took solace in the tone and content of *Gourmet,* a journal born just as the war began. It was an inauspicious time for graceful eating, but *Gourmet* helped its subscribers courageously to transcend the inconveniences they faced. "Imports of European delicacies may dwindle," the first issue admitted, "but America has battalions of good foods to rush to appetite's defense." Wine connoisseurs, for one group, were advised in 1942 to turn their attention to California *vin ordinaire* and by their example to "make it clear to the great non-wine-drinking public that ordinary wines, treated with appreciation . . . constitute a gastronomic resource that is of the first order." War stopped the spice trade, but my lady could instead cultivate an herb garden and make up for lost time, as well as for a lost chef, by adapting favorite recipes to a pressure cooker. "China is helping us," *Gourmet* noted, patriotically, "we must help the Chinese"—by buying almond cookies from United China Relief. The magazine also recommended a Russian cookbook. "The money goes to help Russia," it wrote, "and give Hitler indigestion." And if an issue arrived late because of transportation difficulties, the impatient reader should "blame Japan, not us."

Mostly *Gourmet* led its subscribers to new adventures in gastronomy. "I began with smoked salmon," one epicure wrote in an early prescription, "one of the few delicate meat starters left to us poor gourmets since the end of the caviar era." Another later praised blue cheese, which was produced in increasing quantity in the United States during the war as a substitute for French Roquefort. "Once again," one article observed, "game is in the necessity class, for our increased consumption of game will release domesticated meats for other uses." And game could be good—even rabbit. *Gourmet* published various appropriate recipes along with a wistful verse: "Although it isn't / Our usual habit, / This year we're eating / The Easter Rabbit." And, as a temporary expedient, even margarine, which could serve in place of butter for an infinite variety of hors d'oeuvres.

Bravely the subscribers took *Gourmet's* advice. "I think you are wonderful," one characteristic letter to the journal exclaimed: "If some of us have not the courage and the persistence to hang on to our culture and our way of life, what will become of us?" *Gourmet* knew just what its readers wanted for a future Christmas: a dozen cases of good Scotch whiskey, a large standing roast of the best beef, an oversized steak, five pounds of butter, unsalted perhaps, and a bunch of bananas.

Early in the war, in his acid and penetrating *Generation of Vipers*, Philip Wylie caught the implications of the American mood: "Our war aims remain nebulous, we are told, because nobody has yet hit upon a plan for the postwar world which satisfies the majority of the people on this all-consuming problem of goods. . . . To many, it hardly seems worthwhile fighting to live until they can be assured that their percolators will live, along with their cars, synthetic roofing, and disposable diapers."

Through the war years, advertising exploited the urge to spend, intensified it, and directed it to deferred but glowing postwar possibilities. Some advertisements might have won awards, had any been offered, for bad taste. The city of Miami, inundated with Army Air Corps and Navy personnel at its sundry training stations, nevertheless sought expense-account tourists to overcrowd its filled hotels and night clubs. "Miami's pledge to America at War" offered warmth, relaxation, and salt water as respites from hard work, respites to be won by further burdening the railroads and airlines. The presumed need for respite also inspired the United States Playing Card Company to link its product to victory. For continued strength and vitality, it was alleged, "83 per cent of the people of this nation turn to card playing for inexpensive recreation." War-related themes, like those that associated almond cookies with the Grand Alliance, led Mennen's to brag that 1,300 dermatologists serving soldiers and civilians recommended the company's shaving cream. In a similar vein, Formfit sold its brassieres "for the *support* you need these hectic days of added responsibility."

Even before the cross-channel invasion in 1944, the advertising industry had its largest budget in history and had begun to prepare the public for postwar goods. In 1942 and 1943, industry, exercising some restraints, had emphasized institutional advertising designed to keep the name of a firm in the public eye. Now the switch began to product advertising. The Ford Motor Company, Remington Rand, and W. A. Shaeffer launched campaigns for their cars, typewriters, and pens, while General Electric, continuing to urge consumers to stretch the use of their household hard goods, had copy prepared for a sales effort as soon as steel could become available. The potential market was vast. At the time of Pearl Harbor, the liquid assets of individuals came to $50 billion; by the end of 1944 that figure had reached a record $140 billion. The National Association of Savings Banks conducted a survey of depositors that showed 43 per cent eager to spend their savings for "future needs," 20.6 per cent more precisely for homes and their accoutrements, 9 per cent for automobiles. The Office of Civilian Requirements in June 1944 announced that eleven appliances led the list on Americans' postwar plans for buying, with washing machines first, followed by electric irons, refrigerators, stoves, toasters, radios, vacuum cleaners, electric fans, and hot-water heaters. Those items, advertisers had claimed throughout the war, constituted the American way of life. "Weren't you bragging just a little, Yamamoto," *The Saturday Evening Post* had asked. "Your people are giving their lives in useless sacrifice. Ours are fighting for a glorious future of mass employment, mass production and mass distribution and ownership." "Some day Johnny, front line observer, will climb out of his foxhole," Western Electric predicted, "into a world freed from fear of dictators. When that day comes, the telephone . . . will help to place all peoples . . . on friendly speaking terms." Johnston and Murphy promised that "when our boys come home . . . among the finer things of life they will find ready to enjoy will be Johnston and Murphy shoes. Quality unchanged." The advertisement that probably most bothered Bill Mauldin, an advertisement neither better nor worse than dozens like it, displayed a Nash Kelvinator refrigerator as the emblem of a sentimental homecoming. The lady in the picture, chin up, spoke out for an unaltered world in which she and her heroic soldier-spouse would pick up where they had left off: "I know you'll come back to me. . . . And when you do . . . you'll find . . . everything your letters tell me you hold dear. I will be wearing the same blue dress I wore the day you went away. And on my arm the silver bracelet you gave me . . . on our anniversary. . . . Everything will be here, just as you left it" and withal a new refrigerator in the kitchen.

That new refrigerator, like other new products and services of whatever make, as Americans learned from advertisements and public-relations stories in 1944, would surpass all their previous expectations. It would manufacture and eject ice cubes, quick-freeze fruits and vegetables, contain freezing compartments and revolving shelves, use compact motors to save space, and come in many colors to blend with any decor. Radios, too, would incorporate wartime developments to produce better tone and receive frequency modulation, and television would provide glorious new at-home entertainment. The happy housewife would save time with a new, small dishwasher, a new automatic toaster and an electric coffeemaker. Her stove would gain sparkle with Revere ware and improved copper cooking utensils. Her husband, when he was not preoccupied with his new 8 mm movie camera the size of a man's pipe, would be able to convert his basement into a playroom because his heating plant would be small enough to fit into a closet. He could even leave home comfortably. Pullman illustrated double-decker postwar sleeping cars with the legend: "Goodbye standing room only," and Pan American promised "in the post war period world travel will be within the reach of the average man and his family." But the postwar man, the probabilities were, wanted to remain near his wife, his conveniences, and his toys. Happily, the seasons of the year would not deter him. Air conditioning would make his house livable the year through: *"Now*—we'll be glad to put your name down for earliest available data on postwar air conditioning and refrigeration equipment," General Electric proposed in June 1944 while American troops struggled through the mud of Normandy and Guam.

The house and all that went into it, "the American home," best symbolized of all things material a brave new world of worldly goods. The vision was in part a fantasy woven by advertising, in part a romanticizing of desires born of depression circumstances and wartime deprivations. After the decade of the 1930's, during which new building dropped 61 per cent below construction in the preceding ten years, housing in 1940 had already become scarce and grim, especially for low-income groups. Of all available housing units, more than 14 per cent needed major repairs and improvements. "Not less than 416,000 families," according to the Bureau of Labor statistics, had "established households in the backs of stores, in public buildings, warehouses, and garages, and in shacks, houseboats, barns, tents, boxcars, caves, dugouts. . . . More than four-fifths of . . . new properties . . . were beyond the reach of more than four-fifths of the families in non-farm areas."

Wartime dislocations, not least the mobilization of 13 million men, many with wives who traveled with them, further strained conditions. Over 4 million workers—with their families, some 9 million people—left their homes for employment in war plants. "Scarcely a section of the country," one federal report noted, "or a community of any size escaped the impact of this great migration." Yet the shortage of building materials forced the War Production Board in April 1942 to ban all nondefense construction and put stringent limitations on the alteration or improvement of existing structures. By June 1945, over 98 per cent of American cities reported a shortage of single-family houses, over 90 per cent a shortage of apartments. Various estimates judged 75 per cent of all plumbing and electrical equipment and 68 per cent of all interiors to be below par. Conservation orders had also necessarily prohibited production of sinks, furniture, bedding, and electric appliances, and curtailed production of electric wiring and of plumbing and heating fixtures.

In the circumstance, *Better Homes and Gardens,* a periodical addressed to the middle class, instructed its readers in the reuphol-

stering of old furniture, the conservation of fuel, and the eradication
of mice and of "white water rings" on wood. In the suburbs, according to the New York *Times,* "under the pinch of war, the majority
and the minority are one. We all bear the badge of the great frater-
nity: a bandage on one finger and barked knuckles on all the others.
. . . We . . . have cotter pins and bent nails in with our change.
. . . Perhaps the best way to launch this whole subject would be in
a huge mural, with . . . a panel called Man and Moloch (man with
his head in the furnace, studying a jammed gauge) . . . [and] Man
Cutting a pane of glass for the First Time in His Life (lots of glowing
red in that one)."

Outside the suburbs, in cities overrun by war workers, conditions
were much worse. Near Detroit, at Henry Ford's Willow Run bomber
plant, designed to employ 100,000 workers, there was "no housing
whatsoever, excepting a trailer camp and a scant handful of over-
priced dwellings." In Portland, Oregon; Hartford, Connecticut; and
Richmond, Virginia, "furnished rooms, trailers . . . makeshift
houses . . . had to be utilized." Trailer camps, in the view of the
head of the Federal Security Administration, constituted "a new kind
of 'slum on wheels' that offers almost unparalleled hazards of over-
crowding and insanitation." Just as bad was one typical old Victorian
house in a Connecticut boom town, a house with five second-floor
bedrooms. "Three of them," one reporter wrote, "held two cots
apiece, the two others held three cots. The twelve cots were all occupied by workers at the aircraft plants. . . . 'This is beginning to
sound like big business,' I said.

" 'Sure it is,' the landlord said. . . . 'But the third floor is
where we pick up the velvet. . . . We rent to workers in different
shifts . . . three shifts a day . . . seven bucks a week apiece.' "

Those victimized by the "hot bunk" system, like those with scars
from cutting glass, looked forward avidly to postwar space and conveniences. In 1943, with the end of the war still not in sight, residential real estate buying reached proportions unknown since the 1920's.
In boom cities like Detroit and Washington, the well-to-do were buying houses partly because there were none to rent. In many suburbs,
including Westchester County, New York, others bought real estate
with full confidence that rising prices would assure postwar sales at
large profits. The cost of small homes had mounted 15 to 30 per cent
since Pearl Harbor. Within the following year they rose another 10 to
30 per cent.

Columns of letters from young couples to *Better Homes and*

Gardens described the homes and furnishings and refrigerators for which the writers were saving. "I can see our house going up," one wife wrote, "stamp by stamp, bond by bond, joist by joist." Another correspondent believed "children should have the security of a home and the pride of a home that belongs to their family . . . a neighborhood where they'll be included in things because they *belong* . . . a place to which they can bring their friends." The Celotex Corporation encouraged the chorus. "As America drives under war's incentive," it advertised, "the products of our future greatness are being shaped. New wonders are coming from the men of science and industry. . . . Housing will undergo tremendous change. . . . Out of undreamed-of progress . . . will emerge your 'Miracle House' of tomorrow." It was to be "within the reach of the *average* family" but nonetheless fantastic—luminescent panels instead of incandescent bulbs, furniture and upholstery that would never wear out—the proper equivalent in a house of the new, static-free radios, new and more powerful automobiles, new tickets to a consumer's heaven. "Is This Worth Fighting For?" another advertisement asked. It then depicted an old-fashioned living room complete with fireplace and walls of western pine, the latter the product of the manufacturer who paid for the ad.

To most of the American people at home, that advertisement symbolized much of what was worth fighting for and, well before the fighting ended, what was worth spending for, whether to satisfy yearnings long unfulfilled or whether just for the fun of spending. The postwar world as most people visualized it, the Office of War Information told President Roosevelt, was "compounded largely of 1929 values and the economics of the 1920's, leavened with a hangover from . . . makeshift controls of the war." The American way of living returned, during the new prosperity of the war years, to patterns that Americans liked to believe had marked national life before the Depression, patterns they wanted to preserve and to project into the postwar period.

That way of living, they believed with sufficient cause, would also satisfy Americans in arms. After all, even Bill Mauldin did expect Willie and Joe to settle down. Where more likely, if the civilian consumer could arrange it, than in a little white house in the suburbs where a young wife wearing a pretty blue dress and an anniversary bracelet would greet her veteran husband every evening with slippers to change into from his Johnston and Murphy shoes, with a Scotch whiskey highball and a sirloin steak, all within the efficient antiseptic environment created by electric appliances that cooked the meals, did the housework, kept out the cold or heat, and left the couple free to listen to their radio. With modest alterations to suit individual tastes, with allowances for Mom's blueberry pie for dinner on alternate Sundays, that was the decent, uninspired picture, for thousands of Americans, of their postwar world.

3. Enterprisers at War

In 1942 Fred Gimbel persuaded his brother, Bernard, to try to sell, on a 10- to 20-per-cent commission, the works of art of which William Randolph Hearst was eager to dispose. It was, as *Fortune* said, "simply a question of risking the floor space," and the Gimbels took the chance with 80,000 square feet. Aside from the free advertising they received, during 1942–43 they sold about $4,225,000 worth of Hearst's hoard. Fred Gimbel calculated that the venture brought to the New York store about 50,000 curious customers, many of them itching to spend. One New Jersey syndicate bought a complete, 10,000-ton Spanish monastery, boxed, that had cost Hearst $500,000, for $19,000. The advantages of the art business in the affluent war years moved the brothers to absorb the Kende Galleries into their emporium, where daily auctions became an attractive feature at "plain old Gimbels," now adapted to wartime merchandizing.

The Gimbels, like other successful businessmen whose operations were by no means essential to the war, displayed a remarkable capacity to adapt to the wartime mood and wants of American civilians. Gimbel Brothers owned eleven large department stores. In 1942, the centenary of the family's first American venture, Bernard Gimbel was "taking full advantage" of the war boom. He had decided earlier, against advice of the Harvard Business School and various economists, that the war would be a long one. Anticipating severe shortages in consumer goods, in 1942 he borrowed $21 million to build up his inventories. Gimbel's $3.5 million stock of nylon and silk hose lasted well into 1943. For Christmas of that year the New York store had 500 electric trains to offer. Still later it featured beach robes. Hose, trains, beach robes—all were made of critical materials and available almost nowhere else. Earlier and later, Gimbel's converted Army and Navy surpluses into briskly moving frills—100,000 small flare parachutes sold as toys; 1,000 field telephones brought $29.95 apiece. In 1944, for the first time, the Gimbels' network of stores exceeded Macy's in gross volume, $194.5 million from which the stockholders shared $4 million of profits. Bernard Gimbel liked to quote Francis Bacon, one of whose axioms he had followed: "In all negotiations of difficulty, a man may not look to sow and reap at once; but must prepare business, and so ripen it by degrees."

Macy's also flourished during the war years, though some merchants thought its top management was "suffering from profundity." That quality was attributed primarily to Beardsley Ruml, sometime foundation officer and dean at the University of Chicago, treasurer of Macy's, and chairman of the New York Federal Reserve Bank. Ruml, whose wartime activities focused more upon public fiscal and financial problems than on the operational details of Macy's, nevertheless helped his store sustain its well-being. In 1944, Gimbel's banner year, Macy's still showed a slightly larger net profit. That owed much to innovations Ruml had sponsored, among them an inexpensive form of installment buying that especially fit the needs of war workers, and a $12 million issue of 2½-per-cent debentures that provided easy cash to sustain inventories. Ruml also completed plans for an aggressive postwar expansion of Macy's outside New York City. Furthermore, his concern about Macy employees who could not afford to retire on pensions because of the taxes due on their previous year's earnings moved him to the advocacy of a major change in federal income taxation, the forgiveness and pay-as-you-go scheme that Congress adopted in 1944 over the Treasury's objections. Whatever the inequities in that program, it did solve the retirement problems at Macy's. Beardsley Ruml, for all his preoccupation with public issues, remained, as *Fortune* put it, "one of the most enthusiastic . . . and capacious consumers in the United States," a fitting role for the guiding intelligence of a prosperous store in a time of heavy buying. There was a great deal that Macy's could still tell Gimbel's during the war years.

The Parker Pen Company, another adaptive firm, in those years was "proud of its fountain pens, but . . . even more proud of the way it . . . advertised them." The company had ridden to glory on advertising during the 1930's when its conspicuous red Duofold pen marked any man who owned one as rich. Besides snob appeal, Parker had introduced "look appeal" by manufacturing pens in bright colors and identifying them with current fashions. The "51," the top of Parker's line when the war began, symbolized prestige. By late 1944 supplies were so short (the pen was made of Lucite, the plastic also used in bomber noses) and demand so high for the "51" that it drew black-market prices up to $200 in China and India. But Parker had more profitable markets elsewhere. The War Production Board allowed pen companies about 60 per cent of their prewar production, with about half of their output designated for the Army and Navy. In spite of those restrictions, Parker increased its dollar volume by concentrating production on its three most expensive pens. It also increased its profits, which in 1943, even after wartime taxes, were up 30 per cent over 1941. Parker extended its domination of foreign markets, primarily by turning to air freight, and utilized its idle plant space by accepting war contracts on a no-profit basis, but also under conditions that cost the company nothing, for it leased the expensive new equipment it used at a nominal rate from the Army, which had bought and still owned it. Parker also expanded its hold on the domestic market for ink. For its Quink, a new product in 1941, Parker spent $2.5 million in advertising. Parker had, as *Fortune* put it, "a good war angle. . . . By helping the government advertise the morale-building value of mail for servicemen, Parker naturally created a blue-sky demand for ink." Happily, no restrictions impeded ink production. During the two years preceding October 1944, ink sales trebled but Quink sales rose 800 per cent. By that time, Parker was spending more than ever before on advertising to prime the postwar market for a new model pen, more prestigious, more stylish, more expensive even than the "51."

Robert W. Woodruff of Coca-Cola, manufacturer of a five-cent drink, proved as adaptable as had the Parker Company with its fifteen-dollar pen. The most widely distributed mass-produced item when the war began was not an automobile or a razor blade but Coca-Cola. Woodruff kept it so, sugar rationing to the contrary notwithstanding, and also assured its expansive postwar future. He accomplished his purpose early in the war by persuading the Army and Navy that Coca-Cola was an essential product because it exactly suited the wants of soldiers and sailors. That message brought the armed services to encourage the company to carry overseas the bottles, the syrup, in time the bottling and syrup plants, that made possible a continuing, broadening manufacture and distribution of the drink, particularly since sugar for the armed services did not come from the limited domestic allotment. The operation, expensive and sometimes nonprofitable, permitted Coca-Cola everywhere to follow the flag, and Woodruff kept the price at a nickel. Before the war ended, he had used it to establish a worldwide taste for his beverage, by then more American than sweet corn or turkey giblets. He had also moved most of his management units from Atlanta to New York in preparation for the coming decade of international growth, which he expected, correctly, would be the greatest in the company's history.

At the least Woodruff's peer in shrewdness, Philip K. Wrigley also thrived by identifying his product, chewing gum, as an essential wartime commodity. The war seemed at first to doom chewing gum to temporary extinction. Gum was half sugar, which was rationed to between 70 and 80 per cent of 1941 consumption. More important, the rubbery saps of exotic trees that made gum chewy came from Malaya and Borneo, which Japan held, and Central and South America, where supplies were tantalizingly plentiful but subject to the dire shortage of shipping space. Further, Wrigley's plant could not be converted to war production because his workers were largely unskilled and his machinery adaptable for little except making and packing chewing gum. Nevertheless, he came out ahead.

Wrigley preserved his access to the raw materials he needed first by directing his gum-tree tappers in South America also to tap the rubber trees that grew in the same places and were easily harvested concurrently by the same men. With the raw rubber, the gum sap could then be moved to ports from which, now and then, a ship carrying rubber found space for a little of Wrigley's indispensable chicle. He did not need much—the entire gum industry used only 12,000 tons a year. But he did need sugar, too, and to obtain it he managed, as Woodruff had, to make his product seem essential. To that end, at the request of the Army, he supplied a stick of gum for each package of "K" or combat rations. The Subsistence Research Laboratory of the Chicago Quartermaster Depot believed the gum would help relieve soldiers' thirst, keep their mouths moist, substitute for tobacco when smoking was prohibited as a release from tension, even in some measure keep teeth clean. Charmed by that reasoning, Wrigley also undertook to pack the "K" rations that included his gum. The operation filled one of his plants and overflowed into another he acquired for the purpose. There was little if any profit in it, but gum was safe for the duration.

The profit rose during the war, and prospectively in cascades for the postwar period, from Wrigley's success in marketing chewing gum to civilians as "a war material." His tactic was extraordinary. Before the war, Wrigley and his rivals competed not by cutting price but by aggressive promotion—the use of premiums for dealers, the distribution of free samples, and, especially, advertising. Wrigley spent some 25 per cent of his revenues on advertising, a proportion beyond that even of cigarette companies. "Doublemint Gum," Americans learned to believe, was "healthful, delicious." With the war, Wrigley dropped that message. He devoted all the radio time he continued to buy—over $2 million—to war-related compaigns, to the recruitment of Sea Bees for the Navy or the publicizing of the Army's mission and progress. He then added a new program glorifying war workers. On that show, the featured performer, Ben Bernie, sometimes told his audience to "chew gum, any kind of gum," adding "personally, I prefer Wrigley's Spearmint." So, increasingly, did the thousands of war workers who listened to a program about themselves. Meanwhile, Wrigley had gone to work on their employers. To them, he stressed the benefits of chewing gum. Demand for gum, he maintained, had always soared under conditions of stress. Experiments he financed suggested that chewing gum provided direct physiological relief from tension, as well as diminishing thirst and the need for tobacco. War workers under pressure, it followed, would relax with gum, and also make fewer trips to the water fountain or a smoking area. "To help your workers feel better—work better," Phil Wrigley concluded, "just see that they get five sticks of chewing gum every day." He would distribute this gum—an essential war material, he ruled in September 1942—only to essential war industries and only when an order was accompanied by a letter from an official on the firm's letterhead "stating the need for chewing gum in that particular plant."

One-third saliva, one-third sweetener, one-third hokum, perhaps, but the new message got across. Wrigley took no chances. He had his division managers listen to a fifteen-minute "mind conditioner," beside which any mere soap opera paled. As *Fortune* reported, the scene was laid in a factory; the characters were the workers, the president, an insinuating voice called Monotony, two temptresses called Thirst and Nicotine, and Adolf Hitler himself at his snarling nastiest. Hitler started the action by saying he had friends slowing down production in every factory in America; the temptresses then did their bits, followed by the narrator: "Monotony . . . fatigue . . . false Thirst . . . nervous tension. Yes—these are the agents of the Axis." After more of the same dire script, the president, distressed, asked a foreman why his division produced more than the others. The foreman admitted that chewing gum drove away false

Thirst and the rest of the saboteurs. The president, just before the curtain, gave the crucial order: "Make *chewing gum* available to *every person* . . . employed in this plant." Wrigley tried to, and also looked forward to selling "Essential Gum . . . in peacetime not as a minor pleasure but as a major necessity."

There were, of course, real necessities—not pens or soft drinks or gum—not available in department stores, but vital for the conduct of the war. Gimbel, Ruml, Woodruff, and Wrigley accommodated to circumstances inherently disadvantageous to their enterprises. Equally shrewd and shrewder men seized the opportunity to make the necessities and the resulting profits, and often to do so by utilizing techniques of finance and production adapted precisely to the special conditions of government procurement. Some companies simply responded effectively to new, wartime demand. For one, Walter Kidde and Company, manufacturers of carbon-dioxide fire-detecting and extinguishing systems for tanks, planes, and ships, lifted its sales from $2 million in 1938 to $60 million in 1943, its number of employees from 450 to 5,000. For another, Lights, Inc., producers of aviation and automotive lighting equipment, in the same period moved sales from $150,000 to $10 million, while its payroll grew from 25 to 280. For a third, D. W. Onan and Sons, makers of generator engines and generators, saw sales rise from $300,000 to $50 million, the payroll from 60 to 2,100. There were hundreds like those three, sprung from the brow of Mars. There were none that surpassed, in success or reputation, the new giants of the aircraft and shipbuilding industries, and none of them more celebrated than the enterprises of Andrew Jackson Higgins and Henry J. Kaiser.

Andrew Jackson Higgins seemed to incarnate the self-made, self-confident, swashbuckling industrialist of American folklore. *Fortune* considered him a happy cross between Huey Long and Henry Ford. Like the Kingfish, he enjoyed sharp suits (he leaned to dark gabardines), colorful stories, and rich profanity. Like Ford, he was a master of the techniques of mass production and a suspicious critic of the eastern financial establishment and its alleged associates in Washington, D.C. ("District of Confusion," Higgins called it). Unlike Ford, he got along well with labor unions and had a reputation for popularity with his workers, whom he addressed frequently over a loudspeaker system. But he also pushed them. "Don't relax," Higgins preached. To reinforce that sermon, he posted a picture of Hitler, Mussolini, and Hirohito sitting on water closets in the men's washroom. "Come on in, brother," the caption said. "Take it easy. Every minute you loaf here helps us plenty." He also urged his workers to report shirking among their fellows, and on occasion he implored them to quicken their pace. "Whenever there are some of these emergency calls asking for apparently the impossible," Higgins explained to a congressional committee, ". . . I have a loudspeaker system and the labor bands . . . play a few stirring pieces, including the *Star Spangled Banner*, and somebody gets up and makes a speech. When the tears are running down their eyes . . . I ask them how they are going to do it . . . and we get along fine."

He got along fine with the Navy, too, for his Higgins Industries Inc. excelled in producing motorboats, especially power landing barges, tank lighters, and PT's, the speedy patrol torpedo craft that won early fame in the Pacific war. That production, rather than his extravagant manner, was the key to Higgins's wartime success, and it in turn grew out of his prewar experience first in the lumber industry and then in boatbuilding. Ten years before Pearl Harbor his boats were breaking speed records on the Mississippi. His designs allowed those boats to navigate waters too shallow and treacherous for their competitors. Six years before Pearl Harbor he had also designed a tank landing boat, a model that, with refinements, proved to be superior to an alternative developed by the Navy's Bureau of Ships. Two years before Pearl Harbor, when Hitler invaded Poland, Higgins had purchased an enormous stockpile of high-grade lumber. And after Pearl Harbor he enlisted the Navy, the mayor of New Orleans, and through them senior executives in the steel and railroad industries, to provide his yards with the right kind of steel for his purposes. The prewar experience, his access to raw materials, the staff of inventive engineers he employed, his own skill in production, and the easy availability of war contracts and of financing for them (advances and progress payments from the Navy, plentiful reserves against taxes) opened the road for the stunning expansion of Higgins Industries. As one critic observed, the concern was more a production department than a modern corporation, but while production counted most, it prospered, and its salty management and home city prospered with it. In 1935 Higgins's total sales came to $400,000; in 1943, to over $120 million. By 1943, Higgins had become a local hero in New Orleans and a national symbol of wartime industrial success second only to Henry J. Kaiser.

In the emergency production of war materials, Franklin Roosevelt was said to believe, energy was more efficient than efficiency. Furthermore, speed was often as important as quality, and costs mattered less than results. The prevalence of those standards gave Henry J. Kaiser the levers for which he had been grasping. Kaiser exuded energy, much of which he spent in spurring his senior executives to harder work. "He pounds you," one of them said. His rages and threats equaled his appetite, which accounted for his pronounced jowls, paunch, and duckwaddle of a walk. He made a fortune before the war on government contracts, and on government contracts during the war he raised the edifices of his ambitions. As his son Edgar put it proudly: "We are building an empire."

The Kaiser empire rested on sand—sand and gravel, the first business in which Kaiser had excelled. As a prominent West Coast sand-and-gravel man he became one of a consortium, the Six Companies, that won the federal contract to build Boulder Dam, "the most spectacular single construction job" of the Depression years, also one with profits after taxes exceeding $10 million. The Six Companies, the *Wall Street Journal* commented, "evolved an organization that combines the merits of a Chinese tong, a Highland clan and a Renaissance commercial syndicate with all the flexibility and legal safeguards of the modern corporation." While the work went forward during the years 1931–36, Kaiser revealed a flair for borrowing necessary capital against still unrealized income, for acquiring needed raw materials according to a precise timetable, for organizing groups of workers in transitory communities to which he provided better services than usual and from which he evoked strong loyalties and steady performance. He also proved his skill in negotiations with government officials. From his suite at the Shoreham Hotel in Washington, he and his agents made the connections without which contractors do not receive public contracts. They learned, too, to be adroit in public relations and quick in soliciting tips from the top about pending government plans and effective procedures for abetting them. Between 1933 and 1942, the Six Companies won the contracts and completed the work on Bonneville and Grand Coulee dams on the Columbia River and for sinking the piers for the Golden Gate Bridge. In those years, Kaiser established an office at Rockefeller Center in New York, rented a permanent suite at the Waldorf, and ran his long-distance telephone bill to its steady wartime level of $250,000 a year. He managed, too, to enlist, as one Washington agent, that incomparable lobbyist, the former New Deal wonder boy, Thomas G. Corcoran. It was Corcoran who led Kaiser to other useful acquaintances, among them Lauchlin Currie, an economist on Roosevelt's White House staff, William Knudsen, the General Motors executive who had joint command of the War Production Board, and Jesse Jones, the head of the Reconstruction Finance Corporation and its complex of related offices, the largest aggregate of lending agencies ever put together in the history of the world.

Kaiser and the Six Companies brought their customary acquisitive speed to the gathering and execution of war contracts, available in magnitude on a compatible and remunerative cost-plus basis. They built cantonments, military roads, boats, shelters. Above all, they built ships. At ten different yards, eight of them on the West Coast, they found and held the labor (not least because of Kaiser's inventive scheme for group medicine), procured the machinery and steel (not least because of Kaiser's direct lines to the government desks setting priorities for scarce equipment), and devised the assembly-line methods (not least because they were uninhibited by conventional practices) that produced the tonnage the country needed. They built Liberty ships, small aircraft carriers, tankers, troop ships, destroyer escorts, landing ships—in all, 30 per cent of the total national program in 1943—for contracts exceeding $3 billion. Some of their ships were less sturdy than those manufactured by traditional means and in older yards, but during 1942 and 1943 enemy submarines were sinking American vessels so rapidly that sturdiness and longevity seemed less important than sheer volume of production. Kaiser got the volume by adopting the techniques of the construction business to his new task, particularly by prefabricating as much as possible. The delivery of a Liberty ship, the basic cargo carrier of the war, consumed, on the average, 355 days in 1941, with total deliveries that year just over 1 million tons. Roosevelt called for eight times that volume in 1942. Kaiser's yards responded by cutting average delivery time to fifty-six days, and even completing one Liberty ship in fourteen. One of his ships foundered on the pier before even sailing, but his productivity had made him something of a national hero and assured his occasional access to the White House. There on one occasion he strengthened the President's resolve to overrule the Navy and to put flight decks on cargo ship hulls in order to create a fleet of baby flat-tops from which to combat German submarines in the Atlantic and to escort Navy task units in the Pacific. On both oceans, the baby flat-tops succeeded splendidly in their missions. Kaiser's yards built a large proportion of them.

Indeed, the Six Companies built so much so fast that their older competitors, United States Steel and Bethlehem Steel among them, became jealously constrictive about allocating steel to their rival.

Kaiser's previous undertakings led him to a resolution of that problem. He and his associates, aware of the additional profits that would accrue if they could verticalize their construction business, had before the war built the world's largest cement plant. Its operations broke the cement monopoly that had long inhibited building on the West Coast. Now Kaiser, though his associates declined to take the chance of joining him, shot off to build his own steel plant at Fontana, California. He solicited the necessary endorsement from the WPB and borrowed $106 million from the RFC against future, yet unrealized shipyard earnings. As he said at the time, he risked nothing but the taxes he would otherwise have to pay on those earnings. The Fontana mill, moreover, would, he thought, provoke United States Steel into expanding its capacity on the West Coast, and thereby help to develop the region that held Kaiser's affection. Expecting, as he did, to acquire government-subsidized wartime facilities for about ten cents on the dollar after the war, Kaiser considered Fontana cheap at twice his loan.

A possessive man, Kaiser tended to think about Six Company properties as if they were his own. Though he held only 7.5 per cent of the equities in the Joshua Hendy Iron Works, a Six Company property that built marine engines, he called it "my engine company." That habit, as well as his adventurous spirit, contributed to the reluctance of the others to join the steel venture. They were also dissatisfied with the profits from the Permanente Metals Corporation, set up on Kaiser's initiative, financed by a $20 million RFC loan against future profits at two shipyards, and engaged in manufacturing magnesium. Unlike his partners, Kaiser cared less for immediate profits in magnesium, a scarce light metal, than for its potentialities in the aircraft industry. Corcoran, with his own expansive vision of the future, had prompted Kaiser to get into the magnesium business. Kaiser soon thereafter entered into an agreement with another ebullient promoter, Howard Hughes, to produce a huge, light, postwar seaplane to carry heavy cargoes. That dream collapsed, but not before Kaiser had begun to make plans for postwar adventures (built on wartime profits, by wartime methods, and in part in wartime plants) in automobiles, prefabricated housing, and helicopters. Without perceived limits of time or space or possibility, he was forever building his empire.

In the process, Kaiser saw himself, according to *Fortune*, as "at least a joint savior of the free-enterprise system." That delusion was as paunchy as the man. Of enterprise Kaiser had an imperial measure, but it was free, if at all, only for him. Certainly his operation bore little resemblance to the classic models of the economic philosophers of capitalism. Energy, ability, ambition, those things Kaiser had; but government supplied his capital, furnished his market, and guaranteed his solvency on the cost-plus formula—and so spared him the need for cost efficiency, rewarded speed at any price, and came close to guaranteeing his profits. Kaiser's enterprises, from Boulder Dam through the Richmond shipyards and on to Fontana Steel, partook in extraordinary degree of the characteristics of a huge federal public-works agency, liberally funded by taxpayers' dollars. In the circumstances, the United States needed that kind of agency. The loose rhetoric of American business culture described it as free enterprise.

Kaiser's delusion was by no means his alone. During the war Republic Steel thrived on business the government sponsored according to priorities and regulations the government prescribed, and the government underwrote the profits. In some ways Republic had never had it better. Yet in one of the company's reflective advertisements, Old Joe told Young Joe what he was fighting for—"your right to live your own life . . . without being pushed around by some bright young bureaucrat who wants to do all your planning for you." Reynolds Aluminum, the beneficiary of a huge federal loan, boasted of its "astonishing increase in production" of aluminum. Gulf South, another war baby, expected the world to be rebuilt by "Individual Initiative and Free Enterprise." Warner and Swasey, makers of turret lathes for war industry, warned against the foreign doctrine that government planning should include control of production. That would ruin the nation and everyone in it. "If Free Enterprise had not flourished here," *The Saturday Evening Post* informed an imaginary housewife in Hamburg, "the cause of world freedom might now be lost for centuries." The *Post* went on to attribute the stature of the United States, "the last bulwark of civilization," to the American cycle of *"mass employment, mass production, mass advertising, mass distribution and mass ownership of the products of industry."*

That cycle, during the war, began with government spending. Higgins and Kaiser, working for the government, achieved mass production, and with others like them furnished a mass employment unknown since before the Great Depression. The wages paid directly by war industries, and indirectly from government disbursements, supplied the mass purchasing power that permitted ownership of available civilian goods, with their attractions amplified by mass advertising. Gimbel's and Macy's, as well as other mercantile establishments, accommodated skillfully to the affluence of the war years, as in different ways did manufacturers of such nonessentials as soft drinks, chewing gum, and fancy fountain pens. The process encouraged Kaiser to dream of empire, Gimbel's and Parker to dream of profits, and Sadie the welder to dream of some immediate and more postwar comforts. In the context of those and similar expectations it was a lovely war after all, with the fighting remote and prosperity returned. In that context, for many Americans much of the time, it was difficult to remember that the demand for the guns had fostered the production and consumption of the butter. Free Enterprise, so claimed its champions, produced the guns, while government, for its constrictive part, rationed the butter. Government also, as some among those champions knew from their own beneficial experience, allocated its order for the materials of war in a manner that restored the vitality and assured the future strength of the largest and most powerful of the industrial institutions in the nation.

4 / War Lords and Vassals

1. Industry and Government

The accomplishments of American industry during the war restored much of the prestige that businessmen had lost after 1929. Their reputations gained just as much, perhaps, from the reliance of the federal government on their experience in management. Business executives and their legal and financial associates had the skills that Washington needed for the massive wartime programs of procurement and economic controls. Those who were recruited to high administrative posts knew the satisfactions of command and the flush of publicity that carried rewards commensurate with glittering battle stars. So at least it was by the time the war ended, but it was not so during the months before the United States became a belligerent.

Before Pearl Harbor, Roosevelt moved warily toward the mobilization of the economy. He held back partly because he sensed that greater speed would alarm those millions of Americans who were willing to support the enemies of the Axis but remained unwilling to join the fight. Furthermore, well into 1941 the recovery of the economy from the Depression had not yet proceeded so far as to strain the capacity of American industry. Conversion from civilian to military production remained erratic and inadequate. In addition, the booming civilian market made some industrialists, Detroit's automobilers especially, reluctant to divert their attention from conventional pursuits. Those in their turn were absorbing machines, manpower, and scarce raw materials already necessary for defense purposes. Still the President delayed. He was long loath to invite into his administration those "economic royalists" who had led the opposition to the New Deal, those men of business and finance of whom he had earlier said "they are unanimous in their hatred for me—and I welcome their hatred." Characteristically, too, he resisted centralizing authority over the economy in the hands of one subordinate or even one group of subordinates. As ever, he preferred to divide responsibility among several associates so that he could decide the main questions himself—so that his agents would, as he told them to when they had a problem, "bring it to Papa."

Circumstances overcame his hesitancies. When the Nazis swept across western Europe in the spring of 1940, the President, who was eager to aid the British, confronted the bankruptcy of the American armed forces. He confronted, too, the need for greater national unity at the very moment of a national political campaign. Beginning then, moving more rapidly after his re-election and the ensuing passage of the Lend-Lease Act, continuing thereafter as American involvement in the war in the Atlantic grew, he created one defense agency after another, usually with overlapping but, in the aggregate, growing authority. He pushed on during the year after Pearl Harbor until by May 1943 he had delegated sufficient authority and centralized sufficient responsibility to permit the effective mobilization of the economy. Though he remained the ultimate boss, though his subordinates squabbled over many small and some large issues, though there was never really a czar for domestic affairs, Roosevelt had built the essential apparatus, and it functioned without his guidance and ordinarily without his personal intrusion.

His selection of subordinates for both the War and Navy departments and the various civilian agencies revealed his understanding of the necessity for bipartisanship, at least on the surface, and his willingness during the emergency of war to accept counsel and assistance from businessmen and their spokesmen. Furthermore, the decisions he made when those wartime subordinates disagreed demonstrated his overriding concern for the nurture of the armed services according to prescriptions they wrote. That concern, given their predilections, made him often not just the expedient ally but the outright champion of big business, so recently the target of his prewar attacks on "price-rigging, unfair competition . . . and monopolistic practices . . . that flow from undue concentration of economic power." When the end of the war was at last in sight, the President indicated that he had not entirely abandoned that attack, but during the war he suspended it.

He believed he had no choice. The need he felt to destroy the enemy—a feeling spurred by American military setbacks in 1942, by the dangers to the United States and her partners from German military power and military technology, by the ghastly character of German policies in occupied Europe, and by the growing casualties that arose from prolonged battle—that feeling made total victory, to be achieved as rapidly as possible, almost an obsession with the President. To its achievement he surrendered principles to expedients, as in Italy to the distress of Robert Sherwood, and as on the home front to the equal distress of men who believed that he yielded more to big business than circumstances required. Though he never acquired a benign view of business, he developed a necessitarian view of the war.

That was the attitude also of Henry L. Stimson, the lifelong Republican and extraordinary American whom Roosevelt persuaded to accept appointment in 1940 as Secretary of War. As Stimson at once discovered, he had inherited a "situation . . . that is perfectly horrible." His predecessor, Harry H. Woodring, a weak and petulant administrator, had devised no useful plan and spent no effective energy to place the Army on a wartime footing. He had hoped and expected that the United States would remain a nonbelligerent. The airplane industry, for example, still starved by Congress, had received essential funds for expansion only because of orders placed by the French and British, who had had to conduct their business through the extraordinary agency of the Treasury Department. Stimson, understandably appalled, rapidly brought order, direction, and purpose to his domain. In the months before Pearl Harbor he recruited talented civilian assistants and with them worked out a solid basis of relations with the Army's top command, especially General George C. Marshall. A pillar of integrity, Marshall made it his first business to bring the Army, as a balanced force, into a condition of readiness for war. His senior subordinate for assuring supply for that mission, General Brehon Somervell, a dogged man, gave military considerations top priority in procurement. Marshall and Stimson supported that policy, whatever the resulting liabilities for small contractors less efficient than the industrial giants.

First in preparing for war, later in prosecuting it, the Army and the War Department worked with a resolution rare in Washington. More even than Roosevelt, Stimson was obsessed by a sense of urgency and a determination for victory. His sympathetic biographer put it exactly:

> In his desire to prepare the body and spirit of the country in a time of danger Stimson was obviously less sensitive than many of his colleagues to the possible side effects of the mobilization of national energy. He was not much concerned by the thought that some of the New Deal social gains would have to be set aside for a season. He was not much disturbed by the idea that the War Department might

become a great Bourse. He was not much afraid that . . . an over-whelming display of strength by the Army against a resistant industry would subvert the democratic process. . . . He did fear that the country might be left powerless. . . . He was single-minded in the task of preparing . . . the country to cope with the conditions of the world.

So in comparable measure were his senior subordinates, all of them from backgrounds like that of Stimson himself. He was a part-ner in Winthrop and Stimson, a leading New York law firm that catered to corporate clients. His close advisers included Harvey Bundy, of Choate, Hall and Stewart, a premier Boston law office; John J. McCloy, of the eminent Cravath law firm in New York; Robert A. Lovett, of Brown Brothers, Harriman, a major New York financial house; and Judge Robert P. Patterson, a New York lawyer before he reached the bench and again after he left office. A man of similar past, Under Secretary (later Secretary) of the Navy James Forrestal had emerged from the Wall Street investment house of Dil-lon, Reed. None of these men had ever been New Dealers. Like Stimson, as lawyers or financiers they had for many years instructed corporate clients in how to advance their private interests. The transi-tion from influential counselor to prescriptive customer was easy to make. They made it with ebullient confidence, a mark of their eastern patrician standing, of education in prestigious boys' schools and at Harvard or Yale or a similar elite institution, of companionship on boards of trustees of great museums and charities, of easy access to powerful agents in public life. That background generated among them a common sense of noblesse oblige, a selfless but not uncritical patriotism, and an unimpeachable personal morality.

Accustomed to command, they were at ease with the power they now exercised along familiar lines of personal and business acquaint-anceship, lines outside the New Deal's networks. For his part, Roosevelt became again accustomed to their faces, the likes of which he had known throughout his own patrician life. He was also as gladly grateful for their presence as they were to serve. Still, no more than other men were they infallible, and just as much as other men were they captives of their past. That past precluded them from conspiring to create some combination of military and industrial in-terests that would dominate national policy. It did not preclude their sponsorship of policies that successively mobilized big business, aggrandized it, and linked it to the military establishment.

From the outset, the senior civilians in the War Department were impatient with most other civilians whom Roosevelt appointed to run first the defense and later the war agencies. The Office of Production Management, established in January 1941, had co-directors, the labor leader Sidney Hillman and the former General Motors executive William Knudsen. Uncomfortable in their artificial partnership, they were also solicitous of their former constituents. Neither their agency nor its prewar successor, the Supply Priorities and Allocations Board, had the authority or the unity to move the country to the degree of industrial mobilization deemed vital not only by the military but also by such ardent advocates of preparedness as Harry Hopkins, Roosevelt's intimate and the head of lend-lease, Secretary of the Interior Harold Ickes, Secretary of the Treasury Henry Morgenthau, Jr., and Vice President Henry Wallace. New Dealers all, they supported Stimson's repeated requests for improved federal organization for war. Roosevelt stepped toward their objective by creating the War Production Board in January 1942. It was to "exercise general responsibility" over the economy, particularly all aspects of production. But the WPB and the War Department soon clashed.

Donald Nelson, the head of the new agency, had been recruited to Washington from Sears, Roebuck and Company, and had participated in defense planning since 1940. He had revealed both strengths and weaknesses as an executive. Jovial, warm-hearted, and gregarious, Nelson developed a growing sympathy for the New Dealers in government and a happy sensitivity to the Democrats in Congress. As they came to like him, he came to identify with a concern important to many of them, the preservation of small business in the United States. That goal, along with his easy manner, won him, too, the loyalty of most of his staff, a loyalty touched in the end by pity. For Nelson lacked force and decisiveness. He tolerated insubordination to the point, invariably too late, where his pride provoked him to get tough. He acquiesced in Roosevelt's appointment of a special office to oversee the synthetic rubber program, a program that should have fallen to him had he in time identified and begun to remedy the problems that created a rubber shortage. Most crippling, he let the Army and Navy control military procurement. As he knew, they had experienced purchasing officers already at work, and they were doubtless less vulnerable than his agency could be to charges of favoritism or collusion in letting contracts and defining profits. Their independence, however, left him without the authority to define and enforce over-all production priorities which affected not the military only but also lend-lease, other war-related and essential civilian production, and the health of businesses, large and small.

In order to speed procurement, the War and Navy departments

turned naturally to big business. It was easier to deal with one contractor instead of many. Further, big business had agents seeking contracts in Washington, and strong lines of private credit. Big business had much of the plant and most of the experience to handle enormous orders for military material, and big business had the stock of executives and engineers capable of undertaking the management of new, technologically difficult programs. The need to induce industry to co-operate during 1940 and 1941 made the military especially eager thereafter to sustain that co-operation. The inducements for established firms were precisely those from which newcomers like the Higgins and Kaiser enterprises also profited: subsidies or low-interest federal loans to enlarge plants and build new machinery, fast tax write-offs for expansion and retooling, generous contracts negotiated on the basis of costs plus a fixed fee, the assurance that facilities for war production financed by government funds would be available at bargain-basement prices for postwar use. Those and other devices guaranteed large profits without risk. That was Stimson's intention. "If you are going to . . . go to war . . . in a capitalist country," he wrote, "you have to let business make money out of the process or business won't work."

Besides profits, industry needed a reliable flow of materials. During much of 1942, contractors battled each other for shares of such scarce basics as steel, copper, and aluminum. Nelson failed to solve the problem until he turned it over to Ferdinand Eberstadt, a New York investment banker, who in the fall of that year contrived the Controlled Materials Plan. Under that arrangement, operative during 1943 and thereafter, the armed services, the Maritime Commission, the Lend-Lease Administration, and the Office of Civilian Supplies sent to the War Production Board periodic estimates of their needs. The WPB then allotted to each claimant a proportion of available stocks, and each in turn distributed its share to its prime contractors. The system restrained federal agencies from ordering more than industry could provide within the defined allotments, but left unsettled the ability of Nelson to hold allocations for the War and Navy departments to optimal levels, and the ability of the WPB to direct the contracts let by those departments to small as well as big business.

As with risk-free profits, so with essential materials, the War and Navy departments naturally favored the large, established firms that they had induced into war work and found so convenient, as well as so capable, in executing that task. Though convenience and capability determined the pattern, it was reinforced by the compatibility of Stimson and his associates with the premier industrialists who were their wartime partners. In 1940, when the defense program began, approximately 175,000 companies were providing some 70 per cent of the manufacturing output of the United States, and one hundred companies produced the remaining 30 per cent. By March 1943, even though twice as much was being produced, that ratio had been reversed. The one hundred companies previously holding only 30 per cent now held 70 per cent of war and civilian contracts, and were still gaining in proportion to the others. That development reflected the decisions of the War and Navy departments about contracts, materials, and priorities. The great bulk of federal funds expended on new industrial construction had gone to the privileged one hundred companies.

That development may have been necessary, as Stimson and his men indisputably believed, but Donald Nelson was less certain. He had begun to understand how broadly the War and Navy departments were exercising the authority that appeared to reside in the War Production Board. He was endeavoring therefore to establish a dominant role for the board. The Eberstadt plan seemed to leave the board with ultimate control over the distribution of raw materials, but actually it did not. Eberstadt had not intended that it should. A quiet, remote, extraordinarily gifted intelligence, Ferdinand Eberstadt had first gone to Washington in 1942 as chairman of the Army and Navy Munitions Board, an appendage of the armed services. When it was absorbed by WPB, he continued to believe that authority over priorities, which the Munitions Board had had, properly belonged to the military. His Controlled Materials Plan did give WPB the power to allocate stocks of raw materials to various claimants, but it left to those claimants decisions about how to use what they received. Eberstadt, as Nelson's public-relations officer perceived, wanted WPB to be "a materials-control agency, pure and simple." In contrast, Nelson intended his agency to exercise "the basic control over the economy of a nation at war."

To that end, early in 1943 Nelson persuaded Roosevelt to recruit for the WPB Charles E. Wilson, president of the General Electric Company, an expert on production who was renowned for his personal drive. Nelson charged Wilson with supervising the production of the end products to be fabricated from the materials WPB allocated. He also made Wilson his number-two man. The War Department attempted to get Roosevelt to fire Nelson, to put Bernard

Baruch in his place, and to make Eberstadt Baruch's operating deputy. But before Roosevelt could fire Nelson, Nelson fired Eberstadt. That victory had no substance. The supervision of production, assigned to a group of which Wilson was chairman, remained dominated by the War and Navy departments, to which Wilson looked for guidance. Nelson, unable to provide that guidance, was understandably also unable to control his subordinate. The bureaucratic infighting, an entertainment for official Washington, left intact the arrangements of power that predated the episode.

2. Pains for Small Business

Never during the war were the guardians of small business successful in their mission. Though they were alert, persistent, and even influential in Congress, they had little impact on the pattern of procurement that the military established.

Through the Senate Special Committee to Investigate the National Defense Program, Harry S Truman, of Missouri, the chairman, established his personal reputation as an ardent foe of waste and collusion. The committee's hearings, which Truman structured by antecedent staff work, exposed the involvement of American corporations in international cartels that had operated to the detriment of the national interest and domestic business competition. Yet those disclosures cost the guilty corporations no war contracts. The hearings also explored the connections to industry of the hundreds of business executives who had accepted part-time or full-time posts in war agencies either without compensation or for a nominal dollar a year. Those men, the committee's counsel concluded, had had to create the organizations for national mobilization. They "very naturally believed that the largest and best organized business concerns . . . would be best suited to take care of the defense program." Indeed, "large companies simply offered their managerial skill, their willingness to plan new facilities, retool . . . get together . . . key personnel and train . . . new labor, so that in the last analysis there would be an entirely new plant and industry. This, of course, was done at government expense."

As one result, small business was confined largely to civilian work, soon subject to severe shortages of materials. The hearings alerted executives participating in federal programs, efficient and patriotic men, to the dangers of inadvertent collusion with their business friends, but again without effect on the flow of orders to big business. Primarily, the committee's work identified Truman as an available candidate for the Democratic nomination for Vice President in 1944.

Others in Congress adopted a different strategy in behalf of small business. They were an odd group, held together by a common dedication to open competition and a common suspicion of the great eastern banks, investment houses, and corporations—a collectivity that antagonistic Americans had labeled long ago as "Wall Street." Stimson and his associates personified that Wall Street. It seemed to its wartime critics to be engaged, as ever, in infiltrating government and directing policy against the interests of the folk and their hometown companies in the hinterland. Foremost of those critics in the House of Representatives was Wright Patman, a scrappy, dogmatic Democrat on the important Banking and Currency Committee, a Texan suffused in the suppositions of the populist past, and a self-appointed scourge of eastern banking. He picked up the program initiated by the Senate Committee on Small Business, on which two major voices were those of James E. Murray, a Montana Democrat, and Robert A. Taft, Ohio's "Mr. Republican." Murray had earned impressive wealth as the successful manager of small enterprises. As one of his admirers observed, his was "a world of light and darkness with the powers of evil symbolized by big business," for which he blamed the crash of 1929 and the Great Depression. A persistent champion of New Deal social legislation, Murray had as much enthusiasm for small business as did Taft. The most talented conservative in public life, a proud, learned, cerebral man, Taft detested Franklin Roosevelt, distrusted Wall Street, and believed with evangelical fervor in the free market and its preservation. Anxious about big government and the big military as well as big business, Taft suspected the War Department and the war agencies of conspiring, under Roosevelt's inspiration and in the guise of wartime necessity, to destroy capitalism in America. He attacked federal spending no more ardently than he promoted the bill Murray sponsored to establish a Smaller War Plants Corporation. The measure had one sentimental advantage. In Congress, as in most of the country, small business was as American an emblem as grits or apple pie.

The Senate committee began hearings on small business only

eight days after Pearl Harbor. The defense program to that date, as the hearings revealed, had awarded 75 per cent of all contracts to only fifty-six large companies. The committee, complaining that procurement officers had followed the easiest path, concluded that those officers had "failed to recognize that small business—now in grave danger—represents a phase of American life worthy of preservation." In its first version, the bill to accomplish that preservation established a division for small business within the War Production Board as well as a special lending agency to assist small concerns. After Nelson objected to an independent office within his board, a substitute bill left him with authority over the new office, an authority he used with his characteristically good intentions and administrative clumsiness. More significant opposition to the bill came from Secretary of Commerce Jesse H. Jones, who was also Federal Loan Administrator. A rich Texan with an outsized ego and an incomparable jealousy of his presumed prerogatives, Jones fought to prevent the creation of any lending agency outside his domain. The rigged statistics he presented to support his case damaged it as much as did Taft's sharp and telling inquiries.

The Senate Banking and Currency Committee reported out a measure that preserved the independent lending authority the Small Business Committee had recommended. By a vote of 82 to 0 the Senate approved the bill. Patman, co-operating closely, had meanwhile begun to direct a similar measure through the House, which, on May 26, 1942, after incorporating a few amendments, also voted unanimously, 244 to 0. The Murray-Patman Act required Nelson to appoint a "special deputy . . . to assume the responsibility for the welfare of . . . small concerns" and set up a Smaller War Plants Corporation with a revolving capital fund of $150 million to finance the conversion of small plants to war or essential civilian work. The new corporation could also take prime contracts from the procurement divisions of the armed services and subcontract them to small enterprises. "Our committee," Senator Murray had said during debate, "visualizes the Smaller War Plants Corporation as an aggressive agency with capital and facilities . . . and staff . . . to provide both finances and contracts . . . with the smaller manufacturing plants."

The vision faltered in execution, partly because of administrative difficulties within the Smaller War Plants Corporation, largely because wartime priorities conflicted with the needs of crippled manufacturing firms. In July 1942 Nelson appointed Lou E. Holland deputy chairman of WPB for smaller war plants and a member of the board of directors of the SWPC, the lending office for the program. The head of a corporation that manufactured sprinklers and the former president of the Kansas City Chamber of Commerce, Holland transferred to his public office the attitudes of small enterprisers, including a disdain for the federal bureaucracy. He was determined to make a new start for his special constituency. On that account he rejected Nelson's offer to turn over to the SWPC the Contract Distribution Division of the WPB. That competent office had over one hundred branches throughout the country ready to help small business to solicit subcontracts. Unable to build a substitute organization, Holland concentrated instead on negotiations in Washington with senior procurement authorities, especially in the War and Navy departments. His voluble and defensive manner consumed time grossly disproportionate to any visible results. Constantly homesick, Holland returned to Kansas City at the end of 1942 with his prejudices confirmed by the experience he had shaped.

Nelson had meanwhile helped to get the Smaller War Plants Division started in contracting and subcontracting. In October 1942 he reported some progress to Roosevelt, especially in procurement of such items as incendiary bombs, fuel pumps, cathode-ray tubes, and ammunition boxes. Yet, as he admitted, the judgment of the prime contractor, who had to be sure about the promptness and quality of deliveries, continued to be the determining factor in subcontracting. SWPC could only continue to try to help small plants to convert their operations to a state attractive to prime contractors. While "zealous to prevent discrimination . . . against small plants," Nelson would not endorse loans for their conversion or allocations of material for their use except on the basis of wartime needs. Materials especially, he warned, "shall not be doled out as relief currency to any plant to enable it to continue a non-essential activity."

The application of that criterion left some small plants without sufficient earned income to cover the fixed costs on their facilities and tools, costs particularly for interest on mortgage debts and for obsolescence. Those distressed firms caught the special attention of Holland's successor at SWPC, General Robert Wood Johnson, chairman of the board of Johnson and Johnson, the large manufacturer of surgical materials. Johnson, a big-business executive with experience in procurement for the Army, had a genuine interest in small business, a good instinct for organization, and a bemused intolerance for Washington's bureaucratic labyrinths. Civil-service procedures, he complained, were interminable. In spite of them, he reorganized his shop and focused its purpose, as he put it, "to save small industry in distress." That focus ran counter to Nelson's policy of supporting only essential work. It also muddied the image that small business men had of themselves. They disliked what they took to be Johnson's patronizing air. They believed they had the means to play a legitimate and patriotic role in which they could take pride. They wanted prime contracts as well as subcontracts, "a full meal," not just "crumbs from the table." Their friends in Congress agreed. Wearily giving in to their criticism and to bureaucratic impedance, Johnson resigned in December 1943. As he informed Nelson, he had managed to obtain more prime contracts for small business and he had enlisted some co-operation from the Army. He had also been urbane where Holland was parochial. Still, neither man was sufficiently astute or aggressive to direct the policies of SWPC toward the stimulation of small enterprise in the postwar period when necessitarian conditions would no longer obtain.

That objective, important to Nelson, moved him in January 1944 to the appointment as Johnson's successor of Maury Maverick, the stormy Texas Democrat who had impeccable credentials as a New Dealer and a deserved reputation for operating in the manner his surname described. Maverick, who had at one time been in the lumber and hardware business, had served two terms in the House of Representatives and also as mayor of San Antonio. In those roles he had attacked monopoly, machine politics, and social decay. He had also been a top administrator within the WPB since 1942 and an advocate of organizing scientific research in the United States so as to use federal funds to make the fruits of invention available to small business. While co-operating with his friends in Congress, Murray and Patman especially, Maverick restructured SWPC and recruited a new senior staff. He defined his agency's purpose as dual: to assure the efficient use of the full capacity of small plants for war production, and to prevent the wartime economy from "needlessly or accidentally" injuring small producers "and the American free enterprise system."

The latter goal involved enhancing the competitive position of small business for the postwar period. Maverick therefore sought not only prime and subcontracts, but also technical and managerial advice for small enterprises, more extensive business loans and insurance, priority for small business to reconvert for civilian production, and priority for small business in the distribution of surplus tools and war plants. He made imaginative proposals for multiple tenancy in large plants, for financial help to small business for the acquisition of surplus tools, and for the protection of small subcontractors from the shock of abrupt terminations of procurement contracts. He also stirred Nelson to new zeal in the battle over reconversion.

It became a grim fight. Nelson recognized the "need for on-schedule production," the special demand for tanks, the shortages of manpower in some significant sectors of industry, the galaxy of difficulties that remained unresolved at the time of the cross-channel invasion of Europe. Yet he also insisted on the necessity of making immediate adjustments "to permit the resumption of civilian production wherever . . . no interference with the war effort will result." That resumption would ease the transition to a peacetime economy, help to prevent unemployment, and, as Maverick's proposals indicated, benefit small firms that had been unable to convert fully to war production. The last prospect particularly antagonized big business. The giant firms had already begun to advertise the postwar products they intended to sell in the predictable boom market for civilian goods. Nelson's plans might give small firms a head start in reaching that market. In the judgment of Bruce Catton, then on Nelson's staff, the response from "the war lords of Washington" was rather like that of an elephant frightened by a mouse.

No less real on that account, the anxieties of big business coincided with the fears of the military that Nelson's turn toward reconversion would dampen wartime morale. The War Department influenced the Office of War Information to stress the danger of shortages and the need for hard work. More important, the Joint Chiefs of Staff contended privately that war production would have to increase or force a revision of strategic planning. That conclusion rested on slim evidence, which Nelson attacked in July 1944 in letters to Roosevelt and to the Joint Chiefs. He lost his case. In some measure the President shared the Army's worry about morale. More significant, he left the decision about the timing of reconversion to James F. Byrnes, chief of the Office of War Mobilization. Byrnes, who had never trusted Nelson, had confidence both in big business and in the War Department. The issue of reconversion, moreover, had split Nelson and his second-in-command, Charles Wilson. Long attuned to accepting the advice of the military, Wilson adopted their view of reconversion. "I am opposed," he wrote Roosevelt, "to any interference with war production as its necessities are outlined . . . by . . . the Joint Chiefs of Staff." He therefore, as he added, considered subversive the plans and the conduct of Nelson and Maverick and their friends.

His plans jettisoned, his reputation shredded, his influence spent, Nelson in August 1944 accepted from Roosevelt a mission to China, an exile the President had designed for removing from Washington those whom he intended soon to drop. Wilson resigned, eager to return to General Electric and unwilling to take over at WPB where so many of Nelson's men remained. On his return from China, Nelson also resigned. Roosevelt then appointed in his place Julius A. Krug, a veteran bureaucrat without the spunk to tilt with the War and Navy departments.

Spokesmen for small business had protested against Nelson's exile. One typical letter complained that Roosevelt had sold out to the War Department. In principle that contention was perhaps correct, but in practice the defeat of Nelson and Maverick occurred too late to make much difference. Big business, by virtue of its contracts for conversion from civilian to war production and for expansion thereafter—contracts spread out from 1940 through 1944—already had a first claim to the facilities the government had financed. At the time of reconversion, small business received more surplus property than it might have without Maverick's intercession, but far less support than he or Nelson had thought appropriate.

In essence the defeat of Nelson had more significance; conversion to and reconversion from war production, along with related activities dominated by the War and Navy departments, did threaten free private enterprise in the United States as Maverick, Murray, Patman, and Taft understood that concept. Congress had tried to legislate a vision of society, a vision in which small enterprises would compete fairly against large in a market uncontrolled by government. It had been many decades since that vision accorded with circumstances in the United States, but Congress had voted to preserve a dream. It was a futile and a sentimental but not an unpopular gesture. Even in a time of war, a people and its representatives had a right to dream.

For their contrasting part, the Joint Chiefs—brave, intelligent, dedicated men—suffered from a "professional deformation" that discouraged those doubts about national policy that deserved more attention than they received from the President and his advisers. The necessitarian direction of procurement gave up the whole of the vision which many Americans would have liked to preserve. The postwar hopes of small business shrank before the wartime gains of industrial giants. "Free private enterprise" was transformed into a cliché that big business glibly used to sanction its size and wealth, and to arrogate to corporate management credit for the enormous growth of plant and productivity that government had underwritten. In fact, "free private enterprise" was translated while the war went on to mean immunity for business from federal efforts to free the market from the controls that business had conspired to impose upon it.

3. Holiday for the Trusts

Big business had had an unusual history in the United States. The Sherman Antitrust Act of 1890 reflected, as one of its interpreters later observed, an economic ideal that "industrial progress can best be obtained in a free market, where prices are fixed by competition and where success depends on efficiency rather than market control." The act itself declared illegal "every contract, combination . . . or conspiracy, in restraint of trade or commerce among the several States, or with foreign nations." The government was to prosecute violators in the federal courts. That law, the same analyst wrote, was "historically unique. . . . No other nation had any legislation like it." Between 1890 and 1941, although enforcement of the Sherman Act was irregular and big business opposed it, the ideal behind the act persisted. "The antitrust laws," as their interpreter wrote elsewhere, "remained as a most important symbol."

Those were the words of Thurman Arnold, whose most important and celebrated book, *The Folklore of Capitalism* (1937), seemed to contrast with his energetic efforts to enforce the Sherman Act immediately before and during the war. The book suggested that the antitrust laws served as a ritual that symbolized the ideal of a free market while industrial combination advanced apace. Crusades for the enforcement of the act, Arnold wrote, "were entirely futile but enormously picturesque, and . . . paid big dividends in terms of personal prestige." Furthermore, the crusaders "were sincere in thinking that the moral answer was also the practical answer. Thus, by virtue of the very crusade against them, the great corporations grew bigger and bigger, and more and more respectable."

Much in history supported that conclusion. Early in the twentieth century, Theodore Roosevelt, despite his resolve to vitalize the Sherman Act, had a larger concern with creating federal agencies to oversee the practices of big business. The consolidation movement in American industry, as he knew, had arisen largely to overcome the economic debilitation of unfettered competition. That movement, unarrested by regulation, gained momentum during World War I and especially during the 1920's, when large American firms undertook to join cartels with their European and British competitors in order to divide markets and fix prices. The National Industrial Recovery Act of 1933, the first of the New Deal's designs for overcoming the Depression, suspended the Sherman Act to permit industry-wide agreements on production, prices, and wages. Influenced by many of his advisers, Franklin Roosevelt in 1933, like Theodore before him, believed that the efficiencies of bigness, the economics of scale, had to be preserved, but that the federal government should monitor the behavior of industrial and commercial giants. After the Supreme Court struck down the NRA, after business proved hostile to New Deal regulatory measures, after economic recovery faltered, Roosevelt, influenced now by another group of advisers, in 1938 became for a season a champion of vigorous enforcement of the antitrust laws.

Of a similar mind, a majority of the Congress supported the creation of the Temporary National Economic Committee to undertake a searching investigation of monopoly in the United States. Roosevelt then appointed Thurman Arnold Assistant Attorney General in charge of the Antitrust Division of the Justice Department. TNEC hearings exposed the extent of the monopoly problem, and Arnold and his assistants presented, as he recalled, "a chamber of horrors of current business practices." They also initiated prosecutions under the Sherman Act and the statutes supplementing it. Congress underwrote that effort with appropriations for the Antitrust Division that rose from $413,000 in 1938 to $2,325,000 in 1942. In that time, moreover, the Truman Committee, digging in where the TNEC had left off, focused on monopoly and cartels in defense industries.

One case especially caught public attention. In investigating the Standard Oil Company of New Jersey, Arnold's division discovered that that corporation had agreed not to develop processes for the manufacture of artificial rubber, in exchange for a promise from I. G. Farbenindustrie, the large German petrochemical firm, not to compete within the United States in petroleum products. That agreement seriously delayed the development of artificial rubber in the United States, a delay that hurt the war effort for many months. Jersey Standard, faced with the damning evidence, pleaded *nolo* and paid a modest fine. But in the absence of a trial the American public generally ignored the episode, though Arnold believed that Jersey Standard would renew the cartel agreement once the war ended. He therefore responded eagerly to Truman's request to present the case on synthetic rubber before the Senator's committee. Those public hearings provoked just the outraged response that Arnold and Truman had intended.

Arnold had pursued that course partly to dramatize his conviction that the Antitrust Division had an important wartime role. "The existence of war," he believed, did not "require management by private monopoly. . . . The vast government spending for war production . . . created a great opportunity for conspiratorial agreements between businessmen with respect to prices, bidding, consolidations, and mergers." Determined to prevent unnecessary restraints on trade and thereby to preserve as much as possible of a free market, equally determined to divorce American corporations from the cartels they had joined, Arnold was fighting to prolong the prewar mandate that the President had given his division. In that fight he had the support of his immediate superior, Attorney General Francis Biddle.

Early in 1942 Arnold and Biddle had compiled commanding evidence of questionable corporate behavior. For two examples, the Du Pont Company and the American Lead Company had pooled patents in their industries to restrict production and prevent other American firms from doing business in the United States. They had also entered agreements with foreign competitors to divide international markets. One co-conspirator with Du Pont was Imperial Chemical Industries of Great Britain. Its head, Lord McGowan, had baldly asserted that it was necessary for the security of business to give his cartel the power to eliminate competition and stabilize production and prices throughout the world. In other cases the major proponent of that kind of stabilization had been I. G. Farben, whose business practices had the effect, whether inadvertently or not, of enhancing Nazi military strength while penalizing American. Pan American Airways was already attempting, both surreptitiously within government circles and openly through company propaganda, to establish itself as the sole instrument for postwar American transoceanic aviation.

Those developments, past or pending, involved corporations on which the War and Navy departments were relying for war production and services. To the senior civilian officials of those departments, men comfortable in the corporate world, Thurman Arnold appeared a scourge. "He had frightened business," Henry Stimson wrote, ". . . making a very great deterrent effect upon our munitions production." Arnold, he said elsewhere, was a "self-seeking fanatic" whose campaign was "playing havoc with the National Defense." Stimson believed, as did Donald Nelson, and as the National Defense Advisory Committee had put it as early as July 1940, "that industrial combinations deemed necessary for the effectiveness of the defense program" should be immune from prosecution under the antitrust laws.

The resulting attack on Arnold took several forms. How could he, one of his critics asked, advocate a course of action that his *Folklore* had earlier described as an ineffectual ritual? "The reason I claimed the antitrust laws were futile," Arnold answered, "was that there was never sufficient effort and money to enforce them." Now he had the money and was eager to make the effort. Arnold's popularity on the Hill persuaded Stimson of the futility of trying to have Congress suspend the antitrust laws. Instead, the service departments influenced the War Production Board to grant immunity under those laws in some six hundred cases of collusion supposedly necessary for efficient military procurement.

A third attack outflanked Arnold. Taking their troubles to the White House, Stimson and Robert Patterson, for the War Department, and James Forrestal, for the Navy Department, argued that the government could not enlist companies like Du Pont, indispensable in the war effort, if Arnold simultaneously investigated or prosecuted them for violations of the antitrust laws. The senior executives of such companies, so the argument went, could not give proper attention to organizing war work if they had to divert their energies to preparing defenses against Arnold's suits. Roosevelt referred the issue to Samuel Rosenman, his personal counsel, who agreed with Stimson about the need to postpone antitrust activities for the duration of the war, an agreement that accorded with the President's own bias. Whatever his enthusiasm for antitrust in 1938, Roosevelt in 1942 had enthusiasm only for victory.

Arnold and Biddle deferred, as they believed they had to, to that attitude. On March 20, 1942, they joined Secretaries Stimson and Knox in signing a letter to the President, made public eight days later, that in effect suspended antitrust activities. The text, which Roosevelt approved, gave all the spades and trumps to the service departments:

> In the present all-out effort to produce quickly and uninterruptedly a maximum amount of weapons of warfare . . . court investigations, suits, and prosecutions unavoidably consume the time of executives and employees . . . engaged in war work. . . . We believe that continuing such prosecutions at this time will be contrary

to the national interest and security. It is therefore something which we seek to obviate. . . .

On the other hand we all wish to make sure: 1. That no one who has committed a violation of law shall escape ultimate . . . prosecution; 2. That no such person shall even now be permitted to postpone investigation or prosecution under a false pretext . . . ; 3. That no one who has sought actually to defraud the government shall obtain any postponement. . . .

Each pending and future Federal court investigation, prosecution or suit under the anti-trust laws will be carefully studied . . . by the Attorney General, and the Secretary of War or the Secretary of the Navy. . . . If [they] . . . come to the conclusion that the . . . prosecution . . . will not seriously interfere with the all-out prosecution of war, the Attorney General will proceed. If they agree that it will interfere; or if after study . . . they disagree, then . . . the Attorney General will . . . defer his activity . . . providing . . . that he shall have the right . . . to lay all the facts before the President whose determination . . . shall be final. . . .

The deferment . . . of the investigation, suit, or prosecution shall not mean . . . the exoneration of the individual or corporation. . . . As soon as it appears that it will no longer interfere with war production, the Attorney General will proceed. . . .

Predictably, almost all outstanding investigations and prosecutions were deferred. Arnold's resulting irritation made him delighted to give the facts on Jersey Standard to the Truman Committee, but the successful development of the American synthetic-rubber program took the sting out of the initial public response. Now that his office had become essentially a sinecure, Arnold took a bitter view of his defeat. It had, as he later observed, facilitated the War Department's grant of war contracts to "a handful of giant empires, many of them formerly linked by strong ties with the corporations of the Reich." Because the government tolerated restrictive practices, those few giants would have extraordinary advantages for controlling postwar markets. "The representatives of big business," Arnold contended, "filtered into the War and Navy Departments and the War Production Board. And F.D.R., recognizing that he could have only one war at a time, was content to declare a truce in the fight against monopoly. He was to have his foreign war; monopoly was to give him patriotic support—on its own terms."

Roosevelt neutralized Arnold early in 1943 by appointing him to the Court of Appeals for the District of Columbia, a position from which he did not resign until after the war had ended. In spite of his energy and the generosity of appropriations for his division, Arnold's battle of 1942 for wartime antitrust had in the end confirmed the insights of *The Folklore*. While the giants grew, the crusader, "enormously picturesque," made a reputation of a kind by engaging, like so many of his predecessors, in a diverting ritual.

Yet the irony of Arnold's failure was tempered by the victories of his successor. Even *Fortune* had kind words for both the free market and Wendell Berge, the new head of the Antitrust Division. Businessmen might condemn a competitor who cut prices, *Fortune* allowed, but "their spine tingles" when English industrialists and Soviet commissars are told that "competitive enterprise is the American way and don't forget it." Most Americans also applauded "the competitive system" even though a free market no longer determined either farm prices or union wages. The old ideal, alive and well, was nurtured by Berge, whose policies avoided, as *Fortune* saw it, Thurman Arnold's "windmill technique. In a world where an excess of competition can be as socially disturbing as a complete lack of it, Mr. Berge knows antitrust enforcement has to be selective."

Berge had enunciated his criteria for action: "We try to concentrate on monopolies that shelter inefficient or obsolescent industry or suppress new products or technologies needed to supply jobs and expand the economy. We believe profits are justified only as a reward for risk taking; we have no use for a corporation that entrenches itself behind a Siegfried line of illegal patent or monopoly weapons and still expects to draw down rewards appropriate to venture capital." Like Maury Maverick, Berge supported a bill proposed by Senator Harvey Kilgore to set up a government-sponsored research agency whose discoveries would be free to all users. And like Arnold, Berge had drawn a bead on international cartels, on which he concentrated, with Biddle's encouragement, during his time in office. The Webb-Pomerene Act of 1918, Berge held, did not authorize American firms to participate in cartels. It merely permitted them to form associations to pool their sales abroad. He disagreed with Lord McGowan, who had urged repeal of the Sherman Act; he agreed instead with the London *Economist*, which considered cartels "a buttress . . . for the high-profits-and-low-turnover mentality."

Though *Fortune* approved, the War and Navy departments had yet to relent. In the face of their protests, between April and August 1944, Berge and Biddle brought the antitrust issue to a new resolution, not the least because they employed a new tactic. Arnold had always stressed the ideal of free enterprise. In dealing with the White House, Berge and Biddle now emphasized the insidiousness of cartels, particularly the liabilities for American industry and national defense that had been arranged by cartels controlled by leading German firms. In the postwar period, Berge contended, the United States could not afford to invite a similar risk. A prudent government would act at once to break up cartels, those involving the British as well as the Germans, before they could get a grip on patents Americans would need for both military and industrial development.

The revived antitrust campaign selected as its first target E. I. Du Pont de Nemours and Company, and Imperial Chemical Industries, Ltd., along with their senior corporate directors and officers. The complaint charged those companies with conspiring to create a worldwide division of chemical markets, with continuing that illegal practice during the war, and with planning secretly to return after the war the markets previously allocated to their German fellow conspirators. Secretary of War Stimson, citing the letter of March 20, 1942, immediately asked for a postponement of hearings. The Du Pont Company, he wrote Biddle on April 25, 1944, had prime contracts with the Army totaling over $1.4 billion and was operating for the government nine ordnance plants constructed at a cost of $580 million. ICI was the largest supplier to the British government of various essentials, including munitions, small arms, aviation fuel, and paints and plastics. "The most important officers in each of the companies," Stimson went on, "have been named as defendants." The scope of the inquiry was vast. It would tie up the executives and inescapably interfere with the war effort; counsel for the defendants estimated it would take at least four months just to file answers, let alone to participate in further proceedings. Secretary Forrestal wrote a similar letter for the Navy Department. In it he quoted Lord Halifax, British Ambassador to the United States, whose views were presented, too, to Secretary of State Hull. On instructions from the Foreign Office, Halifax also requested a stay "on the grounds that the burden of preparing their defense would seriously impede the operations of Imperial Chemical Industries on vital war production and research."

Against that formidable counterattack Biddle held his ground, with the foreseeable result that Forrestal and Undersecretary of War Patterson referred the controversy to James F. Byrnes, in 1944 Roosevelt's deputy on matters of this kind. The Justice Department, Patterson wrote Byrnes, had instituted the civil suit against the defendants by a bill of complaint of eighty-seven pages, alleging a number of violations of the antitrust laws covering a period since 1897. Even the Justice Department, Patterson believed, recognized that the case could not be tried on its merits without interfering with the war work of Du Pont and ICI. The War Department argued that those companies could not even prepare an answer to the bill of complaints except with unacceptable diversions of crucial executive energies. Ignoring Byrnes, Biddle wrote directly to Roosevelt. The Attorney General, opposing a postponement, did not see "how it is possible to determine whether a trial of this case will seriously interfere with the war effort until the nature of the defense is known." The preparations of pleadings and motions, Biddle continued, "consumes the time of private counsel. It does not consume the time of executives . . . who might be engaged in war production." The defendants had had their time for filing answers extended on three occasions. Now Biddle wanted to press ahead.

Aware of the sensitivity of the issue, Byrnes suggested to Roosevelt that Rosenman settle the controversy, since Rosenman had helped to construct the letter of 1942. Rosenman, like Byrnes eager to avoid unnecessary risks, told Roosevelt that he would, of course, intercede if the President so wished, but the executive order creating the Office of War Mobilization charged Byrnes with the responsibility. Roosevelt agreed, but Byrnes demurred. With two reluctant kittens under his throne, Roosevelt settled the matter himself by telling Stimson at a Cabinet meeting that he had decided in favor of Biddle.

That decision set an important precedent. In June the Justice Department moved to reopen its case against the Bendix Aviation Corporation for participating in worldwide cartels in aircraft instruments. On request of the War and Navy departments the case had been postponed in February 1943, but in February 1944, Biddle charged, Bendix had attempted to continue the cartel. The corporation had then written a letter trying to transmit information about aviation de-icer equipment to a Spanish subsidiary of the German Robert Bosch firm. Incensed by that and other developments, Biddle, as he wrote Roosevelt, wanted to proceed at once to free the "industry now of artificial, uneconomic and unlawful limitations in order to insure efficient preparations for the postwar development of the aircraft industry in this country." As ever, the War and Navy departments requested postponement, but Biddle had made the long-run necessitarian argument—postwar development of aircraft—his own. He won.

The Attorney General advanced a similar case for reopening proceedings against an international cartel that included Rohn and Haas Company of Germany, Du Pont, and I. G. Farbenindustrie. They had allegedly restricted production, fixed prices, and allocated markets for acrylic products used in the preparation of artificial rubber, pharmaceuticals, dyestuffs, and photographic articles. One affected product was plexiglass, essential in bomber noses and gun turrets on combat aircraft, which Du Pont and the Rohn and Haas Company of the United States supplied to the armed services. The War and Navy departments therefore again asked for postponement, but again they lost. That was the outcome, too, in the reopening of a case against the National Lead Company and Du Pont, whom Berge charged with collusion with ICI, I. G. Farben, and others to control titanium pigments. In spite of characteristic objections from Patterson and Forrestal, Roosevelt wrote Biddle on August 14, 1944: "You are authorized to proceed. . . . Further delay should not occur in the interest of justice and the enforcement of the law."

The change in the President's position between March 1942 and August 1944 reflected the change in the national situation. Early in 1942 American, British, and Soviet forces were falling back as Germany and Japan advanced. By the summer of 1944 the Russians were marching through eastern Europe, the Anglo-American invasion was bursting out into the interior of France, and American forces in the Pacific were making the Mariannas and the Philippines steppingstones to Japan. Now that victory was in sight, though still elusive, Roosevelt confronted new necessities. In the economic realm they were precisely those that Berge and Biddle pointed out. Furthermore, the national political campaign was under way and the President had reasons for stirring the sentiments of New Dealers worried about his choice of a running mate, who were heartened by his support of Berge's program. The change in his position proved significant, for the successful prosecution of the cases that Biddle reopened in 1944, though not completed for some years, did strike at the cartels. Had Stimson, Patterson, and Forrestal prevailed, the postponements of 1942 and 1943 might have become indefinite; for when Harry Truman became President in April 1945, he quickly fired Biddle, and succeeding attorneys general systematically diminished the Antitrust Division. Biddle and Berge had reached Roosevelt in time to win some important cases.

In some measure they also preserved the ideal of antitrust, though their actual victories left the symbol by and large a substitute for reality. Reinterpretation of the Webb-Pomerene Act, although it was damaging to cartels, little affected big business within the United States.

During the war there was no free market, only a controlled market. With exceptions, to be sure, the Office of Price Administration controlled prices; the use of cost-plus contracts controlled profits, though not expenses; and the War Labor Board controlled wages, with some slippage. Thurman Arnold, concerned though he was with the preservation of competition, admitted that the most effective form of wartime antitrust arose from the watchfulness of John Lord O'Brian, general counsel of WPB, who "never wavered in his purpose to prevent unnecessary combinations and restraints of trade." But the war, as the armed services saw it, made necessary the very combinations of economic power that Arnold and Maverick condemned. As market libertarians, they had no effective recourse. "The little guys," as one of Berge's associates later recalled, "needed subsidies to function at all; the big guys could deliver, and got bigger. When reconversion came, the big guys simply outbid the little guys and kept the market power they had obtained." They had, of course, obtained most of it long before the war began. Moreover, as the War and Navy departments so often asserted, the war was an unpropitious time to attempt a redistribution of economic power. The most that Arnold, Berge, and Biddle should have expected they in fact achieved. Along with corporate bigness, antitrust, that most important symbol challenging the corporate appetite for power, also remained.

4. Of Size and Abundance

Size, like abundance, characterized economic life in the United States during World War II. Not just big business, but big labor and big agriculture—strong before the war—were stronger still when it ended. Despite management's hostility, unions grew dramatically between 1941 and 1945, a period during which total membership rose from 10.5 million to 14.75 million. That growth depended in part on the decision of the War Labor Board in 1942 to adopt a "maintenance of membership" policy. It gave every worker hired in a shop with a union contract fifteen days in which to resign from the union. Thereafter he had to pay union dues for the duration of the contract between the union and the firm. Labor leaders had wanted a closed shop, which would have forced all workers into unions; management had advocated the open shop, which would have permitted all workers the continuing choice of remaining outside the union. The compromise, which reduced the likelihood of strikes, also much assisted union recruitment. For their part, the unions used their bargaining power to extract higher wages within the limits the government set, and to negotiate for more compensation in the form of unprecedented fringe benefits, particularly pensions that would require payment only after the war's end.

The War Labor Board's formula to limit wage increases so as to control inflation covered hourly rates rather than over-all take-home pay. Partly because overtime wages were common, average weekly earnings for labor rose 70 per cent during the war. Profits rose less steeply. Furthermore, wartime federal tax schedules permitted an advance in disposable income for labor larger than for management and ownership groups. In the whole period after 1919, a marked redistribution of national income occurred only during the years of World War II. The labor movement remained divided, but although the American Federation of Labor and the Congress of Industrial Organizations were unrelenting in their rivalry, the national affiliates of both federations emerged from the war with more members, more money, more bargaining power, and more prestige than they had ever had before.

Agribusiness—the business of large-scale commercial farming, often undertaken by corporations rather than private individuals— did almost as well. Its influential lobby, the Farm Bureau Federation, and the pertinacious farm bloc in Congress, exacted from the government a considerable ransom in return for votes needed to establish controls over nonagricultural prices and wages. Farmers had prospered memorably during the period 1910–14, so much so that the ratio between agricultural and industrial prices for those years defined "parity," the goal for which agricultural interests had striven since 1921. Early In 1942 the farm bloc In Congress managed to set ceilings on farm prices at 110 per cent of parity, 110 per cent of heaven. Roosevelt fought to keep agriculture in line, but commodity prices, swollen by wartime demands, stayed close to or above parity for the duration, and only federal subsidies and rationing programs prevented consumers from having to buy food at extortionate costs. The large commercial farmers profited from the war just as big business and union labor did. And when the fighting stopped, the Farm Bureau Federation was still one of the most powerful lobbies in Washington.

The hardening pattern of bigness in American life came during the war to affect even colleges and universities. By the spring of 1942 enlistments and conscription were reducing enrollments, especially in graduate and law schools. In the colleges, undergraduates began to prepare themselves for the more interesting and remunerative military skills. There occurred a mushrooming of courses in mathematics, astronomy, navigation, cartography, electronics, meteorology, even in Japanese and Russian. Only the larger and wealthier institutions had the staff to offer all those subjects, which grew to the detriment of conventional courses in the liberal arts. The resulting distortions of liberal education worried faculty members less than did their feelings of irrelevance in the national emergency. By October 1942, according to one federal survey, teachers remaining in the colleges were "suffering a terrific loss of morale and feel that they are giving little or nothing to the war effort. They are jealous of their colleagues now in Washington . . . and also seriously considering enlistment in the armed forces . . . ; the government has been vague and contradictory over what it expects from the schools. Some college people believe that a good many schools will be closed completely. All of these fears add up to a general sense of frustration and uselessness."

That sense drew confirmation from financial problems that had plagued higher education during the Depression and now intensified with war. Decreased enrollments obviously caused decreased income from tuition. Fund raising from private sources could not compete against war-related appeals. Income from endowment rose more slowly than costs for food, maintenance, and equipment, especially equipment for the sciences, which were much more expensive to teach than were the humanities. Smaller institutions suffered most severely from those conditions because they did not have the programs in science and engineering for which students could receive temporary draft deferments, and they ordinarily lacked the residential plant to accommodate Army and Navy training programs. In 1942 the Massachusetts Institute of Technology expected to hold its regular enrollment, but Dartmouth, Amherst, Williams, and Wesleyan— four excellent eastern colleges—expected a decrease of at least 17 per cent. Smith, another small institution, housed a training camp for the WAVES, the Navy's women's corps, but most Army and Navy programs, larger in number, moved into more commodious universities like Harvard, Columbia, Northwestern, and Notre Dame. By the end of 1942, Harvard was almost bereft of law students but had 4,500 servicemen in special programs in various schools, as well as a full complement of medical students training for a first professional assignment in uniform.

The smaller colleges survived in part because the Army and Navy contrived to assist them. During 1942 and 1943 the immediate need for troops in the field or crews at sea had yet to reach the peak that would arise in 1944 with the invasion of France and the accelerated push in the Pacific. In those earlier years the smaller colleges could offer basic educational training, most of it quasi-scientific, to college-age draftees who lacked important skills. The Navy's V-12 program and the Army Specialized Training Program sent qualified young men to the campuses, the Army for nine months, the Navy for as much as two years, and later drew them out for combat duty. By the end of 1943 ASTP included some 140,000 student-soldiers, of whom 20,000 were in medical, dental, or veterinarian courses, 73,000 in pre-science and pre-engineering programs, and about 13,000 in special foreign-language courses. Like ASTP, V-12 put most of its student-sailors in those subject areas, and required those below the professional level to devote between a fourth and a third of their academic time to English, in the use of which most draftees were deficient, and American history, which was usually interpreted as if the national past had been a pageant of triumphant democracy. The shortcomings of the ASTP and V-12 curricula troubled college teachers whose age or family responsibilities trapped them on the campuses with static salaries and outsized teaching loads, but the very existence of the programs kept alive colleges that might otherwise have had to close, perhaps never to reopen.

Those colleges were threatened by the manpower demands of 1944. The Congress preferred to move young men out of training programs before conscripting fathers, previously draft-exempt. The Army and Navy needed to draw upon both sources, though General Marshall properly emphasized the necessity for young men for combat duty. Accordingly, ASTP fell off to about 35,000 students, mostly those near their M.D. degrees, and V-12 was ordered to trim proportionately. The dean of one small college in the South begged the President to adjust plans to protect his institution and others similar to it. Several deans of medical schools warned that the mobilization of premedical and first- and second-year medical students would produce a doctor shortage within three or four years. Those considerations and others, which Harvard President James B. Conant called to the attention of the White House, persuaded Samuel Rosenman, who was handling the problem for Roosevelt, to work out a new scheme with the Army and Navy. It permitted the enrollment of 100,000 seventeen-year-olds, still a year below draft age, in voluntary pre-induction programs that would send them to college during the summer. Many of those volunteers would have time for two or three terms of college before they went into active service.

That new development, like V-12 and ASTP, gave American colleges a role in the war effort and an essential subsidy. It accomplished for small colleges what the SWPC achieved for small business. It also provided an emblem of the concern of the federal government for higher education. Like the antitrust laws, the emblem had more symbolic than actual effect, for other wartime policies had meanwhile enhanced the strength and prestige primarily of the largest and richest American universities.

Those policies flowed from the needs of the Office of Scientific Research and Development, the federal wartime contracting agency for scientific and medical research and related activities. OSRD worked closely with the Army and Navy, and co-ordinated the efforts of the National Defense Research Committee, the National Academy of Sciences, and the National Advisory Committee on Aeronautics. Dr. Vannevar Bush, the head of the agency, a former vice-president of MIT, and his two closest associates, Presidents Karl T. Compton of MIT and James B. Conant of Harvard, all outstanding organizers of American science, understood the urgency of winning the race against the enemy for new offensive and defensive weapons. OSRD contracts made possible the rapid wartime development most dramatically of the atom bomb but also of radar, antisubmarine gear, the proximity fuse, flamethrowers, penicillin, rockets, and various insecticides. Without those and other vital products, the United States would have suffered more casualties, fought more months, and perhaps even lost the war. Anxious always about those possibilities, Bush pushed for speed and efficiency. OSRD contracts, given the necessities he defined, went to those universities and corporations with the facilities to handle them, and American scientists, engineers, and technicians moved to the posts where the work went forward.

Strength, in those circumstances, begat more strength. Twenty universities received the bulk of OSRD contracts; the heaviest concentration was at MIT, followed by the California Institute of Technology, Harvard, and Columbia. The contracts were made on a cost basis, with reimbursement only for actual expenditures. Still, at the participating universities OSRD funds underwrote the construction of new and the utilization of older laboratories, which remained serviceable after the war ended. Consequently, during the war the most powerful scientific and engineering establishments of 1940 gained an even greater lead over their rivals.

No federal officer planned it that way. The experience of the academy merely reflected, as did the experience of industry, the manner in which wartime needs reinforced institutional patterns of prewar society, and in so doing stamped postwar conditions. Small colleges, like small businesses, had to solicit federal subsidies to survive the war. In contrast, the War and Navy departments had to mobilize the great universities, like the giant corporations, to produce the ingredients of victory. Only a deliberate decision to reform American social institutions while fighting the war, a decision to impose an older vision of America upon the country that would enjoy the peace, could have changed the outcome. That kind of decision was precluded by the necessitarian view of the war that imbued both Washington and the hinterland.

In the consequences, as the leaders of labor unions and agribusiness recognized, the future of their constituencies depended upon their cultivation of sufficient size and power to counterbalance the strength of their industrial competitors in the struggle over total national resources. Wartime conditions also abetted that process. Before the war ended there had already emerged the profile of power that John Kenneth Galbraith described in 1952 in his *American Capitalism,* a shrewd description of the manner in which big government, big agriculture, big labor, and big business exercised a continuing, reciprocal countervailance—a balance of forces—in national political and economic life.

If no one planned it that way, few objected; the confirmation of size during the war occurred simultaneously with the return of abundance and the satisfactions, or at least the seductions, of affluence. For most Americans it was easy to associate the pleasures of steady employment and current or anticipated consumption with the existence of giantism. Working men and women, familiar with their unions but innocent of the macroeconomic theory that explained the return of prosperity, tended to attribute their wartime good fortune to those unions and to friendly politicians rather than to the impact of war-inspired deficit spending. So, too, farmers with large landholdings, wartime recipients of unprecedented income, tendered their grateful loyalty to the Farm Bureau Federation and the farm bloc. Both those enterprisers like Kaiser who adapted to wartime circumstances and those corporate managers who enlarged their domains by satisfying military procurement needs attributed their success to their own business acumen and their organizational talents. As they saw it, their prodigies of production confirmed the value of free private enterprise. There were, to be sure, exceptions in each of those groups, men who understood Keynesian economics and perceived, without yet stating, Galbraithian institutionalism. But, for the majority of Americans, by 1945 size and abundance appeared to march hand in hand.

For that majority, expectations for postwar life grew out of their perceptions of wartime gains. The worker or the farmer, eager to consume, counted upon his union or his lobby to maximize his earnings. He was already hungry to acquire the shiny new stoves and shoes that industry was dangling before him in advertisements that promised astonishing new comforts once free enterprise could reconvert. Henry Kaiser was as eager as Henry Ford to profit from the demand for automobiles instead of tanks, and to operate without government controls. By 1944–45, corporate and personal affluence beguiled Americans, as did selfish identifications with their own interest groups.

That mood worried those who had struggled to build up war production and technology, especially those who believed that preparedness for war was necessary for the eternal future. Secretary of the Navy Forrestal wanted to preserve the close relationship of the War and Navy departments with industrial suppliers. Vannevar Bush proposed various schemes to link subsidized postwar military research with the National Academy of Sciences, and through it, with scientists in universities. Wartime advocates of compulsory national service now urged peacetime legislation to construct a reserve of soldiers, sailors, and technicians.

In contrast, Wendell Berge, Francis Biddle, Robert Nathan (once of the WPB), and Senator Harley Kilgore, who had assumed Truman's watchdog role, continued to attack monopoly in business. Others on the Hill addressed their concern against the influence of the farm bloc and the unions. The hopes of the War and Navy departments contradicted those of the opponents of bigness, while the incipient return of "normalcy" bode to engulf both groups.

Much of the resolution would lie in political decisions, some made by 1944 and 1945, some still in the making. Political decisions, affected perforce by the culture of a new "normalcy," had also to determine the immediate future of groups of Americans whose color or religion or national origin had provoked their government and their fellow countrymen, in the name of wartime necessity, to deny them their hopes and even their Constitutional rights. In the processes of politics, economic and ethnic issues alike had continuing influence, sometimes to conflicting partisan ends. And both influences were growing, for, like size and affluence, prejudice in America reached new dimensions during the years after the Japanese attacked Pearl Harbor.

5 / Outsiders

1. Little Italy

In those little places, so one reporter observed, where Italian was spoken and pasta was served, an attentive diner might on occasion overhear another patron say to his neighbor: *"Roberto vincerà."* Perhaps just "Roberto," an acrostic consisting sequentially of the first letters of Rome, Berlin, and Tokyo, the capital cities of the Axis powers. It was a salute of a kind among those Italian-Americans, never many in number, who were caught in the cult of Mussolini. An attentive diner might have heard the word before December 8, 1941. Thereafter Roberto, with exceptions too few to matter, gave way to Kilroy, that omnipresent symbol of the American GI.

Suspicions of Italian-Americans receded slowly during World War II, as still more slowly, if at all, did American prejudices against aliens or citizens who were yellow or black. Racial and ethnic prejudices had characterized American attitudes for centuries before 1941. No four-year experience could have dispelled those biases, certainly not a quadrennium of war that, like all wars, imposed high premiums on loyalty and conformity, real or cosmetic. Yet World War II posed a special test of the ability of American culture to accommodate to its inherent pluralism. Both official doctrine and the sentiments of decent men defined the war as one to erase Nazism, with its vicious racial creed. But the war began for the United States with a surprise attack by brown men whom most Americans came to view as subhuman beasts. Informed at once by that powerful hatred and by a contrasting but more abstract, less palpable commitment to expunge racism, the wartime culture contained an ambivalence that Americans had always felt, usually less starkly.

Kilroy was a polyglot. The great waves of migration had brought to the United States aliens from most parts of the world; most recently, in the early twentieth century, men and women of yellow or swarthy complexions, of religions other than Protestantism, of cultures remote from the British or American. All aliens, including those who had become naturalized and those fluent in English, were perceived as people apart. Federal studies during World War II located instances of discrimination even against immigrants from Great Britain. The newcomers themselves were conscious of and often ambivalent about their own origins. Race prejudices, virulent against blacks, penalized Asian and Chicano Americans, too, and in lesser measure religious prejudices affected Catholics and Jews. The latest of the white arrivals, those from eastern and southern Europe, felt the disdain of descendants of onetime western Europeans who were now, at least by their own definition, 100 per cent American.

Americans of good will, eager to combat prejudice, sometimes inadvertently revealed their own condescension. So it was, for one example, with the editors of *Fortune*. "Europe," that journal noted in 1942, ". . . is inhabited by people who have uncles in America. Every fourth white American knows of blood relations in Europe." Immigrants and their children, the article went on, retained their ethnic identity more stubbornly than the metaphor of the melting pot admitted. But they were not radical or essentially foreign: "Actually the European immigrant and his children have the same motives in politics as Americans: distinct personal stakes, more or less nebulous desires to 'change the world,' the gregarious urge to be part of a potent social force." That sympathetic statement immediately preceded a lament: except for the British, Europeans had no "prolonged experience of a two-party system." Further, the children of immigrants, divested while growing up of a strange accent, were nevertheless conscious of the hostilities within "an Anglo-Saxon environment." (*Fortune* still equated "American" with "Anglo-Saxon," as did its founder, Henry Luce, and as did Winston Churchill.) Those children, once adults, sensitive to their differences, sometimes became "passionately interested in the Old Country." Those foreign identifications among men and women naïve about the two-party system, so *Fortune* implied, threatened the political stability essential in wartime.

Fear of the alien and the colored, which *Fortune* intended to combat, assumed a special focus after the Japanese attack on Hawaii and the German and Italian declarations of war against the United States. The nationals of the Axis countries were now not just aliens but enemy aliens. An accompanying stigma touched their children. Their problems, though special, had much in common with those of every minority group, white, brown, black, or yellow. All had a stake in the way in which the culture absorbed the impact of war. That stake involved the entire nation, citizens as well as aliens, the privileged as well as the disadvantaged. As one congressman put it early in 1942 in a letter to Archibald MacLeish, a letter that could have been more inclusive than it was: "The courage and imagination with which we meet the problem of an alien population will determine in large measure the pattern of our national life both now and in the postwar period."

The onset of war disturbed the previous pattern of Italian-American life. It had not been an easy one. Not yet fully accepted by American society, the almost 5 million Italian-Americans in the United States were nearly all immigrants or children of immigrants. In 1942 some 600,000 of them were not naturalized. Some of the older members of that group planned one day to return to Italy. Many more, regardless of age or citizenship, retained an affection for and often a pride in Italy, an attitude reinforced by the rejection they felt in American society. Like other recent immigrant groups, the Italian-Americans, largely unskilled and unlettered, ordinarily commanded only the worst kinds of jobs, and even those had been hard to find during the Depression. Again like the other groups, they tended to cluster within communities of their own, where the church, fraternal organizations, and social and political clubs helped them at once to preserve many of the comforting old ways they had known and to find networks of friendships and sources of self-esteem.

The quest for self-esteem accounted in large part for the considerable satisfaction that the militant Italian-language press had expressed in the successes of Mussolini. One such newspaper extolled the conquest of Ethiopia as "another beautiful page" in Italian history, an episode marked by the "boldness and ability" of Italian soldiers. Its readers, it urged, as a "sacred duty" should "help the Patria financially, morally, and if it should happen, with our lives." Its campaigns raised nearly a million dollars for the Italian Red Cross. One subscriber, a college student, in 1939 welcomed the Nazi invasion of Poland in a letter predicting that "as the Roman legions did under Caesar, the New Italy will go forth and conquer."

In 1940, during the months after the Italian declaration of war against Great Britain and invasions of France, Egypt, and Greece, about 80 per cent of Italian-language newspapers were profascist, about 12 per cent middle of the road, and only about 8 per cent antifascist, much of the last group being socialist, anarchist, or communist. The profascist papers stressed both the "disastrous consequences" of any American participation in the war and the legitimacy of Mussolini's aspirations. "The declaration, often repeated by Il Duce, that Italy wants a place in the sun is nothing but the vindication of a poor and prolific people"—so wrote the daily *La Notizia* of Boston. "Italy was forced into her holy war of liberation by her enemies. . . . Italy fights in defense of her down-trodden rights"— so said the weekly *Corriere del Connecticut* of New Haven. "The pact of alliance between Italy, Germany, and Japan marks the decline of an epoch, the rise of another. . . . It is the foundation on which future mankind will be organized"—so judged the biweekly Cleveland *L'Araldo.* "It would be a crime if our passionate politicians and our people repeat the history of 1917. Isn't being the sucker once enough?"—so asked the weekly *Il Nuovo Americano* of Paterson, New Jersey.

All those views postdated the speech of June 10, 1940, in which Franklin Roosevelt had condemned Mussolini for his "stab in the back" of France. Both the Italian attack and the President's criticism of it spurred American suspicions of Italians, as well as Italian trepidations about their place in American society. One Connecticut congressman reported "a wave of sentiment in New England . . . against Italians." The Assistant Attorney General of the United States found "some sections of the population hysterical." That hysteria had already led, here and there, to "the wholesale denunciation of all aliens" and in some instances to mob violence. An Italian-language newspaper complained about a "barrage of propaganda . . . aimed at destroying pride in Italian ancestry." In a different but equally agitated mood, a college undergraduate wrote: "I hate Mussolini. . . . I am sick and tired of being picked on for being an Italian. . . . Damn it, I'm just as American as any of the rest of my class." Yet in June 1941, in the Italian districts of Boston 72 per cent of the residents opposed American entry into the war, 65 per cent disapproved of Roosevelt's foreign policy, and only 10 per cent believed the United States had any proper part to play in combating fascism. Those attitudes had accounted in November 1940 for a discernible drift away from Roosevelt and the Democrats in Italian-American voting.

Two Harvard psychologists, specialists on public opinion, summarized the situation in 1941. The Italian-Americans in Boston, they concluded, were dissatisfied on many counts: "They report discrimination in job seeking, they feel neglected, apart. Children in the community do not like to go . . . to playgrounds outside the area because 'all the kids there are Americans.' The militant Italians . . . are . . . paranoid about their status as marginal Americans. In Italy they see the gratification of desires never fulfilled here: power, command, and 'glory.' "

Even the militants, except for an infinitesimal minority who turned to sabotage, could not retain that vision after the Italian declaration of war against the United States. The vast majority of Italian-Americans, according to one government analysis, disapproved of that declaration. They disliked even more the executive order of December 8, 1941, that designated as enemy aliens all Italian (as well as German and Japanese) noncitizens. Accompanied by several hundred arrests of suspected Italian agents, the order suspended naturalization proceedings and restricted physical movement and the opportunities for employment of those to whom it pertained. Most other Italian-Americans were relatives, usually children, or friends of those whose lives were directly affected. At first let down, they soon felt confused by their equivocal position. The older Italians, especially those still aliens, were bewildered. As one psychologist put it, "there is no single, well-defined outlet for their emotions in this war." Their adult children shared their resentment over what they construed to be unjust discrimination.

The mood of the Italian-Americans reflected the dissatisfactions they had long felt, especially about their occupational status. Their middle class, slow in emerging, had yet to acquire enough members, visibility, or authority to provide either commanding symbols of achievement or adequate sources of favor and advancement of their less privileged fellows. There were exceptions, acknowledged leaders in politics and the law, especially renowned professional entertainers and athletes. There was also a cadre of Italian émigré intellectuals, liberal refugees from fascism; but in spite of their talent and intrepidity, they had made little impact on their former countrymen within the United States. A majority of Italian-Americans, uncertain about their possibilities, were apprehensive about their futures. They need not have been, for the Fair Employment Practices Commission reported no special prejudice against them as compared to other white ethnic groups, and the Office of Facts and Figures discovered that a mere 2 per cent of Americans considered the Italians the nation's most dangerous enemy. Those data, even if they had been publicized, as they were not, would not much have diminished an apprehension rooted in a prolonged experience of alienation and an immediate distress about the enemy-alien order.

The order, and even more the attack on Pearl Harbor, moved Italian-Americans immediately to express unequivocally their loyalty to the United States. The United Italian-American League in a wire to Roosevelt urged the defeat of all enemy forces. The Mazzini Society in another telegram pledged its support of the war, as did Italian-language newspapers. On Mulberry Street, the New York *Times* reported on December 12, 1941, store fronts carried photographs of Roosevelt's smile where earlier Mussolini's scowl had appeared, a change that contrasted with the persistence of German symbols in Yorkville. The previously militant New York Italian paper suddenly stressed the homogeneity of its constituency with the national culture. An Italian food store in Boston put up a sign, "We're 100 per cent for the U.S.A.," and a photograph of Private First Class Joe d'Agostino. Throughout the country, thousands of young Italian-Americans, praised by their press, enlisted in the armed forces. During 1942 and later their achievements received the kind of plaudits common for all valorous GI's.

Still, the persisting ethnicity of all immigrant groups in 1942 bothered the Italian-Americans particularly because, as one study concluded, they were eager for a kind of miracle, "American victory without Italian defeat." Federal officials, alert to that condition, moved deliberately to mollify the Italian-Americans, whose political support Roosevelt had always solicited. As he knew, their votes would much influence the outcome of the 1942 congressional elections in states like New York, Massachusetts, and Connecticut. At the President's direction, Attorney General Biddle combated discrimination in employment against Italian-American citizens and administered the enemy-alien laws, as they applied to Italians, with restraint and compassion. He appointed Edward Corsi, an eminent Italian-American, head of the alien review board in New York, a choice that even former militants applauded.

In California, where Japanese-Americans were harassed, Italian-Americans soon enjoyed reasonable treatment. A former Assistant United States Attorney there held that Italian aliens had demonstrated their loyalty by the successful way in which they had brought up their children. The clinching evidence, he said, was the record of the DiMaggios, who had raised nine lovely children, three of them— Joseph, Dominic, and Vincent—major-league baseball stars. Secretary of War Henry Stimson, a proponent of repression of Japanese-Americans, took a contrasting position on the Italians. "I desire," he wrote the commanding general on the West Coast, ". . . that you do not disturb . . . Italian aliens and persons of Italian lineage except where they . . . constitute a definite danger."

Other government officials were pressing for clarification and liberalization of federal policy toward all enemy aliens. At the least, Archibald MacLeish proposed in May 1942, the Justice Department should establish boards to determine the loyalty of individual aliens and to remove all restrictions from those they cleared. The department considered that task too large to undertake, but MacLeish's advisers, already prepared to go further, advocated the outright removal of Italians from the enemy-alien classification. That step, unlike any short of it, would at once satisfy Italian-Americans at home and assist ongoing propaganda efforts in Italy and Latin America. Officials within the War Manpower Commission and the War Production Board, moreover, were eager to mobilize Italian-American labor without restrictive enemy-alien regulations. But the FBI was dubious about the proposal, and the Navy characteristically confused its fears of Italian-Americans with threats to national security. In New York City especially, Italian-Americans worried that they might be subjected to the evacuation policy the government had initiated against the Japanese-Americans in California. *Il Progresso,* once but no longer a militant paper, published several articles about the potential injustices of evacuation, a point also made in a much publicized speech by Luigi Antonini, the vigorously liberal General Secretary of Local 89 of the International Ladies' Garment Workers' Union. In those statements Naval Intelligence discovered a propaganda campaign designed to "create doubt, unrest, and confusion concerning . . . the defense command recently created in the area." Misinterpretations like that confirmed the very anxieties Antonini had expressed. Fortunately Roosevelt had already instructed the Secretary of War that the military were to take no action against American citizens bearing Italian or German names without express, previous Presidential approval.

The Bureau of Intelligence within MacLeish's office corrected its naval counterpart. The great majority of Italian aliens, the bureau reported, had lived in the United States more than twenty years; most were more than forty-four years old and 10 per cent more than sixty-four; most had American children, and many also had grandchildren. Some still identified with their place of birth, but "most of them

consider that their interests as well as those of their children lie here."
Before Pearl Harbor 40 per cent of Italian aliens had applied for
naturalization. They were an industrious, peaceful, law-abiding folk.
Though they made up 63 per cent of all enemy aliens, in May 1942
they constituted less than 14 per cent of enemy aliens in federal
custody or suspected as threats to American security. A removal of
Italians from the enemy-alien classification, the report concluded,
would dispel their insecurities, neutralize propaganda addressed to
them from the Axis nations, ease the administrative burden of the
Department of Justice, and assist American propaganda in Italy.

The report, especially persuasive in the light of the prewar experience of Italian-Americans, which it reviewed, convinced the State
Department but not the Army, the Navy, or the FBI. Their objections
prevailed until continuing manifestations of Italian-American loyalty
and the approach of the November elections added weight to the
arguments that Biddle and MacLeish had been advancing since May.
Democratic congressional candidates in the Northeast needed the
Italian vote, which the Republicans had been cultivating. Aware of
all this, and in any case sympathetic to the Italians, Roosevelt on
Columbus Day 1942, lifted the enemy-alien designation from them
and established simplified procedures for their rapid naturalization.
Attorney General Biddle announced the President's decision to a
cheering crowd at Carnegie Hall. American Italians, Biddle said,
loved not Mussolini's Italy but "the older Italy of freedom." The
announcement came too late to help the Democrats, who in November suffered again from Italian-American defections, but not too late
to accomplish the purposes MacLeish's counselors had defined.

More important, the President's order marked the beginning of
the end of Italian-American separatism. As in the past, Italian-
Americans, like Irish or Polish or Scandinavian Americans, preserved
much of their ethnicity, but henceforth, increasingly, they did so from
choice rather than from necessity. As Max Ascoli, a liberal Italian-
American intellectual then in government, concluded in 1942, "The
war has given the final blow to the segregation of the Italian communities in America." That development, he wrote, involved a two-way
traffic, an understanding by the Italians of their role within American
culture and its network of institutions, and a willingness by other
citizens to accept Italians as Americans.

Before the end of the war, *Seventeen* was expressing that willingness to its nubile feminine audience. In a speculation of November
1944, entitled "If I Were Seventeen," Joseph Cotten mused: "I am a

youth again, back in the splendid town of Petersburg, Virginia. . . .
Life is too cut and dried in the nice part of Petersburg, so right away
I hike myself over to the back streets. . . . I see just as much of that
sweet Italian family as I did the last time I was seventeen. After all, I
know by this time that they all turned out well."

Patronizing, to be sure, but at least *Seventeen* promised romance
to Roberto's loyal American daughter. Federal wartime policy, after
an uncertain start, had helped to make that promise feasible. In the
case of enemy aliens who unlike Italians were not white and who
lacked political punch—and where American prejudice became hysterical and the federal government did not reverse its course—wartime policy with its imperative on loyalty produced antithetical
results.

2. "A Jap's a Jap"

In April 1943, a year after the federal government had begun to
incarcerate thousands of guiltless Japanese-Americans in internment
camps that the Army preferred to call "relocation centers," the President had apparently forgotten who was in charge of those centers.
Responding to Harold Ickes, who had protested against the wretched
conditions in the camps, Roosevelt asked Milton Eisenhower to draft
a proper reply. But Eisenhower had resigned from the directorship of
the War Relocation Authority nine months earlier. Roosevelt's lapse
of memory suggested his lack of concern about the most blatant mass
violation of civil liberties in American history. Even Milton Eisenhower, who was more sensitive to that issue than was his chief,
interpreted protests against the treatment of the Japanese-Americans
as expressions of "liberal groups and kind-hearted people" who did
not understand military necessities. The misunderstanding was that of
Roosevelt, Eisenhower, Henry Stimson, and those generals who let
sheer racial prejudice prevail in the name of necessity. They did so in
the spirit that had motivated Westbrook Pegler, in his syndicated
column curiously entitled "Fair Enough," to write in February 1942
that "the Japanese in California should be under armed guard . . .
and to hell with habeas corpus until the danger is over."

Unlike the Italian-Americans, the Japanese-Americans were not
white and had no political importance. At the time of Pearl Harbor

there were only about 127,000 of them in the United States, a mere .1 per cent of the total population. Almost all of them, some 112,000, lived on the West Coast, where for years they had experienced all kinds of racial discrimination—laws against intermarriage with whites, legal exclusion from swimming pools and dance halls, extralegal bars to employment and to middle-class-housing districts. The Immigration Act of 1924, prohibiting further immigration from Japan, also made 47,000 Japanese already in the United States, the Issei, ineligible for naturalization. Various state laws had long denied them the right to vote and to own land. Besides the Issei, by definition enemy aliens once the war began, there were about 80,000 Nisei, Japanese-Americans born in the United States and therefore American citizens, and their third-generation children, Sansei, almost all of them very young. The Nisei themselves were young, because their immigrant parents, struggling against poverty, had tended to postpone marriage. As a consequence there was a missing generation among Japanese-Americans. In 1942 the median age of the male Issei was fifty-six, of the female Issei forty-seven, but of the Nisei, only eighteen.

More than the Italians, the Japanese-Americans had remained bicultural. The exclusion of further Japanese immigration, the ineligibility of the Issei for citizenship, and their inability to own land, made them wary of American society. They had no reason to absorb American concepts about civil liberties, for their own liberties were circumscribed. They had no opportunity to learn to use the ballot. Rejected by Americans, they clung to many of their old traditions, and often learned English badly if at all. Yet they worked hard, rooted themselves in the United States, sometimes prospered, and regularly urged their children to pursue American educational opportunities vigorously and to establish for themselves both a place and a stake in American society.

The ambiguity of Issei culture persisted in the lives of the Nisei. They learned Japanese, which they spoke with their parents though not among their peers. They celebrated Japanese festivals as well as American holidays. They ate Japanese food at least as commonly as American, usually married according to arrangements made by Japanese intermediaries, and subscribed more often to Buddhism than to Christianity. A minority of them, the Kibei, who had gone to Japan for part of their education, especially felt a cultural and linguistic displacement after returning to the United States. Another minority, those Nisei who had had exceptional social and economic advantages, integrated the two cultures comfortably; but the majority were far less privileged, and far more torn between the ways of their parents and an Americanism distorted by the constricted opportunities open to them on the West Coast.

That majority, like the Issei, were employed largely as agricultural tenants or sharecroppers, especially in truck farming, or as fishermen, or in retail trade or domestic or commercial services. Few had the chance to become mechanics, fewer still to enter professions. In varying measure, all of them felt the sting of white racial prejudice. All had more reason for developing the paranoia of apartheid than had the Boston Italians. Worse, whatever hysteria developed against the Italians and Germans was trivial in contrast to the waves of hatred that West Coast Americans, and others, too, came to discharge against the Issei and the Nisei during the months following the raid on Pearl Harbor.

At first the reaction seemed temperate. When adult Nisei expressed their horror over the attack and their loyalty to the United States, West Coast Americans appeared to accept them at their word. The Issei, as enemy aliens, suffered various restrictions on travel and employment, which were annoying rather than oppressive. General John L. DeWitt, head of the West Coast Defense Command, and Secretary of War Stimson opposed, for the time being, "discrimination . . . on the ground of race."

Soon, however, rumor began to feed the prejudices of both the military and the West Coast whites. Early reports from Hawaii, later proved to be false, attributed sabotage to Japanese-Americans in Oahu. Though martial law prevailed there, the government never took special measures against Japanese-Americans on the islands, partly because there were proportionately so many of them—one-third of the population—largely because they were manifestly loyal and harmless. The latter condition also obtained on the West Coast; but there the very absence of sabotage came to be regarded as evidence that some terrible Japanese plot was brewing.

That fantasy gained some credibility when in December a Japanese submarine torpedoed an American freighter off the California coast; but any danger to the United States was imaginary. "Vinegar Joe" Stilwell, then a major, soon to be a general, found the Fourth Army "kind of jittery" on December 8, 1941, when it sounded an alarm for an air raid in San Francisco. The next day word of a Japanese fleet sailing between that city and Los Angeles quickly proved "not authentic." Two days thereafter, Stilwell—"like a damn fool," in his own assessment—briefly believed another report that the Japanese fleet was nearby. "Not content with that . . . blah," Stilwell wrote in his diary for December 13, ". . . Army pulled another . . . today. 'Reliable information that an attack on Los Angeles is imminent. . . .' The Army G-2 is just another amateur, like the rest of the staff."

The interpretation of innocent circumstances as ominous affected the Japanese-American fishermen and their families who had long lived on Terminal Island in San Pedro harbor. They resided in shacks, earned meager wages, and had little contact with anyone outside their own group. Still, the accident of their proximity to a naval base struck the suspicious as a planned exercise in sabotage. On December 7, 1941, the Justice Department detained some of the adult males. As rumors of espionage grew, though no case was ever authenticated, pressure against the Terminal Islanders mounted, until on February 1, 1942, a mass raid preceded the apprehension, without cause, of the remaining men.

The six to eight weeks after Pearl Harbor provided professional agitators, the Native Sons and Daughters of the Golden West especially, with the opportunity they had always sought to rid the West Coast of the Japanese, Issei and Nisei alike. The Japanese advances in the Pacific, particularly in the Philippines, provoked nationwide anxiety and with it a general hatred of the Japanese that frightened Japanese-Americans as much as did the propaganda directed against them by hardened racists. That propaganda was vile. Shortly before Pearl Harbor, one journalist had predicted that Japanese-American fishing boats "will sow mines across the entrance of our ports. . . . Sierra passes and tunnels will be blocked. . . . Japanese farmers . . . will send their peas and potatoes and squash full of arsenic to the markets." After Pearl Harbor, a California barbershop offered "free shaves for Japs" but was "not responsible for accidents." A funeral parlor advertised: "I'd rather do business with a Jap than with an American." The Los Angeles *Examiner* reported "armed Japanese in lower California ready for invasion." "Herd 'em up, pack 'em off . . . let 'em be pinched, hurt, hungry and dead up against it," wrote columnist Henry McLemore, adding, for those who might have missed his point, ". . . I hate the Japanese."

Demands for their evacuation flowed from most of the California congressmen, sundry West Coast mayors, California Attorney General Earl Warren (who later deeply regretted the position he had taken), the American Legion, and various farm, labor, and business groups whose members were eager to profit by taking over Japanese-American enterprises or at least by eliminating their competition. That agitation convinced the Japanese-Americans that the developing policies of the federal government grew out of prejudice rather than military precautions, a conviction wholly persuasive in the light of the generous conclusion that General DeWitt had reached by February 1942: "A Jap's a Jap. . . . It makes no difference whether he is an American citizen or not. . . . I don't want any of them. . . . There is no way to determine their loyalty."

By February, with hostility toward the Japanese still rising, De-Witt's earlier scruples about justice had crumbled, as had the restraint of most other public men, civilian as well as military. A crisis had arisen between the Justice Department and the Army. In January 1942 DeWitt designated broad areas along the West Coast, including some whole cities, from which enemy aliens were to be evacuated. The Justice Department, loath to remove over 10,000 people, also lacked the means to do so. Attorney General Biddle, in the view of Henry McLemore, was handling the problem "with all the severity of Lord Fauntleroy playing squat tag with his maiden aunt." Actually the Attorney General quickly yielded to the military. At the President's direction, which it had solicited, the War Department drew up Executive Order 9066 of February 19, 1942, authorizing the department to designate military areas and to exclude any or all persons from them. Secretary Stimson utilized the order only on the West Coast and only against Japanese-Americans. There he delegated his authority to DeWitt. "If Stimson had stood firm," Biddle later wrote, if Stimson "had insisted, as he apparently suspected, that this wholesale evacuation was needless, the President would have followed his advice. And if . . . I had urged the Secretary to resist the pressure of his subordinates, the result might have been different. But I was new to the Cabinet, and disinclined to insist on my view to an elder statesman whose wisdom and integrity I greatly respected."

Biddle blamed himself too much. Roosevelt was susceptible to the pressure from West Coast politicians and public opinion, as well as to Stimson's advice. Stimson, pleased that the President was "very ·vigorous" about the matter, had always had doubts about yellow, brown, and black peoples. The "racial characteristics" of Japanese-Americans, he held, required identical treatment for aliens and citizens. He expressed no argument with DeWitt's justification for the evacuation of both Issei and Nisei: "The continued pressure of a large, unassimilated, tightly knit racial group, bound to an enemy nation by strong ties of race, culture, custom and religion along a frontier vulnerable to attack, constituted a menace which had to be dealt with."

The evacuation of the Japanese-Americans from the western parts of Washington, Oregon, and California began in March 1942. There was no attempt to distinguish between the loyal and the disloyal, or between citizens and aliens. The evacuees were told not to sacrifice their personal property but to store it if they could not sell it at a fair price. They could do neither. "Commercial buzzards," as one federal official observed, bought up business properties below fair values. The Federal Reserve Bank of San Francisco, which was responsible for personal effects, assumed no liability for stored property and consistently encouraged liquidation. All Japanese-Americans suffered some losses, many of them severe losses, especially the farmers. The government urged them to continue to work their farms until the last moment and threatened punishment for sabotage of crops. The longer the Japanese-American farmer worked, however, the longer he had to retain his agricultural implements, the more rapidly he then had eventually to sell them, and the less he received in return. As he labored, he drained his resources in the costs of seed, fertilizers, and sprays. His capital lay in unharvested crops that some white neighbor would ultimately harvest at an unearned gain. The failure to safeguard the economic interests of the Japanese-Americans transformed many of them, in the analysis of one investigator of their experience, from "a traditionally law-abiding, cooperative minority group into factions . . . protesting and rebelling against . . . the Caucasian majority which was pressing to dispossess . . . and incarcerate them."

There was much else also to persuade them that "they were an unwanted people, rejected by a hostile and prejudiced American public." At first the government simply ordered the Japanese-Americans to leave their homes and settle outside the designated defense area. Evacuation was considered voluntary as families tried to relocate elsewhere. But they were no more welcome inland than on the coast. John Tolan, chairman of the committee of the House of Representatives investigating the situation, concluded that "nobody wants the Japanese." The governor of Arizona had said his state did not "propose to be made a dumping ground for enemy aliens." He recommended placing them in "concentration camps," as did the governor of Arkansas. "The Japs live like rats," the governor of Idaho volunteered, "breed like rats and act like rats. We don't want them."

Neither did the governors of Kansas, Montana, or Colorado. "Voluntary evacuation," Milton Eisenhower later told the President, "was not a feasible solution. Public opinion . . . was bitterly antagonistic to the influx of the Japanese."

On March 27, 1942, General DeWitt ordered a freeze. The Army began moving the Japanese-Americans to receiving stations at race tracks, fair grounds, and similarly depressing facilities. The removal disrupted family life and frightened or angered its victims. As one of them said: "It's enough to drive a man nuts."

So it was. "Racial animosity," as the Japanese American Citizens League put it, lay behind the definition of military necessity. In the name of protective custody, the government now set about removing the Japanese-Americans from temporary receiving stations to internment camps. The policy of evacuation produced the decision for internment, while racial prejudice provoked both. In mid-March, at Tolan's recommendation, Roosevelt had created a new civilian agency, the War Relocation Authority, to set up and administer the permanent camps. By September the WRA had placed more than 100,000 people in ten camps in seven western states. Milton Eisenhower, the first head of the agency, had hoped to establish "small C.C.C. sort of camps" organized to undertake useful projects for which the internees would receive some compensation. He believed in the "essential Americanism of a great majority of the evacuees" and hoped that the "unreasoning bitterness" of 1942 would be replaced by "tolerance and understanding." The Army, however, insisted that it could provide adequate guards only for large centers, perforce inhumane.

The camps the WRA built were fenced in, dotted with tarpapered wooden barracks consisting of several one-room apartments, ordinarily with only partial walls, and invariably furnished only with cots, blankets, and a bare light bulb. A single family or a group of strangers shared each such room, along with communal toilets, laundries, and bathing and dining facilities Those conditions precluded individual privacy or familial intimacy. The camps served unsavory but nutritionally adequate meals, had only marginal schools and recreation areas, and provided only enough medical care to prevent epidemics. They were all in barren, arid areas unsuitable for the development of the economic self-sufficiency that Eisenhower had envisaged.

Furthermore, the effort of the WRA to encourage community government within each camp inverted the traditional relationships of

Japanese-American communities. The deference of the young for their elders could not persist, for the obvious candidates for authority within the camps were the Nisei, the young citizens who were more fluent in English than their parents and better equipped to deal with the camps' officials. The resulting melancholy of many Issei had its counterpart in the unprecedented bravado of a minority of Nisei who became the bully boys of the camps—self-conscious rebels, not without cause, against discipline, familial or official; self-conscious imitators of Shinto ways. Eisenhower and his successor, Dillon Myer, both decent men, had hoped to pursue a principled policy, but it developed predictably that a principled internment, like a principled evacuation, was no more possible than a principled genocide.

Myer also hoped that the relocation camps would be only temporary stations from which the Japanese-Americans would move out to schools, colleges, and employment in friendly communities. But the pervasive national animosity toward Japan, and the equally pervasive fear of disloyalty, made large-scale releases impossible. In order to get out, a Japanese-American had to prove that he was not disloyal to the United States, that he had a position awaiting him, and that it was in a community willing to accept him. Under those and other restrictions, only 17,000, many of them students, could obtain leaves from the camps during 1943. Since they were, by and large, the most acculturated and ablest of the young people in the camps, their departure deprived those who remained of a natural source of intelligent leadership.

While the passing months of a grim and debilitating incarceration generated increasing tensions among the internees, few Americans protested against the barbarous and unconstitutional course of national policy. Representatives of various church groups did try to alleviate conditions in the camps. They also, along with the American Civil Liberties Union, the Fellowship of Reconciliation, and a few similar organizations, continually criticized the abridgments of Japanese-American civil liberties. Even so, most of those protests were tempered by at least a rhetorical concession to military necessity. The problem of the West Coast Japanese, a contributor to *The Nation* wrote in February 1942, was that "his physical characteristics make his position . . . almost unbearable," but "for the sake of war efficiency," surveillance by the FBI would "suppress the ancient Western curse of vigilante rule." Two weeks later *The New Republic*, which was sympathetic to the plight of the Japanese, held that "there is no question that the Japanese colony should never have been tolerated on Terminal Island. . . . The Department of Justice has wisely ordered all Japanese out of the district and . . . has limited the areas within which Japanese aliens may reside." Initially only *The Christian Century* resisted the argument of national security, though Carey McWilliams in *The New Republic* in April 1942 objected to testing loyalty on the basis of nationality or race. "This whole policy of resort to concentration camps," *The Christian Century* warned in June 1942, "is headed . . . toward the destruction of constitutional rights . . . and toward the establishment of racial discrimination as a principle of American government. It is moving in the same direction Germany moved." Only after reports from the internment camps revealed what that brisk warning had forecast, indeed only after the battle of Midway demonstrated that there was no Japanese naval or air threat to the West Coast, did other liberal journals begin unequivocally to attack the federal policy of "Jap Crow." Not until 1944 did *Fortune,* the first of the establishment magazines to speak out plainly, publish an article exposing the outrages to which Japanese-Americans had been subjected. Until then, with few exceptions, the free press of the United States shared the disgrace of American public opinion, of the War Department, and of the White House.

Again with the exception of few of its members, Congress also behaved badly. In 1942 most of the West Coast delegation in the House of Representatives, supported by a majority of their fellows from elsewhere, introduced a bill to deprive the Nisei of their citizenship. For months the irresponsible Martin Dies, chairman of the already notorious House Committee on Un-American Activities, poisoned the atmosphere in Congress with unsubstantiated charges against the Japanese-Americans. Dies focused in part on a strike at the camp at Poston in November 1942 and a riot at the camp at Manazar the next month. Both episodes reflected the hostility between some Kibei, who admitted their antagonism toward Americans, and leaders of the Japanese American Citizens League, who were cooperating with Dillon Myer in a futile attempt to make living and working conditions tolerable. Even those hostilities grew out of circumstances that a Senate committee later accurately defined: the relocation centers, like any concentration camps, were creating disloyalty. Dies characteristically confused resistance to injustice with conspiracy against the nation.

Under the influence of that assumption, the Senate in July 1943 passed a resolution, sponsored by California Democrat Sheridan Downey, directing the WRA to segregate loyal from disloyal internees. The agency had already begun to do so, as the President reported to the Senate. Still not satisfied, congressional hard-liners pressed Roosevelt to fire Myer for his humaneness, which they considered pusillanimity. In January 1944 the West Coast contingent wrote the President to demand the "utter extinction" of the Japanese "military machine," the exclusion from the West Coast for the duration of the war of even demonstrably loyal Japanese-Americans, and the return to Japan "at the earliest possible opportunity" of disloyal Japanese-Americans.

The government's procedures for determining loyalty had already further confused the issue. Once they had been interned, many Japanese came to prefer the routines of the camps to the prospect of confronting the venom of American communities to which they would have to go if they were cleared. The Nisei also worried about leaving behind anxious parents who could not obtain clearances. Consequently responses to interrogations about loyalty were frequently disingenuous. In addition, the interrogations allowed for no subtleties. Examinations began after January 1943 when Secretary of War Stimson announced "the inherent right of every faithful citizen, regardless of ancestry, to bear arms in the nation's battle." Earlier even the Nisei had been barred from military service. Now the Army and the WRA devised a program to process all Japanese-Americans over seventeen. A team of officers and noncoms, none alert to the group psychology at the camps, asked each examinee about his past, his foreign travel, his membership in clubs, his reading habits, and, most directly, whether he would serve the armed forces of the United States on combat duty wherever ordered, as well as whether he "would swear unqualified allegiance to the United States . . . and forswear any form of allegiance . . . to the Japanese emperor." Women had only to promise to volunteer, if qualified, for the Army Nurse Corps or the WAAC. To the Army's surprise, the recruitment produced only 1,200 male volunteers—3,000 had been expected. The process of interrogation, with its presumption of disloyalty, provoked general resentment among the internees. The question about allegiance evoked many negative replies, especially among young male Nisei, of whom 28 per cent refused to swear loyalty to the United States.

Perhaps only the Army of all government agencies could have been surprised. The Nisei had watched their rights abrogated, their property in effect confiscated, their liberties violated, all without a semblance of due process. Now the signal right offered them was that to bear arms. Furthermore, the Issei, understandably frightened, begged their sons not to depart, probably, so they believed, on a journey to death. Both generations wondered about the treatment the Army would offer a Japanese-American. One older Nisei commented that his loyal military service during World War I had spared him no losses after Pearl Harbor. The real surprise was that only a minority of Nisei answered the loyalty question negatively. Those who did enlist, especially the soldiers in the 442 Regimental Combat Team, distinguished themselves for valor in battle in Italy, as did an all-Japanese battalion recruited in Hawaii. Both units, in keeping with the Army's version of democratic doctrine, were segregated.

The loyalty processing created new problems for the War Relocation Authority, for the minority of Japanese-Americans who had refused to swear loyalty became the target of public pressure, including the Senate resolution urging their separation from the loyal. The agency confined 18,000 disloyals in a special camp at Tule Lake. Most of those affected were disloyal only in the technical sense of having replied negatively to the screening questions. Further interviewing showed them to be confused, afraid of white Americans, and as eager to remain with their families as they were resentful about evacuation. Very few expressed allegiance to the Mikado, though some others preferred the prospect of postwar life in the Orient, free from racial discrimination, to remaining in the United States. Yet all disloyals were confined together at Tule Lake, where prolonged exposure to poor food, crowded barracks, and petty authority prompted a series of protests of growing severity. Organized efforts at passive resistance, soon repressed, gave way to riots, and those to retaliation in the form of martial law and severe punishment. Many of the young at Tule Lake moved to overt acts of defiance, including the adoption of Japanese warrior manners and dress. American policies, which had earlier created much equivocal loyalty, at Tule Lake produced alienation and militancy. The Senate resolution of January 1944 proposed as a remedy only further repression.

The public mood also deferred adoption of an equitable policy. By 1944 military necessity could not continue to excuse the internment of the Japanese-American, but Stimson warned Roosevelt that a change of policy "would make a row" on the West Coast, probably lead to riots, and possibly in that event to retaliation in Japan against American prisoners of war. Roosevelt in an election year decided that "the whole problem, for the sake of internal quiet, should be handled gradually." Only after his re-election did he rescind the evacuation order, though even then the West Coast remained closed to 5,000 Japanese-Americans. Gradually the WRA helped the internees to relocate. At the end of 1945, the agency closed the camps. The damage had long since been done, and even in 1946 returning Nisei, veterans included, continued to face racial antagonism, especially west of the Sierras.

Meanwhile in three cases the Supreme Court had upheld government policy. In time of war, the Court ruled in the Hirabayashi case, "residents having ethnic affiliations with an invading enemy may be a greater source of danger than those of different ancestry." It was not for the Court, the decision continued, "to sit in review of the wisdom" of military authorities. In the Korematsu case, the Court on December 18, 1944, upheld the exclusion of the Nisei from the West Coast, though in this instance three justices dissented, with Mr. Justice Frank Murphy opposing automatic endorsement of military decisions and chastising a "legalization of racism." In *Ex parte* Endo the Court on the same day did deny the government's right to detain loyal citizens in relocation centers, but Miss Mitsue Endo had languished in one for two and a half years before the ruling forced her release. Full restitution to the Japanese-Americans for the property they had lost awaited a decision of 1968, twenty-six years after the initial offenses.

The Court's sad record during the war occasioned only one powerful criticism, the stinging analysis of Eugene V. Rostow published in the *Yale Law Journal* in 1945, too late to right the wrongs it identified. As Rostow wrote:

> Our war-time treatment of Japanese aliens and citizens of Japanese descent on the West Coast has been hasty, unnecessary and mistaken. The course of action which we undertook was in no way required or justified by the circumstances of the war. It was calculated to produce both individual injustice and deep-seated social maladjustments of a cumulative and sinister kind. . . . We believe that the German people bear a common political responsibility for outrages secretly committed by the Gestapo and the SS. What are we to think of our own part in a program which violates every democratic social value, yet has been approved by the Congress, the President and the Supreme Court?

In the case of the Italian-Americans, decency begot decency, acceptance encouraged loyalty. In the case of the Japanese, hate produced hate, rejection encouraged repudiation, and mass fear and prejudice resulted in the suspension of traditional Anglo-American civil liberties. So, too, in Great Britain, enemy aliens during World War II were subjected to the repressive policies of panic.

3. Jews in Great Britain

When the question of the treatment of the Japanese-Americans first arose, responsible officials in Washington believed they had nothing to learn from the experience of other countries. They erred, for they could have learned from the excessive precautions of Great Britain, excesses they repeated. Great Britain, where American concepts of civil liberties had originated, had developed policies for dealing with a substantial population of enemy aliens under conditions of national peril that far exceeded any the United States confronted. Those policies and the accompanying expressions of criticism provided lessons in the dangers of yielding to military judgments about civil affairs and in the regrettable limitations of corrective dissent. Because the enemy aliens of Great Britain were white men and women, mostly Germans and Italians, the issue of race did not arise, but after the Nazi victories in Europe in 1940, distrust of foreigners surfaced in England, as did anti-Semitism. The resulting harshness of policy was little relieved even after the possibility of a German invasion had receded.

In the months of the "phony war" following the outbreak of hostilities in 1939, the British government pursued a lenient course toward the 74,000 enemy aliens in the United Kingdom. Tribunals set up to investigate their credentials had interned only 439 of them by the end of March 1940. Most of the others were restricted in their travel and employment. Though some Tories in Parliament favored a tougher line, and though there were occasional "fantastic reports" seemingly designed to foster a "spy scare," the public appeared satisfied with the balance between vigilance and moderation.

That balance collapsed after the Nazis overran Denmark and Norway, the Low Countries and France. The shock of those quick German victories, the anger at Nazi collaborators within the defeated nations, the nakedness of Britain herself, and the air blitz against England during the summer and fall of 1940 stirred a latent xenophobia "We have a large number of German refugees in our midst," a letter to the *Times* of London observed on April 25, 1940, "and while these remain at large it is bound to cause uneasiness and a possible source of danger to our land. Hitler's subtle methods are known to be planned through internal means; we have only to read what has happened in Norway. . . . Surely the right course is to intern all enemy aliens at once."

There were dissenting replies, including one formidable rebuttal in *The Economist*, but even that journal had changed by mid-May, when the full sweep of England's danger had become patent. "There can be no standing on ceremony when the nation is in daily peril," *The Economist* then wrote, "and the Government's new powers to intern all enemy aliens and restrict the movement of other foreigners will be opposed by none. So far they have been applied to 3000 enemy aliens and 10,000 others in eastern and southern coastal counties, but there is no reason at all for not applying them at once to the entire country; sabotage and sedition know no . . . boundaries, and among the 27,000 or so Germans and Austrians still free there must be a cadre of enemies. . . . We can take no chances at all now." A week later *The Economist* urged the confinement of all those of both sexes "who cannot prove innocence to the hilt." That demand, an inversion of traditional criminal practice, made guilt a function of nationality, even among refugees, just as Americans later made guilt a function of race and nationality combined, and just as Hitler had made it a function of religion.

In mid-May the War Office forced the government to proceed against the refugees in England against the advice of the Home Office, which then had to assume administrative responsibility for internments and deportations. The government at that time interned "as a . . . measure of precaution" all male Germans and Austrians over sixteen and under sixty years of age. Some of those affected, mostly early escapees from Nazism, took severe losses of property in businesses that the government had encouraged them to establish during the years since 1936. All the internees suffered from conditions in the camps in England and later at the major internment center on the Isle of Man. Custodial personnel were ordinarily friendly toward their charges, but food and accommodations were crude, genuine refugees and avowed Nazis were herded together, medical attention was inadequate, and as *The Economist* noted, "for quite inexplicable reasons" books and periodicals were forbidden to inmates.

In many instances, German émigrés who had resided in England for twenty or thirty years were confined. As one younger internee, formerly a master at the Rugby School, reported: "The greatest complaint of most . . . of us is the inactivity into which we are forced. Many of us have outstanding technical, scholarly or industrial qualifications; others are young and healthy . . . and very fit for military service or industrial work. It is a most bitter disappointment to us that we have to sit quietly behind barbed wire, and are not allowed to do our bit . . . in this fight against our common mortal enemy."

When the internment of women began, husbands and wives, including many in their sixties or seventies, were often confined in separate camps, unable even to write to each other. So, too, in deporting internees to Canada or Australia, the government frequently separated families and always exposed those who were shipped out to the grave risk of German submarines. In one instance, on July 10, 1940, twenty-two married men interned on the Isle of Man left the camp voluntarily for overseas under a guarantee from the camp's commander that their interned wives in Port Erin would accompany them in the same convoy. The women received an identical promise, which the authorities in both camps then simply broke. As one critic wrote the *Times,* the internment of refugees from Nazi Germany, "besides being cruel and senseless," was destroying the good will, "nay . . . love," which the victims had for Great Britain. "In times of crisis," the Home Secretary replied, setting a precedent for Americans later, "considerations of national security must take precedence over all other considerations."

Alert though they also thoroughly were to threats to national security, the most civilized of Englishmen could not condone government policy. G. M. Trevelyan, the great liberal historian, writing while the Germans were bombing London nightly, found no evidence that the government fully realized "the great harm that is being done to our cause—essentially a moral cause . . . by the continued imprisonment of political refugees." Recalling an earlier English war against an enemy engaged in religious persecution, Trevelyan, like a good don, continued: "I am sure the Prime Minister knows what a fatal injury would have been done to our war effort between 1689 and 1713, if when war broke out, Huguenots who had come over since the Revocation of the Edict of Nantes had been imprisoned for the duration because one in a thousand might be a Jesuit in disguise."

That was the point of view also of the eminent author H. G. Wells and the celebrated classicist Gilbert Murray, among others. The "deplorable management of the refugee question," Murray recognized, was due in part to haste and thoughtlessness born of "deadly peril," but there was a "deeper cause." The demand to "intern all Germans," indeed "to intern all foreigners," struck him as "the reaction of the average ignorant and unthinking man, who can see no difference . . . between one foreigner and another. Oppressor and victim, Fascist and anti-Fascist, are all the same to him. . . . I tremble for any democracy which yields . . . to mob hysteria."

The New Statesman and Nation especially carried the case against official policy, which it attributed to "ignorance, panic and moral cowardice." The employment of skilled refugee artisans and intellectuals, and the recruitment of a European legion, The New Statesman argued, would assist the prosecution of the war. At the least, the Home Office could hasten the identification of possible spies and, while doing so, permit newspapers and personal mail to reach those imprisoned. Merely to have to propose such improvements disclosed "the measure of the barbarism to which we, a free people, have consented."

The New Statesman also condemned the wholesale transportation to the dominions not only of prisoners of war and convicted Nazis, but also of anti-Nazi Germans and anti-Fascist Italians, some of whom were aboard the Andora Star, which was torpedoed at sea. Its sinking had first disclosed to the public the indiscriminate nature of the deportations. Yet within the War Cabinet, only Ernest Bevin, a Labour member, had the courage "to speak strongly" about the general issue.

Statistics bore out strong talk. Of the Germans confined in a typical camp in November 1940, 30 per cent had previously been in Nazi prisons or concentration camps, 70 per cent had been classified in 1939 by British tribunals as "refugees from Nazi oppression," 82 per cent were Jews, 11 per cent were cripples, and 27 per cent were over fifty-five years old. Furthermore, 87 per cent, all the fit men, wanted to volunteer for some form of British national service.

Stung by criticism in the press and Parliament, the Home Office announced in August 1940 procedures by which most refugees could apply for release to a newly created committee. But those procedures proved exasperatingly slow, and the muddle of Home Office administration seemed to The Economist and The New Statesman a disgraceful sequel to internment itself. "There can be no question," Winston Churchill nevertheless asserted, "of releasing persons of enemy nationality merely because they have lived here a long time or have shown no sign of disloyalty or lack of sympathy with the Allied cause."

In November 1941, after Germany had turned its fury largely toward the Soviet Union, the British courts sustained the Prime Minister. The case of *Liversidge* v. *Anderson* tested the authority of the Home Secretary to intern any person if he had "reasonable" grounds to believe the safety of the state so required. Four law lords took the view that the Home Secretary needed only to be satisfied by the evidence placed before him. "With majestic emphasis," Lord Atkin alone dissented. He insisted on the right of a person, in a country ruled by law, to appeal to the courts against executive decisions. Though Parliament had long ruled in Great Britain, Harold J. Laski, a renowned liberal British legal scholar, praised Lord Atkin for his wisdom. "Unless it can be shown," Laski wrote, "that submission to the judiciary by the executive puts the safety of the State in immediate hazard the violation of a supreme tradition is a misfortune which requires instant remedy."

In the aftermath of the decision in *Liversidge,* the government refused to consider any modification of the power of the Home Secretary. The ensuing debate in the House of Commons revealed a growing uneasiness, especially among Labour members, who sponsored an unsuccessful measure that referred to "the grave concern now felt in this House at the abolition of any judicial safeguard for the liberty of the subject." Supporters of that view worried particularly about the indefinite internment of individuals who they had earlier thought would suffer only temporary confinement. But the government was unmoved, and the internees continued only slowly to obtain release from camps, and then only under restrictions imposed on their movement and employment.

There was a further humiliation. Aliens, mostly German Jews, who volunteered and were accepted for service in the British armed forces faced special dangers were they to fall into Nazi hands. Yet the government denied them British nationality, though the grant of citizenship would have had "an immense moral value."

Disturbed by all aspects of the refugee problem, *The New Statesman* concluded early in 1941 that "no country seems capable of eliminating sickening abuses in its treatment of helpless civilians." The editors had one hope: "Cannot the United States move in this matter? Mr. Roosevelt is in a position . . . to ask for a world standard of decent treatment for detained persons everywhere." A further "grand gesture" would open the United States and, through American pressure, Palestine also, as sanctuaries for the Jews of Europe, "the unhappy victims of German racialism."

Those were misplaced expectations. Though exposed to little

peril, the United States under Roosevelt in 1942 treated the Japanese-Americans even worse than the Churchill government treated alien refugees. The American people, press, and Congress were no more openly critical of violations of civil liberties than were their British counterparts. The Supreme Court proved to be as supine as the law lords. Like Churchill, Roosevelt was trapped in illusions about military necessity. And for both the British and American governments, those illusions, along with persisting prejudices, impeded, almost prevented, the provision of safe areas for escape for the Jews of Europe, whom Hitler was systematically murdering.

4. American Jews

During World War I, Jews in Germany remained loyal to the cause of the Fatherland, while German-Americans—Jews and gentiles, citizens and aliens—stood divided in their feelings about United States foreign policy. Many German-Americans opposed the steps that brought the country into the war. Fewer, though still a large number, were unreconciled to the declaration and pursuit of war. During that war other Americans, spurred by official propaganda, gave way to a hysterical rejection of all things German. By 1941, however, the attitudes of 1918 had almost entirely evaporated.

When, after Pearl Harbor, Hitler declared war on the United States, only an insignificant fraction of German-Americans, a tiny minority easily controlled, sympathized with the Nazi regime. Official policy abjured a campaign of hatred against the German people, though it constantly attacked Nazism. Moreover, the German-American aliens in the United States, far fewer than in 1917, were primarily Jews who had fled Germany and other areas under German control. In the United States during World War II, there was no anti-German hysteria. There did persist considerable anti-Semitism, directed against Jews whatever their national origins, and a related indifference to the plight of the Jews of Europe whom Hitler had set out to exterminate.

Wartime public-opinion studies disclosed that Americans distrusted Jews more than any other European people except Italians. Nevertheless, in 1941 Jews within the United States had suffered somewhat less from discrimination than had the Italians, and much less than the Japanese or other non-Europeans. Like the Italians, Jews had used their votes to reward sympathetic politicians. More than the Italians, Jews had received federal appointments, especially appointments of prestige and power, from the Roosevelt administration. Further, American Jews had created strong organizations of their own, many of them well financed by the substantial wealth of the elite of the Jewish community. Jews in the United States, like Jews elsewhere, had struggled, with marked success, to obtain both an advanced education and footholds in the professions. Politically, socially, institutionally, American Jews had the strength to influence government. For their part, anti-Semitic Americans—except for several thousand cranks, most of them uneducated—although they approved of the exclusion of Jews from certain occupational and social groups, did not let their prejudices extend to outright persecution. For the most part the lawyers and bankers of the establishment, many of whom would not happily hire a Jew, were nevertheless appalled by Hitler's doctrines as well as by his ghastly programs.

Still, during the years of war, as in the years preceding them, Jews in America felt the hostility of their countrymen, of whom some were other, more privileged Jews. In what was intended to be a sympathetic analysis, *Fortune* in 1936 revealed the unconscious anti-Semitism of its editors and much of their readership. The apprehensiveness of Jews, *Fortune* argued, contributed to anti-Semitism, whereas the admitted indifference of some 40 per cent of Americans to the fate of Jews in Germany seemed to *Fortune* not "callous" but "the most effective prophylactic against the pestilence of hate." Though *Fortune* recognized that Jews were "not a race in any scientific sense of the term," the magazine proceeded to stereotype Jews. It attributed to Spanish Jews, the aristocrats of American Jewry, for example, "thin features and spare bodies which often take on a typically Yankee look," presumably a desirable distinction. Contrary to anti-Semitic folklore, *Fortune* observed, "Jews are so far from controlling the most characteristic . . . American activities that they are hardly represented in them." There was no basis for the fear that they would dominate the economy. But they did invite prejudice, *Fortune* wrote, because of their "notorious" tendency to "agglomerate," not just in cities but also in self-constituted communities within cities—a curious interpretation of restrictive, anti-Semitic, real-estate covenants. The article concluded by wondering whether Jewish "clannishness" would give way among most Jews, as it had with upper-class Spanish and German Jews, to a willingness to be absorbed in American culture.

That question, like the article itself, underestimated the degree to which American Jews, especially the poorest of them, largely the most recent immigrants and their children, were through no wish of their own denied access to employment, especially in white-collar jobs. Evidence of economic discrimination reached the Fair Employment Practices Commission during World War II. An FEPC hearing in New York City in February 1942 produced compelling evidence of bias against Jews there—in all, 2.5 million people, who constituted about a quarter of the population of the metropolitan area. Defense manufacturers, as also in Detroit, rarely said they would not hire Jews, but application forms for employment usually asked about religion. Approximately 30 per cent of the want ads for employment placed in the New York *Times* and *Herald Tribune* expressed a preference for Protestants or Catholics. Of the graduates of the defense training schools in the area, a significantly larger percentage of gentiles than of Jews had found jobs. Even some of government's own committees on defense employment discriminated against Jews. The FEPC attributed those and similar conditions elsewhere to traditional prejudices against Jews in heavy industry, to the high incidence in aviation and other plants of German and Scandinavian foremen who believed Jews incapable of mastering mechanical problems, and to the conventional, internally contradictory folklore that held that Jews were at once radical and greedily capitalistic.

As the industrial need for manpower grew, so did opportunities for employment for Jews, as also for Italians and blacks, but prejudice did not diminish on that account. Conservative opponents of the New Deal had long considered the President, and even more so Mrs. Roosevelt, too friendly to Jews. During the war the least principled of those opponents attacked Mrs. Roosevelt for her association with allegedly radical Jews, and criticized federal agencies for enlisting in publicity campaigns the services, though without compensation, of such Jewish celebrities as the actor Melvyn Douglas. In spite of specific denials from General Lewis Hershey, the head of Selective Service, rumors persisted at least through 1942 that Jews were evading the draft. One OWI report concluded that an organized campaign to discredit Jews had grown up in Irish and other anti-British communities. Though that was perhaps an exaggeration, there were sporadic episodes of violence against Jews in Irish neighborhoods in Boston where, before the war, the influence of Father Charles Coughlin, the anti-Semitic radio priest, had been endemic. Even within Congress there was a small group that defended the American fascist agitators, invariably anti-Semitic, whom the government was prosecuting (unsuccessfully, as it turned out) for conspiracy to damage the morale of American soldiers. So, too, in anti-Semitic suburbs like Darien and New Canaan, Connecticut, the prosecution of a war against Nazism relieved none of the conditions of restrictive residential covenants, the object of sensitive criticism in the novel *Gentleman's Agreement,* a best seller of the early postwar period.

Nevertheless, wartime prosperity, the increasing geographic mobility of Americans, the homogenizing effects of shared dangers in battle, the essential contributions of Jewish refugee scientists, and the stunned reactions to the Nazi gas chambers, combined to facilitate for American Jews acceptance by the society in which they lived. For many Jews, by 1945 changing conditions within the United States were encouraging. Still, even optimistic Jews could only be dismayed by the anti-Semitism, the bureaucratic inertia, and the ordering of the government's wartime priorities that prevented any effective effort to save the lives of the Jews of Europe until 6 million of them had been subjected to Hitler's "final solution."

No conceivable effort could have saved most of the victims of Nazi savagery, but British and American policies reduced the chances for escape that might otherwise have become available. As the Nazis closed their grip on Europe, Jews and other threatened people could save themselves only by fleeing to havens outside Hitler's domain. Spain and Turkey, fearful of German reprisals, provided only limited sanctuaries. Palestine, which Zionists had wanted to make a Jewish homeland since the time of World War I, was effectively closed to refugees by British policy, which the United States supported. A flood of Jews to Palestine, the British contended, would anger the Arab states, whose wartime co-operation and oil resources were essential. For his part, Roosevelt had never deemed Palestine a politically viable area for Jewish resettlement. Instead, he had harbored fantasies about possible resettlement in the Cameroons or elsewhere in Africa, or in New Guinea, or in one or another of the Latin-American countries, none of which was willing to assume the role he contemplated.

The Immigration Act of 1924 had restricted the number of nationals, Jews and others, of any country who could enter the United States in any one year. Even before 1939, the demand for immigration visas from Jews in European countries far exceeded the available quotas, yet only 8 per cent of the American people were then willing to admit more Jewish refugees from Europe. A solid majority of Congress continued to oppose amendment of the act of 1924. Worse, American consular officers, reflecting if not sharing the anti-Semitism of their seniors in the State Department, consistently resisted Jewish efforts to emigrate, while department officers in Washington, in part because of characteristic bureaucratic inflexibility, impeded any significant emergency relaxation of limitations on quotas.

Though the President might have issued an executive order temporarily to ease the block, on several counts he did not. He never understood how bitterly anti-Semitic was Breckinridge Long, the Assistant Secretary of State responsible for administering immigration policy. Long, for many years an acquaintance though not an intimate of Roosevelt's, had impeccable social credentials, and a patrician manner that hid the passion of hostilities that he confided only to his private diary. To his critics Long's bias seemed transparent, but he denied the charges they brought against him, and Roosevelt attributed his failings not to malice but to "old boy" ineffectuality, which he resignedly considered typical of the Department of State.

More important, subject though Roosevelt was to political pressure, American Jews were divided among themselves about the direction in which they wanted him to move. Until Hitler began to try to exterminate the Jews, a minority of Jews in America feared that an easing of immigration policy and a resulting tide of refugees would exacerbate anti-Semitism to their own disadvantage. In some measure that anxiety persisted even after news of the gas chambers and crematoria reached the United States. The flow and impact of that news, moreover, was reduced by the restrictions on communications from Europe that the State Department imposed. Most American Jews were nevertheless committed to the rescue of their European coreligionists, an objective also of the Quakers and other Christian humanitarians; but the Jewish consensus about the need for rescue dissolved over the question of method. Even after the end of the war, many American Jews were not Zionists, or were even comfortable with Zionist objectives and rhetoric. Dubious about the establishment of a Jewish state in Palestine and about its possible future, they did not think of themselves as having, nor did they desire, Jewish nationality. The whole concept of a Jewish homeland conflicted with their sense of their own Americanism. The intensity of the Zionists, especially of Rabbi Stephen S. Wise, who had occasional access to the President, compensated for the doubts of the anti-Zionists. By 1943, so did the support for Zionism of a majority of American Jews and of several of Roosevelt's advisers, among them Justice Felix Frankfurter and, later, Vice President Henry A. Wallace. Even so, the persisting division of Jewish opinion and influence relieved the pressure Roosevelt might otherwise have felt to combat British policy in Palestine or to modify American restrictions on immigration.

In addition, all proposals for assisting the escape of European Jews clashed with Roosevelt's necessitarian view of the war. On that basis the President accepted the arguments of the Churchill government against any move that might antagonize the Arabs, though Roosevelt did toy with the fruitless thought of trying to bribe King Ibn Saud of Saudi Arabia. The State Department and the British Foreign Office refused to negotiate with either Germany or her satellites about providing ransom, to be frozen until the end of hostilities, for the release of Jews. Roosevelt did warn Germany and her allies that after victory the United Nations would punish all those implicated in political or religious murder, but not until 1944 did he retreat from the doctrine that the most effective means for rescuing Jews was quick and complete victory.

There was substance to that position, as there was also for American collaboration with Darlan in North Africa and Badoglio in Italy. As commander in chief, the President felt a staggering responsibility for the lives and welfare of American troops. A collaboration that promised to spare their lives and to hasten victory had much to commend it. As a dedicated democrat who detested Nazism, the President believed its elimination essential for the postwar world. For good historical reasons, he distrusted Hitler as much as he hated him. From those feelings and from the need to preserve the trust and cooperation of the Soviet Union, there emerged his demand for unconditional surrender, a demand that seemed to preclude negotiations with the ultimate enemy. Yet that demand and its corollary may have cost some American lives, while "rescue through victory," by prolonging the period of inaction, led to the sacrifice of many Jews to Hitler's passion. In this as in other instances, the imperatives of warmaking, as Roosevelt interpreted them, had on occasion inadvertently inhumane results. Roosevelt was not unfeeling. He was harassed, preoccupied with overwhelming matters of state, caught in the web of a mammoth war in which no head of government could wholly escape confusion and uncertainty.

Roosevelt retained his personal popularity among American Jews, however, even after various Jewish groups began in March and April of 1943 to bring some unity to their own belated campaign against his administration for relying upon victory for rescue. Agitation of that issue began with a mass meeting at Madison Square Garden, which put forward a resolution that described the minimum objectives of American Jewish organizations for the balance of the war, and called for negotiations with Germany through neutral agencies for the rescue of Jews. The United States, it also proposed, should prepare sanctuaries for those who could be rescued, explore possibilities for resettlement in Africa, Latin America, and the British dominions, grant special passports for stateless refugees, and guarantee postwar repatriation of refugees who desired it. The resolution urged the United States and Great Britain to accept more refugees, and Great Britain to open Palestine to refugees, especially those who had already reached Spain, Portugal, and the Balkans. The mass meeting turned public attention to the refugee problem, as did new reports about Nazi slaughter of Jews at Treblinka. The House and Senate passed a resolution for the punishment of war crimes, including genocide, and Emanuel Celler led a delegation of Jewish congressmen who protested to Roosevelt against the State Department's refugee policies. The President merely referred them to Breckinridge Long.

Long continued to interpret Jewish agitation as debilitating to the war effort. A relaxation of visa policy, he argued, would permit Nazi agents to ship in among genuine refugees. The "hot headed masses," he wrote privately about the Madison Square Garden rally, "would take the burden and the curse off Hitler." Worse, though the implications of the "final solution" had become public knowledge, Long shaped the proceedings of the Anglo-American Bermuda Conference of 1943 on refugee problems to confirm the doctrine of rescue through victory. As a leading English participant later admitted: "It was . . . a façade for inaction."

By the fall of 1943 some members of Congress and officers of the Treasury Department had penetrated that façade. Identical resolutions in the House and the Senate, sponsored in each case by Protestants, urged the President to create a commission to make plans "to save the surviving Jewish people of Europe from extinction at the hands of Nazi Germany." The hearings that followed the introduction of that resolution revealed Long for just what he was. Meanwhile, Treasury Department lawyers, also concerned Protestants, had begun to probe the extent of the State Department's complicity in withholding information about the plight of European Jews and in resisting efforts to save them by procedures consonant with the national interest.

Secretary of the Treasury Henry Morgenthau, Jr., the only Jew in the Cabinet, had been an intimate of Roosevelt's for many years. Constantly impatient with the State Department, Roosevelt often enlisted Morgenthau for special assignments in foreign policy, which Cordell Hull bitterly resented. In the case of the European Jews, Morgenthau became involved on his own initiative, at first only marginally, later—under prodding by his own subordinates—aggressively. By 1944 his experience had converted him to Zionism, which he had previously opposed. His involvement had also demonstrated, as did most facets of his public career, his ability to recruit and use a first-rate staff and his doggedness as an advocate within the close circle around the President. Until mid-1943 he accepted without argument Roosevelt's deference to British policy in Palestine and his insistence on strict compliance with American immigration laws. But in July 1943 he began to resist. His staff then started to uncover the shocking story that they later presented to him, and through him to Roosevelt, in a fully documented study entitled "Report to the Secretary on the Acquiescence of this Government in the Murder of the Jews."

As that study disclosed, details about Hitler's "final solution" had reached the State Department in 1942. The information came from one Gerhart Riegner, a German Jew who had fled to Geneva, where he represented the World Jewish Congress. Through private channels, Riegner first got his ghastly account to Rabbi Stephen Wise, who asked the State Department for confirmation. The department provided it through an inquiry to the United States minister at Bern. His reply led to Wise's publication of the information and to Roosevelt's declaration that religious and political murder would be punished. A second cable from Riegner, transmitted through the American legation to Wise, reported in January 1943 that the Germans were killing Polish Jews at the rate of 6,000 a day, and deliberately starving to death Jews in Rumania and elsewhere. At that point the State Department, at Long's contrivance, ordered the minister at Bern to cease transmitting messages to private individuals. Though newspapers occasionally circumvented that effort at censorship, Riegner's loss of a channel of communication reduced American knowledge about Nazi genocide, and on that account, as Long had intended, probably tempered protest for several months.

In July 1943 representatives of the American Jewish Congress told the Treasury that for a bribe of $170,000 Rumanian authorities would permit the evacuation of 70,000 Jews. After studying the details of the proposed transaction, Randolph Paul, the Treasury's General Counsel, recommended approval. No dollars or other desirable currencies, he found, could reach Axis representatives until after the war ended. Orally Roosevelt also approved the plan, but for four months Long blocked it in the State Department, first by delaying action, then by contending that the plan needed British clearance, which had been withheld. By late November the Treasury had also developed an interest in providing foreign exchange, again with appropriate safeguards, to purchase the release of some 6,000 Jewish children in France. When on December 6, 1943, Morgenthau wrote Secretary of State Hull a strong letter about both the Rumanian and the French proposals, Hull replied evasively. He referred sympathetically to "the desperate plight of the persecuted Jews" but argued that neither plan had been developed adequately. As the Treasury ascertained, the British Foreign Office and the Ministry of Economic Warfare were reluctant to make even preliminary financial arrangements that might rescue 70,000 refugees from enemy territory only to leave them in need of accommodations in Palestine.

At last, Randolph Paul concluded, the Treasury had got "down to the real issue." He and his associates implored the Secretary to propose to Roosevelt the appointment of a special commission on refugees, a commission that could bypass the State Department. Instead, confident that Hull would co-operate once he learned the facts, Morgenthau took the case again to the Secretary of State, who had become as indignant with the British as was the Treasury group. Now at last Hull issued the necessary instructions and licenses to effect the rescue of the Rumanian and French Jews.

In January 1944 Morgenthau took to Roosevelt Paul's "Report . . . on the Acquiescence . . . in the Murder of the Jews." It contained an indictment of the State Department and a plea for the establishment of a War Refugee Board to assist the rescue and relocation of European Jews. The timing of the report coincided with mounting congressional concern and with a reorganization of the State Department that was transferring Long's responsibilities to the Under Secretary of State, the then newly appointed Edward Stettinius, Jr. With Stettinius's unequivocal support, Roosevelt, by executive order on January 22, 1944, created the War Refugee Board along the lines Paul had recommended. Then and later, American Jews grieved over opportunities lost during the previous eighteen months. The fight, as Morgenthau recalled, had been "long and heartbreaking. The stake was the Jewish population of Nazi-controlled

Europe. The threat was their total obliteration. The hope was to get a few of them out."

That hope rested, during the months remaining before the defeat of Germany, with the War Refugee Board, which functioned from the first with energetic efficacy. Though it could not influence British policy in Palestine, the board, chaired by a former staff member of the treasury, did bring the President to order the construction of temporary camps for refugees in Italy and North Africa, an example the British emulated, and to make available "Emergency Refugee Shelters" in the United States. To those former Army camps refugees could be moved without regard to immigration quotas. A special poll indicated that 70 per cent of Americans approved that decision. Americans nevertheless remained largely oblivious to the terror in Europe. Another poll, taken in December 1944, revealed that they knew Hitler had killed some Jews but could not believe that even the Nazis had methodically murdered millions. Yet the Nazis continued to do exactly that until the very hour of their defeat. During 1944 and the first four months of 1945, relentless genocide destroyed far more human beings than the War Refugee Board assisted.

Jewish leaders in the United States had tried but failed to impress upon their countrymen and their government the terrible urgency of their case. That failure arose from the continuing hostility of the public and the Congress to changes in immigration policy, from Anglo-American caution toward the Arabs, from the bumbling and the biases of the State Department, and from Roosevelt's preoccupation with victory. The insensitivities that at once permitted and reflected those conditions were hardened by anti-Semitism and comparable to the insensitivities toward the Japanese in the United States and the Germans in Great Britain. Even during a war against Nazism, the democratic impulse did not dissolve traditional antipathies toward those whose race or religion or national origin marked them as outsiders.

6 / Black America:
The Rising Wind

1. Jim Crow

Though neither aliens nor enemies, American blacks during World War II suffered, as ever, from white hostility and oppression. That had also been their lot during the years of the New Deal. Federal agricultural policy had resulted in the displacement of thousands of blacks from their tenant farms in the South. The Tennessee Valley Authority, the Civilian Conservation Corps, the Works Progress Administration in the South, and other, less important, emergency agencies had preserved Jim Crow practices, at least in part. Deferring to the sensitivities of Southern congressmen whose votes he needed for his programs, the President refrained from supporting antilynching legislation. North of the Mason-Dixon line, blacks received some help from federal relief and work relief, and some token space in federal housing projects, but private industry ordinarily discriminated against blacks in hiring, wages, and promotion, and the older labor unions continued to exclude them. The President did abandon the lily-white policies of the Hoover administration, desegregate federal buildings in Washington, appoint some black leaders to respectable offices, and abide, though without enthusiasm, his wife's vigorous efforts to advance the status and lift the confidence of blacks. Her compassion and his concessions to it, made as they were in his effervescent style, raised the hopes of blacks for an equitable share in national life. The loyalty of blacks to the United States lay both in that growing expectation and, more important, in their basic Americanness. But that loyalty was ambiguous, for educated blacks particularly resented the poverty of their people, the apartheid of American society, and the obstacles to the correction of those conditions.

During 1940 and 1941, the increasing involvement of the

United States in assisting the enemies of the Axis did not dissolve that ambiguity. In Asia the Japanese claimed to be fighting against the domination of white, European colonialism, not least as it weighed upon the corrupt government of China. Germany's western European enemies had huge imperial holdings within which they subjugated brown and black men to white rule. The battle for freedom, as the Allies fought it, did not promise freedom for dark peoples. That anomaly especially troubled American blacks, and whites, too, in the case of the British in India. Though only a few black fanatics sympathized with the Nazis—for Hitler was so patently a racist that the prospect of his triumph threatened disaster—many blacks felt some affinity for the Japanese, even after Pearl Harbor, and few informed blacks had unequivocal enthusiasm for Great Britain.

Furthermore, during the months of American preparation for war before Pearl Harbor, months characterized by both soaring industrial production and the growth of the armed services, blacks found themselves excluded from the rising prosperity of the nation and from opportunities for advancement and command within the Army and the Navy. "We Americans," Roosevelt had told the European democracies, "are vitally concerned in the defense of your freedom." But black Americans, with ample cause, had an urgent concern for freedoms of their own, so long denied.

During 1940 and the first half òf 1941 the demand for labor in construction, heavy industry, and the aircraft industry sopped up white unemployment while blacks remained without jobs, without training, and deprived of income because of reductions in federal relief programs. In the South, where much defense industry located, the National Youth Administration could not enroll blacks in work-study programs because there were no technical schools for blacks and local practice kept blacks out of schools for whites. Elsewhere the president of North American Aviation admitted that his firm would not employ Negroes, a policy the International Association of Machinists endorsed. The United States Employment Service, a federal agency, continued to fill "white only" requests from employers of defense labor. Its "general policy," the service announced, "with respect to the Negro situation had been to operate according to the social pattern of the local community." That policy reinforced local discrimination against or apathy toward blacks. So did the timidity of Sidney Hillman, the labor leader serving as co-director of the Office of Production Management, who would not consent to antidiscrimination clauses in government defense contracts. According to the calculations of another federal agency, some half-million blacks who could have assisted war production remained unemployed and several million blacks underutilized. "If there is such a thing as God," one of them wrote Roosevelt, "he must be a white person, according to the conditions we colored people are in. . . . Hitler has not done anything to the colored people—it's the people right here in the United States who are keeping us out of work and keeping us down."

Too many people within the federal government were keeping blacks down. Blacks had difficulty finding federal employment, particularly in any but custodial and, less often, clerical positions. Once they had begun to agitate in 1941, the federal government within four years tripled the number of blacks employed. But early and late, the most intractable departments were War and Navy. When in 1940 blacks began to enlist, which they did at a rate 60 per cent above their proportion of the population, the Army assigned them almost exclusively to segregated units and trained them almost entirely for noncombatant tasks as laborers, stevedores, and servants. Blacks could not enlist in the Air Corps or Marine Corps, elite services by self-definition. The Navy accepted blacks only for menial roles as officers' cooks and stewards. Those drafted fared no better.

The resulting "grave apprehensions" that blacks felt had a strong basis in the ill-concealed prejudices of the Secretary of War and of the Army's top command. During World War I, Army tests had examined levels of education rather than intelligence. Deprived as they had been of formal education, black recruits, then largely from the South, performed below the national average. The Army, misinterpreting the results, concluded, as it had expected to, that blacks were inferior to whites, a conclusion that, with singular exceptions, remained military doctrine through World War II. During World War I relatively raw black recruits, serving in segregated units (which later testing demonstrated to have lower morale than did integrated units), on occasion fought badly, just as did some all-white units that were less visible. In 1940 Secretary of War Henry Stimson recalled, though he could have discovered otherwise, only the failures of blacks—"the poor fellows made perfect fools of themselves"—just as he remembered the results only of the skewered tests. Segregation, he also believed, involved sentiments the Army could not change or challenge. It could not operate as "a sociological laboratory" while it prepared for war. Here again the evocation of military necessity confirmed the unfortunate set of prejudice.

"In tactical organization," Stimson said, "in physical location, in human contacts, the Negro soldier is separated from the white soldier as completely as possible." General George Marshall, a Virginian, approved. As he saw it, desegregation of the Army would destroy morale "established by the American people through custom and habit." The segregated blacks, Stimson further argued, again with Marshall's support, had to serve under white officers: "Leadership is not imbedded in the negro race yet and to try to make commissioned officers to lead men into battle—colored men—is only to work a disaster to both."

Later experience exposed the vacuousness of those views, though Stimson and Marshall were never wholly persuaded; but in 1940 blacks had yet to have a chance to prove their martial valor and ability, so their spokesmen could turn only to protest. Beginning with restraint, they appealed through Eleanor Roosevelt for the intercession of the President. Her enthusiasm did not much move him, even in the months before the 1940 election. Confident of the black vote, he conferred with three eminent blacks but agreed only to try to persuade defense contractors to modify their hiring procedures and to token alterations in established military practices. Stimson then appointed William Hastie, Dean of the Howard University Law School, as his Civilian Aide on Negro Affairs "in the . . . development . . . of policies looking to the fair and effective utilization of Negroes." The Army agreed to organize some black combat units and to train some black aviators, and it promoted Colonel Benjamin O. Davis, its senior black officer, to brigadier general; but it still declined "to intermingle colored and white enlisted personnel in the same regimental organizations."

At best, the President and his subordinates had taken a short step toward that typical but inadequate American standard of "separate but equal." As *The Crisis,* the journal of the National Association for the Advancement of Colored People, later put it: "A jim crow army cannot fight for a free world." Nor was a Jim Crow country free. So persuaded, A. Philip Randolph resolved at the end of 1940 to abandon restraint and to substitute action for talk, a course in which he soon enlisted others ordinarily less aggressive than he was.

One of the great leaders in American history, Randolph had lived in Harlem for three decades while that community grew in size, self-consciousness, and, despite its general poverty, reputation as the center of black culture for the United States and much of the rest of the world. By 1940 he had established his own reputation as a forceful, independent man whose quiet talk, courtly manner, and devout Methodism concealed a determination, formulated in his youth, to dedicate himself wholly to "creating unrest among the Negroes." An early champion of black socialism and trade unionism, he had put justice ahead of patriotism during World War I, called for the redress of every black grievance and reprisals against every white assault, and so convinced the attorney general of the United States that he was one of the "most dangerous Negroes" in the country. Socialism, Randolph discovered, did not attract American blacks, but his venture in trade unionism did. An organizer and soon the head of the Brotherhood of Sleeping Car Porters, he battled to obtain for that union full bargaining rights with management, while he also relied on its members to help him "to carry the gospel of unionism in the colored world." New Deal labor policies helped the brotherhood to win its fight, and Randolph's tough reform activities won him growing respect among other black leaders. The ablest of them were ready to support him when he organized the Negro March on Washington Committee, of which he was director.

The committee, as Randolph announced in January 1941, planned to bring "pressure to bear upon . . . the Federal Government to exact their rights in National Defense employment and the armed forces. . . . I suggest that TEN THOUSAND Negroes march on Washington . . . with the slogan: WE LOYAL NEGRO AMERICAN CITIZENS DEMAND THE RIGHT TO WORK AND FIGHT FOR OUR COUNTRY. . . ." To that plan there rallied, among others, Walter White, Secretary and chief executive officer of the NAACP, which historically had eschewed militant tactics; Lester B. Granger of the still more conservative Urban League; Dr. Rayford Logan, a prominent black scholar and the chairman of the National and State Committee for the Participation of Negroes in National Defense—essentially an establishment committee; and the Reverend Adam C. Powell, Jr., then a fiery young Harlem preacher.

Those men and other representative black leaders were committed, as Randolph wrote Franklin Roosevelt on May 29, 1941, to mobilize by July 1 "from ten to fifty thousand Negroes to march on Washington in the interest of securing jobs . . . in national defense and . . . integration into the . . . military and naval forces." The marchers would join in a "great rally" at the Lincoln Memorial, where Randolph invited an address by the President "as the greatest living champion of the cause of democracy and liberty."

Characteristically, Roosevelt interpreted the organization of the March on Washington as a problem in public relations. Sweet talk, he had long believed, would quiet the agitation of the blacks. In part he relied upon his wife's devotion to helping blacks to appease them, for he could always disown the activities of his "missus." In part he expected his personal staff, especially Steve Early, Marvin McIntyre, and "Pa" Watson, to find ways to deflect demands that blacks made on him. That technique had proved efficient in the case of Walter White, whose frequent complaints regularly elicited soothing correspondence and interminable stalling. At the most, Roosevelt was prepared to interrupt his schedule to sweet-talk blacks himself. He had supreme confidence, based upon a remarkable record of success, in the contagion of his charm.

When Randolph first challenged him, the President, although resolved to prevent the march, which would at the least embarrass the federal government, was not ready to concede anything substantial. He had Under Secretary of War Patterson prepare an answer to Randolph for the Army. Confessing uncertainty about procedure, Patterson replied on June 3, 1941, that the War Department had already "done everything feasible to see that no discrimination is practiced against Negroes." The department had urged contractors "to give negroes equal treatment, but obviously we have no measure of adequate control as to their employment policies." On the military side, Patterson cited the promises already made and then being fulfilled. "The complaint," he continued, ". . . is that we have not mingled white troops and negro troops in the same units. Such a mingling was not part of the President's policy, and for practical reasons it would be impossible to put into operation. It would seem that negroes might be inspired to take pride in the efficiency of negro units."

So it did seem to Patterson, who had missed the entire significance of Randolph's demands. Mayor Fiorello La Guardia, of New York City, knew better. Through an aide, Roosevelt appealed to La Guardia "to exercise his persuasive powers" to stop the march. Eleanor Roosevelt was assigned the same mission. Both she and La Guardia recommended that Roosevelt, accompanied by Stimson, Knox, and the heads of the WPB, meet with Randolph at the White House "and thrash it out right then and there." That would stop the march, La Guardia believed, "and nothing else will." Roosevelt accepted that advice and also asked Randolph, pending the meeting, to "stop activities for the mobilization . . . for the march." Randolph declined, "because the hearts of Negroes are greatly disturbed . . . and their eyes and hopes are centered on this march." As he told the President, "we feel as you have wisely said: 'No people will lose their freedom fighting for it.' "

As Randolph later recalled the White House meeting of June 18,

Roosevelt started out to play the raconteur. For his part, Randolph insisted on "something concrete, something tangible . . . and affirmative." He proposed an executive order "making it mandatory that Negroes be permitted to work" in defense plants. Roosevelt hesitated. Supported by Walter White, Randolph predicted that they would have 100,000 blacks in the ranks of the march. At La Guardia's suggestion, the adversaries agreed to appoint a committee to draft an executive order satisfactory to Randolph. He rejected several drafts, until on June 25, in a statement calling off the march, he approved a text that Roosevelt issued as Executive Order 8802. The order declared it to be the policy of the United States "that there shall be no discrimination in the employment of workers in defense industries or government because of race, creed, color, or national origin." To enforce that policy, the President appointed a Fair Employment Practices Committee to "receive and investigate complaints of discrimination" and take "appropriate steps to redress grievances."

The threat of action had resulted in federal intercession, unprecedented since the end of Reconstruction, in behalf of equal rights for blacks. Still, among black spokesmen there lingered dissatisfaction and uncertainty. Roosevelt had taken no step to desegregate the armed services. The effectiveness of FEPC, though it began at once to make some progress toward its assigned goal, was proving limited. Only by continuing their fight could American blacks secure the rights they had never known. The first victory encouraged many of them to press on. Randolph, for instance, kept alive his all-black March-on-Washington organization. The prospect of further mass action challenged the tradition of federal inertia; the expectations of American blacks challenged the customs of their white countrymen, when the attack on Pearl Harbor gave a new urgency at once to preserving tranquillity within the nation and to democratizing its institutions as an example to the world. For black leaders the latter of those objectives, for the President the former, had priority. The resulting tension marked all the years of the war.

2. The Persistence of Prejudice

"We must be particularly vigilant," the President warned the country in his State of the Nation address of January 1942, "against racial discrimination in any of its ugly forms." During the ensuing months, blacks had little occasion to believe that that warning affected their lives. The Office of Facts and Figures and later its successor agency, the Office of War Information, conducted a series of polls of black opinion that consistently confirmed the complaints of the NAACP and of the more outspoken black press. The fundamental patriotism of American blacks prevailed, but against long odds: frustration, pessimism, cynicism, and insecurity characterized the attitudes of blacks toward the war. "It is not a question," one black analyst explained, "of Negro . . . loyalty, but essentially that of lack of enthusiasm for a war which they do not believe is being fought for . . . true democratic principles. They feel that the 'fight for freedom' is meant only for white men and does not include the Negro in America, or dark people anywhere in the world. As in India, Negroes in America want some concrete assurance by direct proof that they will fight not only for others, but that they are fighting for gains of their own."

With justification, most blacks felt that they were getting less than a fair share of "the new jobs created by the war" and receiving lower wages than whites employed at identical tasks. Barriers to employment and earnings ranked first among their grievances. A close second involved the residential conditions in which they had to live, and the continued segregation of restaurants and hotels followed in the scale of expressed discontents.

Those liabilities created less bitterness than did the unchanging policies of the armed services. Angry about military segregation, blacks detested the harassment they suffered from Military Police, the abuses they suffered from civilians in towns near cantonments, especially in the South, and the Army's endorsement of Jim Crow regulations in cities frequented by troops on liberty. Segregation on Southern trains and buses especially irritated northern blacks in training camps in Dixie. Walter White reported to the Office of Defense Transportation about "the bitterness and resentment" which the segregation of railroad coaches had "engendered among Negro soldiers, sailors and civilians." Though he argued that their anger could not "be disregarded," ODT did exactly that.

The War Department created an "intolerable situation . . . on account of . . . practices as to racial segregation," as the director of the National Negro Council wrote Roosevelt in March 1942. A guard employed by a contractor on a War Department job had assaulted a black craftsman who was lining up for a cafeteria luncheon. Police officers in Arlington, Virginia, joined bus drivers there in taunting War Department black workers, men and women alike, commuting from the District of Columbia. Many of the young blacks preferred walking to riding on segregated buses. They had "lived all their lives in . . . metropolitan areas where their civil rights had never before been abridged." That complaint led the White House to direct the War Department thereafter to use no segregated bus lines and to desegregate cafeterias, but the department apparently learned no permanent lesson from the episode.

In another, later, case, the commanding officer of the 25th Infantry, stationed near Walla Walla, Washington—a state that forbade discrimination in public places—subscribed to the wishes of local proprietors who labeled their business premises "off limits" to black GI's. The Army, in the view of one federal official, was catering to prejudice. The War Department characteristically disagreed: "While many people sincerely believe that the force of the War Department should be used in advancing desirable reforms . . . nobody would wish to have the decisions as to just what reforms are . . . desirable made by anything less than a clear democratic majority." That reply simply ignored the statute that the legislature of the state of Washington had enacted in order to protect blacks from the local segregation the Army had condoned.

That kind of military reasoning and the prejudices it protected underlay the open fighting between white and black soldiers at Fort Dix, New Jersey. It also reflected the attitude that produced the extraordinary decision of the Army, endorsed by the American Red Cross, to segregate the blood plasma of whites and blacks. Entirely without genetic or scientific basis, that racist policy naturally infuriated blacks.

Throughout the war humiliation recurred regularly for those in uniform. They knew the disdain of white junior officers, the brutality of local police in Southern towns, the pervasiveness of Jim Crow in America. Years later one black soldier, Lloyd Brown, recalled the greeting he received in Salina, Kansas. With several others, he went one day to a lunchroom on the main street:

> As we entered, the counterman hurried to the rear to get the owner, who hurried out front to tell us with urgent politeness: "You boys know we don't serve colored here."
>
> Of course we knew it. They didn't serve "colored" anywhere in town. . . . The best movie house did not admit Negroes. . . . There

was no room at the inn for any black visitor, and there was no place
. . . where he could get a cup of coffee.
    "You know we don't serve colored here," the man repeated. . . .
    We ignored him, and just stood there inside the door, staring at
what we had come to see—the German prisoners of war who were
having lunch at the counter. . . .
    We continued to stare. This was really happening. It was no jive
talk. The people of Salina would serve these enemy soldiers and turn
away black American G.I.'s. . . .
    If we were *untermenschen* in Nazi Germany, they would break
our bones. As "colored" men in Salina, they only break our hearts. . . .

    In full measure black civilians returned the resentment of work-
ing-class whites as both groups flowed into urban centers newly bulg-
ing with war-related activities. The resulting tension alarmed FEPC
investigators, who in 1943 examined conditions in some fifty such
cities. The area of Beaumont–Port Arthur, Texas, was typical of the
worst communities. In 1940 blacks had constituted nearly a third of
the population there, a population that grew 20 per cent in the next
three years. A resulting shortage of housing, schooling, and recreational opportunities made life uncomfortable for everyone. The influx
of newcomers taxed the resources of the cities while the competition
for jobs, homes, and amenities intensified racial antagonisms. Ultimately an unfounded rumor triggered a raid by a white mob on the
segregated black section. It caused two deaths and seventy-five in-
juries before state police and nearby federal troops could restore an
uneasy order. Though some whites contributed to a fund to restore
damaged property owned by blacks, the cities failed to devise programs to overcome the housing shortage or to provide recreation for
the hundreds of youths who had joined the rioting. Conditions in St.
Louis, Baltimore, Mobile, Indianapolis, Houston, and San Francisco,
among other cities, were little better. In Mobile, for example, white
workers started a riot when a local shipbuilding concern upgraded
some black welders. Federal intercession produced a compromise
that permitted those promotions, but only in segregated shops. Blacks
in Detroit and New York were later to suffer much more.
    The lack of available remedies for their grievances nettled
blacks. Their votes, important in northern cities in 1940, gave them
some potential leverage for 1944, but less than their numbers might
have allowed. Throughout the South blacks remained disfranchised,
while black workers who moved into industrial centers elsewhere
were often ignorant about registration procedures, afraid of their

white neighbors, or indifferent to politics. Politicians in both major parties cultivated the black vote in cities like New York and Chicago where black communities were well established, but even there black political leaders, like their white equivalents, were frequently moved by personal and selfish rather than social considerations.

As before, there was little redress available from the White House. Roosevelt's politically minded staff was less interested in civil rights than in mollifying Southern anxieties. The President himself, while sometimes responsive to criticisms of the Army, continued to regard them as matters not for substantive action but for better public relations. "The Army people," he wrote Elmer Davis after blacks had complained about press releases designating certain units abroad as "for service supply," are "dumb when it comes to . . . information." Davis was to improve the rhetoric in Army announcements. But Roosevelt directed no one to create more black combat teams.

Judge William H. Hastie, the black civilian aide whom Stimson had appointed in 1941, cut to the core of the problem. When the 99th Fighter Squadron, a unit of blacks, was formed, Hastie insisted it should not be segregated from white units in either training or operation. Nevertheless it was. Because the Air Corps considered it expedient to ignore the ethical question of integration, Hastie resigned early in 1943. The Air Corps, he then charged, had failed to train enough civilian technicians, had assigned black enlisted men in the 99th to tasks of common labor, and had proceeded in those and other cases, in spite of the terms of his appointment, without consulting him. Though the Army had decided to organize more Negro squadrons, the training of their technical crews, Hastie said, lagged behind the training of pilots, who were therefore unable to fly with the frequency of their white counterparts. In addition, the Air Corps refused to permit qualified black pilots to take cross-country flights that entailed the use of fields where they would require the facilities of local officers' clubs, unofficially segregated. No recourse to public-relations technique could rebut Hastie, whose persuasive outspokenness Roosevelt disregarded but *Time* commended.

Eleanor Roosevelt, whose criticisms of Army policies Stimson could hardly abide, became the target of venomous gossip in the South, where there circulated "Eleanor stories" about her alleged promiscuous miscegenation. One typical group of ladies from Lynchburg, Virginia, condemned a biracial dance in Washington. "The danger," they wrote the President, ". . . lies not in the degeneration of the girls who participated in this dance, for they were . . . already of the lowest type of female but in the fact that Mrs. Roosevelt lent her presence and dignity to this humiliating affair; that the wife of the president of the United States sanctioned a dance including . . . both races and that her lead might be followed by unthinking whites." So that the President would not miss their point, the ladies, petitioning him to stop mixed dances, continued in their odd spelling: "We are not in favor of sacrificing upon the alter of so called patriotism the white womanhood of this country to boost the morale of the Negro soldier in this or any other war." In a similar vein, a Philadelphia matron had earlier appealed to the White House to send as many black troops to England as possible: "The sooner large numbers are . . . gotten out of the United States the better it will be for American women."

Whites in the rural South, anxious as ever about miscegenation, openly resented wartime increases—partly from industrial wages, partly from soldiers' family allowances—in the incomes of blacks. "Employers of farm labor," a government report told the White House, "complain that . . . the Negroes are becoming too independent. . . . Many employers . . . pay a higher price than they planned to pay. . . . Mrs. Roosevelt is mentioned often as being dangerous because . . . she fraternizes too much with Negroes, and is making them want social equality. . . . Southern employers of farm labor . . . fear that out of the situation may come violence. They are accustomed to having the labor they want, and at the price they decide themselves."

Indictments of federal policies, bland though those policies were, punctuated speeches of Southern congressmen like John Rankin and James Eastland, of Mississippi, and reached a shrill climax in a letter to Roosevelt of August 7, 1942, from Eugene (Bull) Connor—so he labeled his stationery—the Commissioner of Public Safety of Birmingham, Alabama, who became, two decades later, a national symbol of the repression of blacks. "There is no doubt," Connor wrote, "that federal agencies have adopted policies to break down and destroy the segregation laws of . . . the entire South. Unless something is done by you, we are going to . . . witness the Annihilation of the Democratic Party in this section of the Nation, and see a revival of organizations which will . . . destroy the progress made by law abiding white people." A revived Ku Klux Klan would oppose the agencies and developments that Connor went on explicitly to describe: the National Youth Administration for preaching social equality, and the United States Employment Service and FEPC for "causing plenty of trouble when there ought to be unity." The number-one problem of the Negro, Connor believed, was not "Social Equality" but venereal disease. Yet agitation by federal officials made Negroes "impudent, unruly, arrogant, law breaking, violent and insolent." Connor had always voted for Roosevelt. "You have made a fine President," he said. ". . . Don't you think one war in the South . . . is enough? . . . Help us before it is too late."

On that letter, which he sent along to Marvin McIntyre, the President wrote and McIntyre apparently tried to erase: "Mac— What do you think we should do about this? I don't know." Had he cared to, Roosevelt might have replied in a public letter defending the federal agencies and condemning "discrimination in . . . its ugly forms." McIntyre typically proposed an answer that merely urged "the full utilization of manpower" and denied any federal intention to "destroy segregation." Even that timid response worried Roosevelt's counselor Judge Samuel Rosenman, who was, as he put it, "rather inclined to think that a formal acknowledgment and purely innocuous reply should be sent." His advice prevailed. Roosevelt, as he had said, "did not know" how to handle Connor. He also did not care enough about civil rights to risk losing Southern support for his party or to risk civil strife in wartime.

That caution, compatible with the conventional attitudes of white liberals, satisfied neither blacks nor their enemies. The roots of the dissatisfactions of blacks, as one of his advisers wrote Archibald MacLeish early in 1942, lay outside the reach of propaganda: "In every case . . . the Negro discussion of . . . information policy has been overcast by the feeling of the Negro leaders that an adequate information policy cannot be followed until some of the problems themselves are met. . . . Negro leadership . . . is disposed to emphasize these problems at the expense of the larger issue—the Axis threats to all groups in America." As the federal government saw it, from Roosevelt on down, that was indeed the more urgent issue, but since many blacks disagreed, propaganda designed to convert them had negligible effect.

So it was, in a most expensive example, with an illustrated pamphlet, *Negroes and the War,* that Chandler Owen, a prominent black publicist, prepared for broad distribution by the OWI. Owen, stressing the "stake the Negro has in America," emphasized also the racism of Nazism, particularly Hitler's insult during the 1936 Olympic Games in Berlin to Jesse Owens, the great black sprinter, and to other black winners of gold medals. The pamphlet told blacks that, in contrast, the Army had "two full divisions of Negro soldiers, that

Negroes serve in all branches. . . . That there are Negro officers
. . . Negro judges on the bench, Negroes in State legislatures and
in the American Congress." Blacks, Owen argued, had "come a long
way in the last fifty years, if slowly. There is still a long way to go
before equality is attained, but the pace is faster, and never faster
than now." Owen then turned to two heroic symbols of black
achievement, Dorie Miller, a gob who had excelled in bravery during
the attack on Pearl Harbor, and Joe Louis: "Under the lights at
Yankee stadium, *our* champion knocked out the German champion
in one round. Sergeant Joe Louis is now a champion in an army of
champions. Joe Louis doesn't talk much, but he talks truly. He talks
for 13,000,000 Negro Americans, for all American citizens, when he
says: 'We're going to do our part, and we'll win 'cause we're on
God's side.' "

The rest of the pamphlet, in words and pictures, showed black
soldiers poised for combat, blacks at lathes in clean factories, blacks
receiving hospital care and comfortable schooling, blacks in unions
and government posts. Probably inadvertently, the illustrations depicted as many segregated as integrated situations. The pamphlet also
omitted any discussion of the Japanese or of the Africans and Indians. Yet the Pittsburgh *Courier,* a nationally circulated black
weekly, had published an article praising Japanese cleanliness, cour-
tesy, and efficiency, and later an attack on the Chinese as "Uncle
Toms." *The Crisis* had identified British imperialism in India with
Nazi racism. And a survey of May 1942 had reported that for many
blacks "the achievement of democracy at home takes precedence over
the conflict abroad."

Accordingly the senior black adviser to Elmer Davis remarked
the failure of the pamphlet as propaganda in a letter he wrote as he
resigned from OWI: "Any program which attempts to improve Negro
morale within the framework of the status quo without attempting to
eliminate traditional methods of treating Negro citizens will be pallia-
tive, wasteful and ineffective." Lester B. Granger, of the Urban
League, scolded the OWI for the contention that Negroes had come
"a long way," which Granger called "a false argument for the federal
government to be presenting . . . poor grace . . . from a government which has failed in so many essential ways to give forthright
and courageous attention to the problems faced by Negroes."

Elmer Davis then and thereafter ordinarily turned for advice
about black problems to men whom Granger and Walter White considered Uncle Toms, but Southerners in Congress viewed the OWI
and especially Owen's pamphlet as subversive. Blacks themselves

would have condemned the OWI as even more subversive had they known about a letter Archibald MacLeish wrote Eleanor Roosevelt in April 1942 after A. Philip Randolph had raised the threat of demonstrations for civil rights in cities throughout the country. Major demonstrations to support black demands, MacLeish judged, would "play into the hands of . . . Axis . . . elements . . . and certainly put the Administration in a most difficult position. . . . I cannot over-emphasize . . . the seriousness of the entire situation." The demonstrations did not occur, but as Owen's pamphlet and Mac-Leish's letter revealed, information policy rested on principles that the OWI had adopted, without publicity, in full harmony with the policies of the White House. The agency, according to its own definition of those principles, "should realize that the long-range problems of racial and minority-majority antagonisms cannot be settled during the war . . . the war must be won first."

That statement of principles went on to declare that "efforts to improve the Economic position of minorities must be part of the effort to win the war." In pursuing that objective, the FEPC during 1942 and 1943 continued to meet resistance, which it lacked the authority regularly to overcome. Many blacks received war jobs in those years; some, an unprecedented but still small proportion, became skilled operators or even foremen in war plants. But as the FEPC later admitted, the "hard, hot and heavy" tasks still fell largely to blacks. White workers, by and large, rose in status and income, but black workers, entering the labor market, took over the vacated mudsill positions. The FEPC, composed of six part-time members and a small Washington staff, could do little more than expose the problem by holding hearings to examine specific complaints. It had an inadequate budget, only $80,000 in its first year, and its first chairmen were self-consciously gradualists. The agency did delineate a set of policies for fair employment, but, lacking power of enforcement, it succeeded only when black workers were courageous enough to file complaints and corporate offenders decent or embarrassed enough to comply with the intentions of Executive Order 8802.

Especially in the Northeast, where blacks were more conscious of their rights than they were elsewhere in the country, the opportunity to file complaints with the FEPC improved morale. So, too, some industries, aircraft for one, which had excluded blacks before 1941, later made a successful effort to hire and promote them. So did the government's navy yards. But in the heavy industries and transportation, north as well as south, management accepted blacks largely for unskilled and heavy jobs, and powerful unions contrived to exclude blacks from membership and to prevent their advancement. The FEPC could neither order nor persuade those offenders to change their ways. Perhaps worse, the United States Employment Service until July 1942 continued to fill "white only . . . requests from employers," and until September 1943 to do so when employers could not be induced to modify their policies.

In a reorganization of war agencies in July 1942, the President moved the FEPC from within the Office of Production Management to the new War Manpower Commission, which also contained the USES. The limited gains that resulted from exposing USES to the pressure of FEPC were overbalanced by the subordination of the latter agency to the authority of the War Manpower Commission. The commission, with support from the White House, gave top priority to filling manpower needs expeditiously wherever they arose. Insistence upon fair hiring practices would have retarded that process in recalcitrant plants. Accordingly WMC policy so crippled FEPC that while it remained under the commission's control it was able to initiate no hearings at all.

The President was as indifferent to the assigned mission of FEPC as was the WMC. In mid-1942 FEPC sent Roosevelt a summary of its recent hearings on discrimination in defense training. Pending his advice, the agency had not published those hearings because they involved another federal agency. The hearings had confirmed charges that in those states "whose laws require separate schools for Negroes and whites . . . Negroes are systematically denied equal opportunity in defense training programs financed by the Federal Government through the United States Office of Education." The eighteen states were entirely in the South or on its border. As the FEPC interpreted Executive Order 8802, it required the Office of Education to demand those states to stop discrimination. Local officials in the South, however, construed the order to pertain only to hiring practices, not to training, a view that the Office of Education endorsed. Roosevelt asked Under Secretary of War Patterson to evaluate the FEPC report. Characteristically prudential, Patterson replied: "In my opinion it would be inadvisable to publish the summary. . . . It would seem preferable for the Office of Education, by private persuasion or the exercise of the sanctions vested in it by reason of its control over funds and equipment, to secure local compliance." For the President, Marvin McIntyre then wrote Lawrence W. Cramer, the executive secretary of FEPC: "Confidentially, it is felt that it would be inadvisable to publish the summary at this time. If you feel you want to go further into the matter, drop in and see me some day and we can discuss it." As Cramer knew, discussions of black problems with McIntyre never produced remedy.

The evasions of the Office of Education and the caution of the War Department had their equivalents at the Department of State. It, too, easily won White House support in a conflict with the FEPC. As an FEPC study disclosed, industries in California and the Southwest regularly discriminated against Mexicans and Mexican-Americans. Oil companies in Houston and copper companies in Arizona, for example, in both instances supported by unions affiliated with the American Federation of Labor, denied Mexicans chances for training and upgrading. The employers believed that "Mexicans are inferior people; that they cannot assume responsibility, that they are not sufficiently advanced to use additional income wisely, and that they . . . were physically inferior to Anglos and hence did not produce as much. . . . The employment practices of the industries of the Southwest have become so traditional and the Companies . . . have for so long dominated the political as well as the economic life of their communities, that it will be difficult to get them to change their ways. The one thing they seem to fear and the one thing that inspires hope on the part of the employees, is the prospect of a public hearing." The State Department opposed that prospect.

Speaking for the department, Under Secretary Sumner Welles on June 20, 1942, urged Roosevelt to prevent a pending FEPC hearing at El Paso. Discrimination against Latin Americans, Welles wrote, negated the Good Neighbor Policy and therefore "very much concerned" the State Department. Still, the department had concluded "after careful study that the problem can be solved by means of a long-range program embracing the education and improvement of the standard of living of the alien population involved and education of both aliens and citizens to a better understanding of each other." Since racial discrimination abounded in the region, Welles saw no need for hearings to prove it. Precise factual information, he argued, "could best be secured by discreet investigation." More important, Welles cited the belief of the Mexican Under Secretary for Foreign Affairs "that any publicity in connection with our efforts to combat racial discrimination against Mexicans in this country would be most harmful. . . . Axis agents . . . seeking to discredit this country in Mexico . . . are making effective use of reports of racial discrimination. . . . It would be most unfortunate if . . . public hearings conducted by an agency of this Government . . . should afford them further material."

Though Welles was the President's friend and continual adviser on Latin-American affairs, Roosevelt was not always solicitous of the preferences of the Mexican government and not ordinarily swayed by the recommendations of the State Department. In this case he agreed. "For international reasons," he instructed the FEPC, the public hearing "should be stopped." Lawrence Cramer asked the President to reconsider. As he pointed out, since the FEPC had already announced the hearings, cancellation might suggest the "existence of a situation which the United States would not desire to have disclosed" —a possibility that Axis agents could exploit. Further, recent publicity about discrimination against Mexicans in Colorado had proved beneficial. Welles rejected those arguments. The State Department, he wrote, would not have approved, if it had been asked, the publicity afforded the Colorado situation. The announcement of the hearings in El Paso had designated no date, so postponement could be indefinite. The Mexican Ambassador to the United States considered public attention to episodes of discrimination "indiscreet" because it called "them to the attention . . . of individuals who have nothing whatever to do with them." Accordingly the State Department reaffirmed its position against the hearings. Samuel Rosenman drafted a letter to the FEPC that said the President, after again consulting the State Department, supported its request for a cancellation of the hearings and recourse instead to Welles's suggestion for discreet investigations.

Those investigations resulted a year later in indictments against the offending employers. In the interval, those companies, protected from adverse publicity from the FEPC, continued to pay Mexicans less well than Anglos, just as Southern schools permitted blacks only an inequitable share of training for war work. By the end of 1943 the FEPC could barely balance its achievements against its frustrations. The loss was not bureaucratic but substantive. It imposed still another social tax on blacks and other proscribed groups. As blacks realized, the FEPC, however inadvertently, had let them down. And by that time American whites, acting on their own initiative, had turned the poison of racism into violence on the streets.

3. Race Riot

Racial antagonism infected all cities to which war industries attracted millions of workers, white and black. The whites, well paid but crowded and restless, most of them either immigrants, children of immigrants, or destitute Southern and Western farmers, enjoyed a first taste of post-Depression prosperity but carried with them the characteristic psychological impulse to ostracize those still lower on the social scale than they were. As ever, that impulse was discharged mainly against blacks, of whom some 700,000 moved to industrial cities during the war. Competition for scarce housing, in two typical cases, pitted Polish immigrant workers against blacks in Buffalo and Detroit. In several other cities, local and federal authorities were able to control similar competition, but in Buffalo hundreds of blacks employed by Bethlehem Steel were forced to remain in the slums when local white resistance, verging on violence, forced the abandonment of proposed federal housing projects. Detroit, as *Fortune* reported, provided "the cause célèbre."

A major center of war industry, Detroit had a critical housing shortage as early as 1941. In May of that year the President endorsed the expenditure of federal funds for the construction of 1,000 dwelling units, two hundred of them for blacks. The United States Housing Authority authorized the Detroit Housing Commission to act as its agent for the building and management of the project. After considerable exploration, the Detroit commission reluctantly accepted a site recommended by federal officers. That site lay on Nevada Avenue between major concentrations of industry and an existing black residential section, an area close to Hamtramck, a Polish-American community. Announcement of the site provoked a flurry of protests from Polish groups. Those protests, the United States Housing Authority concluded, "were predicated entirely on racial questions." The USHA therefore proceeded to award contracts for the project, which it named "Sojourner Truth Homes" in honor of a gallant and gifted black woman, a poet who had been a slave. As the building progressed, agitation continued, now directed against black occupancy of the homes, but the Detroit commission, undeterred, went ahead with the selection of prospective black tenants.

Polish-Americans then enlisted their congressman, Rudolph G. Tenerowicz, who attacked the project on the floor of the House of Representatives, where he accused Detroit blacks of following Communist leadership. His intercession resulted in a visit to the site by members of the House Committee on Public Buildings. Their reactions, along with mounting tension in Detroit, persuaded Baird Snyder, acting for the Federal Works Agency, and Charles F. Palmer, Coordinator of Defense Housing, in January 1942 to schedule Sojourner Truth for exclusively white occupancy. "Since the policy of the Federal Works Agency," Snyder then wrote Marvin McIntyre, ". . . now is to defer greatly to local recommendations, it was . . . decided to follow the original recommendations of the local housing commission . . . and to develop another site for colored occupancy." Detroit blacks, as one of their congressmen wrote Roosevelt, were "terribly disturbed about this act," as was also Walter White. But speaking for the White House, McIntyre on January 19 supported Snyder. "I think," he wrote, "the important thing is to avoid an open fight at this time."

On this occasion the pusillanimity of the White House could not prevent a fight. Aroused blacks insisted so vehemently on their right to move into Sojourner Truth that the Detroit Common Council and the Detroit Housing Commission adopted resolutions urging the return of the project to black occupancy. Clark Forman, a liberal Southerner who was director of the Housing Division of the Federal Works Agency, accepted that request, which expressed his own views. The anger of Southern congressmen over Forman's decision later cost him his job, but at the time his determination furthered the plans of the Detroit Housing Commission to begin to move blacks into their new homes.

Early on February 28, 1942, the first twenty black families attempted to enter Sojourner Truth according to instructions they had received from the housing commission. The first three of those families, their furniture on trucks, encountered a mob of some two hundred to three hundred whites armed with stones and clubs. The city, expecting trouble, had stationed two hundred policemen at the scene, but the police inspector there told housing officials that his men could not successfully guard the blacks. City officials therefore decided to defer occupancy until they could organize adequate protection. Soon about one hundred blacks gathered in the area, where fighting between blacks and whites broke out and continued through the day. The following day the designated black tenants held a protest meeting, which some Detroit locals of the CIO encouraged. Concurrently some 1,000 whites set up a watch around the project. More fighting occurred. The crowd then dispersed, but a small group of whites, led by the Ku Klux Klan, picketed the area for the next several weeks.

In that interval, Detroit's Mayor Edward Jeffries continually supported the blacks. "If we are one people," he maintained, "the Negroes should go into the project." Meanwhile, several efforts at compromise failed. Finally in April, under instructions from the head of the new National Housing Agency, the Detroit Housing Commission moved the blacks in with protection from several hundred state militiamen. That settled the matter, but the episode, as a Chicago black wrote the President, had demonstrated the need for "vigorous federal action to protect the lives of Negroes." Further developments in Detroit bore him out.

The vacillations of both the city and the federal government during the controversy over Sojourner Truth persuaded the more militant of Detroit's whites and blacks, alike, of the need for private vigilance to advance their mutually incompatible expectations, not only for housing but also for all of the city's strained social facilities. Racial hostility especially affected young people, of whom many in Detroit, white and black, had been rejected as inadequate by both private industry and Selective Service. Idle, insecure, alienated teenagers of both races, reflecting the volatile mood of their elders, formed gangs that taunted each other. Interracial conflict, according to one Justice Department investigation in 1943, had increased "tremendously" within the city's high schools. So, too, on city streets, where the police, their ranks gravely depleted by the draft, were inefficient, partial to whites, and often brutal to blacks. "Detroit," that investigation reported, "is a swashbuckling community, not conspicuous for its social maturity. Negro equality . . . is an issue which . . . very considerable segments of the white community resist. . . . There is a real hoodlum element of Negroes—as there is among whites—which has been responsible for numerous unpleasant and even revolting incidents. . . . White Detroit seems to be a particularly hospitable climate for native fascist type movements. . . . Large segments of the Negro community hate the police, probably not without reason."

The friction "produced by those forces not unnaturally resulted in spontaneous combustion" on the sweltering day of June 20, 1943. That morning more than 100,000 people, some 86 per cent of them black, resorted to Belle Isle, a municipal park on an island in the Detroit River. The police reported several minor complaints: six young blacks robbed a teen-age white couple of two dollars; another group of blacks assaulted some young whites on a playing field; an eighteen-year-old black, recently arrived in Detroit from Alabama, called a white woman on a bus a "mother-fucking son-of-a-bitch," and the black bus driver reprimanded him. About 10:30 A.M. there was a series of fights between young white and black girls on the bridge to the island. Reports of those fights stirred others on the lawn adjacent to the naval armory near Belle Isle. There white sailors joined in. By 11:30 some 5,000 people, 90 per cent white, were milling about the area of the armory and attacking blacks. The fighting continued intermittently for twelve hours, though without serious casualties. Only about 2:00 A.M. the next day did the police conclude that they had the situation under control.

They were wrong. The trouble had barely begun. Soon after midnight a rumor had reached a black night club that whites on Belle Isle had killed three blacks. Similar rumors spread through the black district, where crowds began to gather and then to smash windows, stop streetcars, attack their white occupants, and loot stores, especially liquor and pawn shops owned by Jews, who had the reputation, not entirely undeserved, of fleecing their predominantly black customers. During the night the mob assaulted one white streetcar motorman and beat to death a white milkman and a white doctor making a house call.

A counterreaction began about 4:00 A.M. along Woodward Avenue, which separated the white from the black residential districts. By 10:00 A.M. on June 21, white mobs were attacking and sometimes killing blacks who strayed into the area on foot or in streetcars, and whites and blacks had begun to snipe at each other with rifles. As with the blacks, so with the whites, there was no evidence of organized or planned action, and young men, sixteen to twenty-five years old, played a major part in the rioting. Before it ended, twenty-five blacks and nine whites had been killed, at least four of the blacks and one of the whites by police bullets. Mayor Jeffries, convinced only at 9:00 A.M. on June 21 that the city could not handle the mobs, then asked Michigan Governor Harry F. Kelly for help. He in turn requested federal troops. Almost all of the deaths and casualties occurred before those troops began to arrive between 11:00 A.M. and noon—a whole day after the incident at the naval armory should have alerted the police, and through them the mayor, to the possibilities of a riot.

The tardiness and ineffectuality of the police and city government accounted in large measure for the extent of the rioting, but its cause lay deeper. The presence of federal troops, the imposition of a curfew, the return of superficial quiet, all those dispelled none of the attitudes that had motivated the mobs. Bloodshed had scarcely ended when a sixteen-year-old white boy exulted over his own role: "There were about 200 of us in cars. We killed eight of 'em. . . . I saw knives being stuck through their throats and heads being shot through. . . . It really was some riot." A white college girl recalled that "whites were raving with hate." A white newspaper reporter remained shaken by his experience: "The Negroes were berserk. They were smashing up everything that didn't have a colored sign on it. Then I came downtown. There was a 17-year-old white girl standing on the corner and she said to the boys, '. . . I haven't seen any blood yet. What's the matter with you guys!' " A twenty-year-old black admitted that he overheard several other blacks who had guns: "They were saying, 'If it gets tight, get two whites before they get you.' " Still, he reflected, the police were the worst: "Those police are *murderers.* . . . We didn't have a chance. I hate 'em, oh God how I hate 'em."

In the aftermath of the riot, old Southern hatreds also flared. The Jackson, Mississippi, *Daily News* spoke for the region: "It is blood upon your hands, Mrs. Roosevelt. You have been . . . proclaiming and practicing social equality. . . . In Detroit, a city noted for the growing impudence and insolence of its Negro population, an attempt was made to put your preachments into practice." John Rankin, in the same mood, attributed "this trouble" to "the crazy politics of the so-called Fair Employment Practices Committee in attempting to mix the races." Gerald L. K. Smith, the most infamous American fascist, persuaded Martin Dies to have his House Committee on Un-American Activities investigate the Detroit riots, which Dies, as was his wont, blamed on the influence of Communists on blacks.

For its part, the National Association for the Advancement of Colored People pleaded for a statement from the President. "No lesser voice than yours," Walter White telegraphed Roosevelt, "can arouse public opinion sufficiently against these deliberately provoked attacks. . . . Unless you act, these outbreaks will increase in number and violence." Mayor La Guardia urged Roosevelt to see White as soon as possible. White had rushed to Detroit, La Guardia wrote, "the minute the trouble started." His advice to the governor, advice to delay no longer in requesting federal troops, had saved the situation, which was still so tense that La Guardia begged Roosevelt not to withdraw the troops. Jonathan Daniels, whom the President had recently appointed to his staff, partly to handle racial issues, also advised a "statement of idealism" by his chief. Vice President Henry Wallace, a wartime advocate of civil rights, wondered whether there was "any device by which . . . the Negroes can be assured that the President's heart is where it always has been." Wallace did not realize where that was. As ever, Roosevelt hesitated. He asked for an opinion from Rosenman. He asked the War Department for clarification about the use of federal troops in emergencies like that in Detroit. He directed the OWI "to correlate all available governmental information as to . . . tensions and difficulties involving minority groups in all parts of the country." For the while, the President said nothing.

Mexican-Americans in Los Angeles had already encountered the tensions that Roosevelt had called to the attention of the OWI. Captives still of underemployment and the slums, many Mexicans and Chicanos in California had yet to master written or spoken English. The Hearst newspapers, trumpets of white supremacy, associated Mexicans with sex crimes, knifings, gang violence, marijuana, and a racial torpor that supposedly accounted for their poverty and the resulting burden on relief funds and correctional institutions. By 1943 those biases, common among Anglos, especially working men and women, had come to focus on "zooters," boys and young men who dressed in zoot suits. That fashion had originated in Harlem and spread to the West Coast, where many black and Mexican and some white teen-agers adopted it. The zoot suit consisted of a long jacket and trousers tightly pegged at the cuff, fully draped around the knees, and deeply pleated at the waist. Most zooters also wore their hair long, full, greased, and gathered in a ducktail. Many of them belonged to youth gangs, some of which had wholly innocent social purposes, some organized for delinquent activities. To the antagonistic Anglos, Mexicans and zooters were indistinguishable, and white servicemen stationed in and around Los Angeles shared the prejudice.

There, as later in Detroit, interracial hostility burst out in June 1943. During several days in the first week of the month, sailors from the Chavez Ravine naval base and soldiers on passes to the city congregated in small mobs that roamed through the Mexican districts demanding revenge for alleged attacks by zooters against servicemen. Touring the Mexican section in cars and cabs, the soldiers and sailors crashed into movie houses, stripped Mexican teen-agers of their trousers, cut their hair, and beat them up. The Shore Patrol and the Military Police looked the other way. The city police followed the mobs, watched the beatings, and then often jailed not the attackers but the victims. After two nights of such episodes, the Hearst *Herald and Express* ran the extraordinary headline: ZOOTERS THREATENED L. A. POLICE. After still another and worse night, the newspaper blamed the riots on zooter hoodlums. The immediate blame lay with the lack of recreation for servicemen crowded into a strange city, but the lynch-mob spirit in Los Angeles expressed racial antipathies that had been building for years.

In a letter to the President, Walter White analyzed the Los Angeles riots in a spirit that also animated the observations of *Time*. Zoot-suiters, White pointed out, dressed as they did "to compensate for a sense of being rejected by society. The wearers are almost invariably the victims of poverty, proscription, and segregation." It was essential, White continued, to discipline the offending soldiers and sailors, for "otherwise these members of the armed services will believe the wearing of the uniform gives them the license to act as arbiters of how civilians may dress, speak, act or think." A Citizens Committee in Los Angeles, formed at the instigation of California Governor Earl Warren, also demanded punishment of those guilty of crimes of violence "regardless of what clothes they wear . . . zoot suits, police, army or navy uniforms." That committee noted, too, the inadequate housing and recreation available to Mexicans, and the need for public programs to improve those conditions and "to combat race prejudice in all its forms." The White House also received a letter from Philip Murray, the President of the CIO, who urged the mobilization of OWI, the Office of the Co-ordinator of Inter-American Affairs, and the War and Navy departments to prepare and pursue educational campaigns "to eradicate the misconceptions and prejudices" that had contributed to the events in Los Angeles.

Electing to reply to Murray rather than to White, whose letter Roosevelt had Elmer Davis answer, the President asked Secretary of Labor Frances Perkins to prepare a suitable draft. His request postdated the Detroit as well as the Los Angeles riots, and he postponed sending the letter to Murray until July 14, 1943. "I join you," that short letter concluded, "and all true Americans in condemning mob violence, whatever form it takes and whoever its victims."

Probably no words of the President, certainly not those, could have much reduced the likelihood of violence in other cities. On the hot summer night of August 1, 1943, Harlem exploded with the frustrations of blacks who had heard a false rumor that a white policeman had killed a black soldier. But in New York City the police behaved well, and Mayor La Guardia drove through the streets to quiet the crowds. He also deputized 1,500 black volunteers for police duty. The mobs, after some looting in the Harlem business district, dissipated without entering adjoining white neighborhoods. Even so, the riot led to the injury of some three hundred blacks and the death of six. Black leaders, while describing the riot as a "disgraceful episode," stressed its origin in white repression of blacks and the resulting black attitude of "deep resentment against oppression."

That resentment, hardened by the riots in Detroit and New York, had grown throughout 1942 and 1943 as federal agencies—the Army, the OWI, the FEPC, the White House itself—provided meager remedies for black grievances. Blacks disagreed among themselves about how much the President could accomplish or even attempt. A moderate spokesman, Lester B. Granger, wrote Attorney General Biddle about the need for finding jobs for blacks, for "developing sound public and private housing facilities, providing adequate leisure-time and educational programs, securing the cooperation of the press." Granger did not expect quick results. "The President," he went on, "can do little more at this time than stress the point that racial conflict is a deterrent to victory." Implicitly those words, typical of the Urban League, endorsed Roosevelt's interpretation of all problems as secondary to the winning of the war.

Other black leaders challenged Roosevelt's narrow view of his obligations. The President's reply to Philip Murray particularly offended a black college student. Like others of her age she had expected, after Detroit, at least an impassioned affirmation by Roosevelt of the human rights that blacks had yet to enjoy. No words would have dispelled prejudice, but the right words, as she believed, would have lifted the morale of beleaguered blacks. And whatever the limits of rhetoric, Pauli Murray felt the anger and hatred as well as the sadness of the riot. She said so in a poem, "Mr. Roosevelt Regrets," a poem *The Crisis* published in August 1943:

What'd you get, black boy,
When they knocked you down in the gutter,
And they kicked your teeth out,
And they broke your skull with clubs,
And they bashed your stomach in?
What'd you get when the police shot you in the back,
And they chained you to the beds
While they wiped the blood off?
What'd you get when you cried out to the Top Man?
When you called the man next to God, as you thought,
And you asked him to speak out to save you?
What'd the Top Man say, black boy?
Mr. Roosevelt regrets. . . .

4. Protest and Prognosis

In 1941, in organizing the March on Washington, A. Philip Randolph had demonstrated the effectuality of visible protest to gain for American blacks the rights to which they were entitled. Earlier, during

World War I, the NAACP had accepted the argument of military necessity. "Let us, while this war lasts," W. E. B. Du Bois had then asked his fellow blacks, "forget our special grievances." Du Bois, as he later realized, had erred. Randolph had not. After the attack on Pearl Harbor, Walter White, speaking for the NAACP, warned blacks that "declarations of war do not lessen the obligation to preserve and extend civil liberties here while the fight is being made to restore freedom from dictatorship abroad." As the *Crisis* put it: "Now is the time *not* to be silent." Through its pages, White's activities, and continual litigation in the courts, the NAACP pursued its vocal course. So did the Pittsburgh *Courier,* the black newspaper with the largest circulation, which called for the "Double V"—"victory over our enemies at home and victory over our enemies on the battlefields abroad."

To Virginius Dabney and John Temple Graves, Southern white intellectuals who represented the dominant mood of their region, the *Courier* seemed "radical," "extremist," in trying to use the war "for settling overnight the . . . long, complicated, infinitely delicate racial problem." But to A. Philip Randolph, the *Courier* was the contemptible "spokesman for the petty black bourgeoisie." Randolph used the March-on-Washington organization as his vehicle for planning nonviolent protest. The war, he argued, was not for democracy but for maintaining white supremacy throughout the world. Only direct action by blacks could convert it into a people's revolution. Randolph's plans for civil disobedience struck the *Courier,* for its part, as "dangerous demagoguery" that would expose Southern blacks to fierce white retaliation. After the riots of 1943, that possibility strengthened the case of the moderates among black leaders, but the militants did not cease to press their tactics where they could, often with substantial support from the moderates. To the distress of Dabney and Graves, continual black agitation raised black consciousness and expectations, a condition prerequisite for wartime and postwar progress toward desegregation, and alerted the federal government, as had the riots, to the need during the war to relieve at least some black grievances.

The armed services in 1944, partly because of the manpower shortage, moved hesitatingly toward greater racial equality. In two token steps the Navy integrated the crews of twenty-five ships and rated some five hundred black petty officers for line duties. The Army ordered the desegregation of training camps, of which many in the South nevertheless remained segregated, and sent more blacks into combat overseas, though in segregated units. The utilization of black troops in battle allowed black reporters for the first time to observe and recount the bravery of their fellows. No previous venture in public relations had the impact on blacks of the reports of the war correspondents of Afro-American newspapers.

Those accounts, which spontaneously followed the pattern of white reporting about the GI's, permitted their readers a similar, reassuring exercise in identification with heroic virtues. "Everywhere I go," one black correspondent wrote from France, "are tales of our lads who waded ashore in water up to their knees . . . to take part in the assault that forced Jerry from his strong points." In one sector, where a black howitzer outfit was operating, white infantrymen would not budge "unless these guys are laying down a barrage." That reporter "was amazed at the daring and successful operations of the [black] Motor Transport Corps. . . . It is typically American."

In the Pacific a black correspondent found black troops just like white: "tired, wet, miserable, heartsick." In Italy the all-black 99th Fighter Squadron attracted much attention. "Like football players," a black reporter wrote in a race-blind simile, "bursting into a dining room after a triumph . . . war-weary pilots were jubilant in their description of victories over the Luftwaffe." One correspondent found no color line in the foxholes. Another did not look for one. "The story I have to write," he declared, is "the story of colored troops. in actual combat. . . . Now that I have seen our lads in action on the . . . front, I am both proud and humble."

Like other black reporters, he was also angry about the Army's ways. As one black writer put it: "The only colored officers I have seen since coming to France with rank higher than first lieutenant, have been either chaplains or medical officers. Even the company commanders have been first lieutenants. Such a remarkable coincidence. Like at the big headquarters where we sleep and eat: no colored officers at all happen to be stationed in the immediate vicinity." Another black correspondent apologized to his readers: "If I keep mentioning colors and races . . . I'm sorry. But everywhere I go it pops up. I'm sick of the subject . . . but the army functions on racial lines. Officers talk race, not ability. The men discuss it all the time." That same writer also quoted a white lieutenant, a commander of black troops, who had an unusual way of talking: "I hope like hell that these guys in my outfit get a better break when we get back home. I have seen them go through hell."

An immediate better break for blacks preoccupied Walter White during his tour of several fronts in 1944 as special correspondent for the New York *Post.* In a long cable for the War Department and the White House, he made his usual case against the Army's treatment of black soldiers. "As men approach actual combat and the dangers of death, the tendency becomes more manifest to ignore . . . race prejudice," White wrote. So it had been at Anzio beachhead, so also in North Africa, where the war had now passed. Race prejudice fell off particularly where black combat troops were in the field, as in Italy, but it persisted where blacks were assigned almost exclusively to the service of supply, as in Great Britain. The growing respect for blacks as fighters in Italy was "offset by the regrettable decision to transform the 2nd cavalry division from combat to service duties." As White saw it, the Army should instead have used white troops or Italian prisoners of war. While he was gratified to find fewer anti-Negro stories in Italy than in England, he recommended that the War Department instruct the Army everywhere "to take more prompt and vigorous action against anti-Negro propaganda." The attitude of the War Department, he judged, "lags behind that of much of the personnel of the Army," who would approve "a more courageous and enlightened philosophy." White concluded on a familiar theme:

> The time and field are ripe for progress. Continuation of separation and discrimination is in part responsible for a less determined spirit in the Army than is necessary to meet the very crucial days which lie ahead before the war is won. It must be remembered that our failures of omission and commission are being watched by other colored peoples, who constitute a majority of the peoples of the earth.

The War Department and the Army nevertheless continued to reveal their disdain for that majority of the peoples of the world. Black GI's sent home for two weeks of rest and recreation at the Army's expense reached none of the luxurious resort hotels at Lake Placid or Santa Barbara, Miami Beach or Hot Springs, hotels that white GI's enjoyed. Instead, the blacks were assigned to the Theresa Hotel in Harlem and the Pershing Hotel on Chicago's South Side. Adam Clayton Powell, Jr., Walter White, and other blacks protested in September 1944 in telegrams to Roosevelt. Harlem and the South Side were already overcrowded, and the hotels were unattractive. Worse, "the action of the United States Government in segregating Negro soldiers who also have faced death on foreign battle fronts and in introducing segregation in northern communities where it is both illegal and contrary to public opinion is a reprehensible act and . . . an insult to Negroes."

Unmoved, Secretary of War Stimson on September 20, 1944, reminded the President of the "War Department's long-standing policy not to force the intermingling of the races but to provide equality of treatment." The hotels for blacks on furlough, he wrote, were "the best obtainable for the purpose." As ever, separate but equal proved unequal, but Stimson suggested that at the white resort hotels a black soldier would be subjected "to conditions unfavorable to his mental and physical rehabilitation." The Secretary apparently preferred the humiliation of blacks to stem from public authority rather than private prejudices. Roosevelt, more sensitive, arranged to confer with Walter White and several other blacks, who told him that "any segregation . . . would be objectionable." The Chicago *Defender,* a black newspaper, and the Chicago *Sun,* a white daily, soon thereafter also attacked the Army, as White informed Roosevelt. The President answered merely that he felt satisfied with their earlier conversation about "the question which had disturbed you and about which I was disturbed also." He was not enough disturbed to order the Army to change its offensive policy.

White feared the continuation of segregation in the Veterans Administration, which was charged under the GI Bill of Rights with providing aid to returning servicemen. For the NAACP in September 1944, he urged the appointment not of a special adviser on blacks but of a black as assistant administrator of Veterans Affairs. General Frank T. Hines, the head of the agency, agreed only to combat discrimination. Roosevelt, to whom White then appealed, arranged a conference at the White House, where White met with the President, Hines, and Judge Hastie. Hines, though "most gracious," left the question of the appointment wholly to Roosevelt, who referred it to Jonathan Daniels. Daniels buried the matter, as he did also White's plea to the President for an end to segregation in public housing as well as in housing financed under the GI Bill.

Like the rest of Roosevelt's staff, Daniels, a North Carolina liberal but a gradualist about questions of race, assessed the problems of blacks within a political context. In 1944 he focused upon the Presidential election. In April that year, in *Smith* v. *Allwright,* a case presented by the NAACP, the Supreme Court ruled that the Democratic party in Texas could not exclude blacks from voting in primary elections. That threat to the traditional all-white primary, to "the white man's rule," outraged the deep South. Nevertheless, Attorney General Biddle contemplated, as Daniels informed the President late in September, "moving into Alabama with a criminal prosecution in connection with the denial of the rights of Negroes to vote in the

. . . primary this year." Daniels had just talked about the prospect with Senator Lister Hill, of Alabama, one of Roosevelt's reliable supporters. "I strongly share his sentiments," Daniels wrote, "that any such action by the Federal government at this time might be the fact which would translate impotent rumblings against the New Deal into an actual revolt at the polls. I am sure that any such action . . . would be a very dangerous mistake." The impact of that kind of political judgment, which prevailed in 1944, hurt blacks as much as had the traditional prejudices of the Army.

Only small compensation flowed from the President's decision to concentrate within the FEPC the burden of federal activities in behalf of blacks. To that end he had restructured and strengthened the agency in Executive Order 9346 of May 1943. The order established a new FEPC, placed it again within the Executive Office, so as to give it more autonomy than it had had under the War Manpower Commission, increased its budget, enlarged its authority to conduct hearings and investigations, and required all federal agencies to include, in all contracts they let, provisions obligating the contractors and all subcontractors not to discriminate in their training or hiring practices. Roosevelt also appointed Malcolm Ross chairman of the revived agency. Ross, a longtime New Dealer, had attacked the conventions of his own past in his autobiographical *Death of a Yale Man*. Accepting the President's goals as his own, he admitted that "segregation, per se, is of no concern to my committee." But he was persistently vigorous in combating "demonstrated industrial discrimination" in employment. "We may not be able to wipe out discrimination overnight," he told a *Time* reporter, "but where war manpower needs are at stake we can and shall try." The range of the effort, while short of black hopes, encountered both Southern hostility and judicial resistance.

With continual encouragement from the NAACP, Ross did try harder than his predecessors, and, notwithstanding his defeats, with more success than anyone else in the administration. Under his direction the FEPC aroused the conscience of many white Americans and advanced the employment of minority workers, blacks particularly. Most of his victories were unpublicized. A few, like some of his defeats, made headlines.

On its celebrated cases the reconstituted FEPC broke about even. It won a round previously lost when, in November 1943, the State Department, prodded by Ross, approved hearings in El Paso on discrimination against Mexicans. That decision the Mexican Foreign

Office now approved, perhaps in part because the State Department at least solicited approval, as it had not a year earlier. In December 1943 the FEPC lost a round to the Southern railroads, which refused to comply with the commission's order to stop discriminating against blacks. "Any attempt," the railroads held, with the full support of the all-white Brotherhood of Locomotive Firemen and Engineers, ". . . to promote Negroes to locomotive engineers or train conductors would inevitably disrupt present peaceful relations with employees and . . . antagonize the traveling public." Such delicate problems, the railroads maintained, could not be solved by directives from the FEPC. Furthermore, those directives were "wholly lacking in due process of law. . . . Your committee was and is wholly without Constitutional and legal jurisdiction." That challenge moved the President, whose own authority was under attack, only to order a special investigation of the employment practices of the railroads.

The investigating committee failed to bring about compliance. Its study was still under way when the Supreme Court on December 18, 1944, ruled against the railroad union, but did so in an opinion that *The Nation* described as "turgid, legalistic, and evasive." The Railroad Labor Act had given the brotherhood the right to act as exclusive bargaining agent, according to the decision, but Congress in that act "did not undertake to authorize the bargaining representative to make . . . discriminations" based on race. That ruling, as Mr. Justice Frank Murphy observed, ignored the Fifth Amendment. "To decide the case . . . on legal niceties," he wrote in his concurring opinion, "while remaining mute and placid as to the obvious and oppressive deprivation of constitutional guarantees, is to make the judicial function less than it should be. . . . No statutory interpretation can erase this ugly example of economic cruelty against colored citizens of the United States." Racism, Murphy warned, was "far too virulent . . . to permit the slightest refusal, in the light of a Constitution that abhors it, to expose and condemn it wherever it appears in the course of a statutory interpretation." Yet the Court's decision left FEPC without adequate authority, a condition already exposed by the successful defiance by the Southern railroads.

The FEPC had already lost another round to West Coast ship-builders and the International Brotherhood of Boilermakers, the AFL union that enjoyed exclusive bargaining rights with the shipyards. When the yards first hired blacks, the Boilermakers opposed their employment in any capacity except that of common laborer. After protests from the government, from blacks, and from the managers of the yards, the union, whose ritual excluded blacks, set up auxiliaries for blacks who had the relevant skills. Many black workers, however, declined to join or to maintain membership in the auxiliaries because members, though they paid the same dues as whites, were denied equality of voting and other rights related to employment. The union then demanded that the yards discharge the recalcitrants, and management contended that its contract bound it to comply.

After public hearings in Portland, Oregon, and Los Angeles, hearings to which the unions refused to send representatives, the FEPC found the resort to auxiliaries discriminatory and the employers also guilty because of their complicity in discharging black workers. Two of the five companies involved agreed to follow the ensuing FEPC directives, but the others refused. The unions, harassed by suits that black workers initiated in California courts, contrived substitutes for the auxiliaries that accomplished the same purpose. Again, in spite of FEPC orders and in spite of nondiscriminatory clauses in war contracts, an intractable all-white union prevailed. Howard W. Smith, a Virginia Democrat, chairman of the House Committee on Executive Agencies, invited W. A. Calvin, a vice president of the Boilermakers, to write the committee about the FEPC, which Smith was eager to abolish. To Smith's satisfaction, Calvin identified the FEPC with the Communist party. "Since the date of the creation of the . . . Committee," he wrote, "subversive activities have been conducted . . . by professional agitators . . . whose purpose apparently is to use the negro . . . in the accomplishment of their nefarious objectives. . . . On the Pacific Coast . . . litigation has been instituted against our . . . Brotherhood alleging racial discrimination. Our case . . . was prosecuted by two attorneys who are allegedly notorious pro-Communist." Smith took that testimony seriously in spite of, or perhaps because of, the contradictory letters he received from the NAACP.

Smith's committee also investigated Malcolm Ross's intercessions on behalf of black workers in street-railway systems. Successful in Los Angeles, defied in the District of Columbia, the FEPC won a pyrrhic victory in Philadelphia, where the employment of blacks as motormen provoked a strike that was dispelled only by recourse to federal troops. Ross gained his objective, but the Army, not the FEPC, exercised the indispensable authority.

Indeed, the FEPC was able to resolve successfully only one-third of the 8,000 complaints it received, and only one-fifth of those that originated in the South. Of the committee's forty-five compliance orders, employers or unions defied thirty-five. Even so, antagonistic Southern congressmen succeeded in 1945 in halving the agency's budget and providing for its dissolution within a year.

That record and that response added to the other disappointments of American blacks: the persistence of segregation in the armed forces, in the Veterans Administration, and in public housing; the government's lassitude toward prosecutions to end all-white primaries in the South; the President's evasiveness. Malcolm Ross fought for legislation to establish a strong postwar FEPC, as did the NAACP, but by 1944 the continuing frustrations of the war years had convinced more and more blacks, the educated young especially, that A. Philip Randolph had been correct in preaching that militancy was indispensable.

The most influential advocate of civil disobedience in the first years of the war, Randolph seemed to have become, in the assessment of one of his admirers, "a leader without a movement." Though the Detroit riot had damped the influence of militants, that judgment about Randolph was skewered. Others were already adopting, with modifications, the principles and the strategy he had recommended. They were influenced alike by Randolph and by A. J. Muste, the white head of the biracial Fellowship of Reconciliation. Muste shared Randolph's suspicions of capitalism, experience in labor activism during the 1930's, and commitment to nonviolent protest. Like so many of his black disciples, Muste also identified with the struggle in India for independence from Great Britain and embraced the doctrines of passive resistance associated with Mahatma Gandhi, then in the last phase of his martyrdom in the cause of Hindu equality.

Among the "peace teams" that Muste's Fellowship of Reconciliation organized, one at the University of Chicago had a special interest in racial problems. About a dozen of its members organized the Chicago Committee on Racial Equality, which became the prototype for other, similar, units and for the national Committee (later Congress) of Racial Equality—CORE. From the first, in contrast to the March-on-Washington movement, CORE was biracial. The leaders of the Chicago group included George Houser, the white son of a Methodist minister, and James Farmer, the black son of a college professor. Both, like Muste, were steeped in theology, both were ardent Methodist pacifists, and both became figures of national importance as CORE grew in strength and influence. Among other early CORE leaders who became well known were James R. Robinson, a white, and Bayard Rustin, black and a close associate and courageous disciple of A. Philip Randolph. Dedicated to "Action Discipline," to patience in negotiation as well as intrepidity in purpose, CORE members during the war years methodically studied Shidharani's *War Without Violence*, a Gandhian text adopted as their handbook for combating racial discrimination, as Houser put it, "directly, without violence or hatred, yet without compromise."

CORE gained adherents among both blacks and whites partly because of the success of the demonstrations it led. A first of the persuasive episodes occurred in the spring of 1942 in Baltimore, Maryland, where black members of FOR purchased tickets for *Native Son,* then playing in a segregated theater. They met no opposition except for hostile glances. In Chicago that fall, in Stoner's Restaurant in the Loop, CORE had a harder time. When small groups of blacks were able to find any seats at all, they were served meat covered with egg shell, oversalted food, or sandwiches of garbage. They resorted then to a mass sit-in. An elderly white woman, apparently not one of the demonstrators, invited a black to share her table. As white CORE members followed her example, almost all of the sixty-five blacks found seats, to the "wild applause" of most of the patrons.

So it continued to go outside the South. In Tennessee, police beat Bayard Rustin savagely when he refused to ride in the back of a bus in the section designated for blacks, but in Yellow Springs, Ohio, students from Antioch College and Wilberforce University integrated a theater. In Syracuse, New York, CORE persuaded the manager of a skating rink to admit blacks. During 1943 in Denver, Colorado, CORE effected the integration of a restaurant and of the downtown theaters. Blacks in those theaters took their tickets for the segregated balcony down to the orchestra, where white demonstrators threw off ushers who tried to interfere. James Farmer led a group of CORE and NAACP members into Garfield's Restaurant in Detroit, previously segregated, where the blacks ate, and two months later persuaded the management to abandon segregation permanently. At the federal prison in Danbury, Connecticut, in August 1943, conscientious objectors went on strike to protest segregation in the dining halls. Embarrassed by the resulting publicity, federal authorities integrated all facilities before the end of the year. By that time members of FOR and CORE were planning freedom rides, which did not start until after the war, to force the desegregation of buses operating in interstate commerce.

During the war CORE had yet to be toughened by the kinds of

resistance that characterized its postwar efforts. As James Farmer later observed, looking back from the 1960's:

> . . . I am amazed at our patience and good faith. No action group today would prolong the attempts at negotiation for more than a month before finally deciding to demonstrate. No militant Negro today would dream of trying to persuade a manager to serve him on the grounds that Negro patronage would not be bad for business. We have grown too proud for that. But in those days . . . we regarded the sit-in as the successful culmination of a long campaign to reach the heart of the restaurant owner with the truth. What we took to be his conversion was as important as the fact that the restaurant had indeed been desegregated.

That spirit, as well as the courage and the counsel of Randolph, Rustin, and Farmer himself, set an example for black students at Howard University in Washington, D.C., who persisted in their purpose after the flurry of other demonstrations had temporarily receded late in 1943. Those students had organized a Civil Rights Committee, which in 1943 began "sitting and picketing" segregated restaurants within the black sections of the District of Columbia. After two successes, the students moved downtown to take on Thompson's, a segregated restaurant that served moderately priced food all day long at a site convenient for black government workers. The participants in the pending demonstration, more than two hundred in all, signed a pledge that expressed the intent and endorsed the techniques of non-violent resistance:

> I oppose . . . discrimination . . . as contrary to the principles for which the present World War is being fought. . . . I conceive the effort to eliminate discrimination against any person because of race or color as a patriotic duty. . . . I pledge . . . to serve in whatever capacity I am best fitted. . . . I further pledge . . . to do nothing to antagonize members of the public . . . to look my best whenever I act . . . to use dignity and restraint at all times; to refrain from boisterous conduct, no matter what the provocation.

On April 22, 1944, the first three demonstrators entered Thompson's. Refused service, they sat down quietly and began to read books and newspapers. In a succession of small groups, another sixty-two students followed them. Neither the manager of the restaurant nor the supervisor of the chain could persuade them to leave. Meanwhile other students, many of them women, picketed outside, carrying placards that read: "Are you for Hitler's Way or the American Way? Make Up Your Mind" or "We Die Together. Let's Eat Together." On the street some whites jeered, but a poll of the customers within showed that only 30 per cent of the whites objected to eating with blacks.

The students' plan had not anticipated a further development. Six black soldiers who saw the pickets entered the restaurant, requested service, were refused, and then sat down with the demonstrators. At adjacent tables white soldiers and sailors were eating. White MP's asked the black servicemen to depart, but they remained. An MP lieutenant, alerted by his men, then arrived and requested the black GI's to leave as a personal favor so that the Army would not be embarrassed "in case of an incident." If the Army wanted to avoid embarrassment, several Howard students pointed out, it should have all men in uniform depart. The MP's quickly took that course. Still unable to persuade the management to negotiate, the students sat on for another four hours, until the district supervisor "received orders" from his superiors to serve them. Their tactics had worked, and their continuing efforts led to the reduction, though by no means to the elimination, of segregation in Washington.

The mounting frustration of blacks over wartime conditions pointed to the utility of those tactics, especially when, now that victory abroad was in sight, Southern strength in Congress and influence in the White House made the prospect for victory at home remote. In 1944 Gunnar Myrdal published *An American Dilemma,* his classic study of the conflict between professed national ideals and the circumstances of blacks in the United States. In his preface he stressed "one main conclusion . . . that not since Reconstruction has there been more reason to anticipate fundamental changes in American race relations, changes that will involve a development toward the American ideals." Perceptive blacks realized that those changes would materialize only if they pressed to effect them. Otherwise there was little evidence to sustain Myrdal's optimism. So it was that in 1943, when she composed her poem "Mr. Roosevelt Regrets," Pauli Murray also wrote directly to the President: "It is my conviction . . . that the problem of race, intensified by economic conflict and war nerves . . . will eventually . . . occupy a dominant position as a national domestic problem. It is becoming comparable to the issues of labor and relief which plagued the national government ten years ago." By 1945, Pauli Murray, a disciple of A. J. Muste, seemed prescient.

By that year those considered to be moderates among black

leaders were expressing the anger of A. Philip Randolph and elaborating the warning of Pauli Murray. In that year Walter White published *A Rising Wind,* a book that took its title from a statement of Eleanor Roosevelt: "A wind is rising throughout the world of free men everywhere, and they will not be kept in bondage."

White aptly summarized the impact of the war on American blacks, their alienation, their grievances, and the necessity for satisfying them. "World War II," he wrote, "has immeasurably magnified the Negro's awareness of the disparity between the American profession and practice of democracy. . . . The news that the American Negro soldier has received from back home has been predominantly disheartening. He has heard . . . of the humiliation, beating, and even killing of Negro soldiers in the South. . . . The majority . . . will return home convinced that whatever betterment of their lot is achieved must come largely from their own efforts. They will return determined to use those efforts to the utmost."

The race problem was international as well as domestic. The war had "given the Negro a sense of kinship with other colored—and also oppressed—peoples. . . . He senses that the struggle of the Negro in the United States is part and parcel of the struggle against imperialism and exploitation in India, China, Burma, Africa . . . the West Indies and South America." Consequently a failure by the United States to recognize the claims of the colonial peoples would constitute a "grave peril." The nation and its North Atlantic allies had to make a choice: "to revolutionize their racial concepts and practices, to abolish imperialism and grant full equality to all . . . people, or else prepare for World War III." Further, White predicted, "if Anglo-American practices in China and India are not drastically and immediately revised, it is probable . . . that the people . . . of the Pacific may . . . move into the Russian orbit."

The choice, White concluded, in a message that reflected the substance but intensified the tone of Myrdal's analysis, lay between pursuing "a policy of appeasement of bigots," which the country gave "every indication . . . of following," and thus courting disaster, or living up to American ideals and thereby helping "to avert an early . . . resumption of war. A wind *is* rising—a wind of determination by the havenots of the world to share the benefits of freedom and prosperity which the haves of the world have tried to keep exclusively for themselves. . . . Whether that wind develops into a hurricane is a decision we must make."

Insofar as the wartime attitudes of most American whites and the wartime policies of the federal government portended, the nation, in the light of White's prophecy, was heading for dirty weather. His bitterness flashed the same signal, as did the self-awareness of all blacks, their eagerness for access to the cornucopia of plenty that also beguiled whites, and for access to the opportunities and freedom whites took for granted. Before victory in Europe and Japan, blacks knew that they had lost the battle for victory at home.

While the war heightened the self-consciousness and raised the expectations of American blacks, it had the same effect on white groups—Italians, Jews, Poles, and the other "ethnics." Italian-Americans, increasingly aware that their political strength largely accounted for the relatively easy treatment of their enemy-alien parents, were eager for more recognition and influence in politics, and eager, too, for more income and more and better goods and services. They knew, as did the Jews and the Poles, that they were in competition with blacks and other colored people—Mexican, Japanese, Chinese—for jobs in the economy, which in 1945 was at the start of reconversion, and for housing and schooling and recreation in industrial communities. Their self-interest was threatened by the rising wind. The wartime race riots in the North as much as the demagoguery in the South forebode postwar antagonisms that would affect postwar politics, just as those antagonisms had affected politics during the war. In wartime, indeed, politics reflected as much as ever the many aspects of American culture: the conventions of literature and reportage, the messages of Madison Avenue, the seductions of affluence, the influences of big labor and big business, the necessitarian priorities of the White House and the War Department, and, not the least, the desires and the anguish of the outsiders, especially the blacks.

7 / Congress and the Politics of Comfort

1. Resurgent Republicans

To its vociferous critics on the right, the New Deal had always seemed radical, and Franklin Roosevelt an irresponsible reformer with a mania for power and without scruple in politics or private life. The coming of the war did nothing to change those beliefs. On the contrary, the wartime need for a further centralization and more penetrating exercise of federal authority whetted the suspicions of those who thought "that man" intended to hold the Presidency forever. They were bitter as well as suspicious. However moderate the New Deal would come to appear years later, in 1942 the wealthy still felt, by and large, that Roosevelt had directed the energies of government, as he had said himself, to improve the lives of the poorest third of the nation and to reduce the power and privileges of the "economic royalists," whose hatred he had acknowledged and welcomed.

Whatever their personal wealth, regular Republicans were bitter, too. Their party had lost three consecutive Presidential elections, a sequence of defeats without precedent since the time of the party's creation. Furthermore, the chance to unseat a President in the midst of a great war was remote. The conservative wing of the party, which had opposed prewar aid to Great Britain and the Soviet Union as well as the reforms of the New Deal, blamed the Eastern liberals and their internationalist candidate of 1940, Wendell Willkie, for the GOP's most recent failure. Those conservatives, mostly from the Middle West, were resolved to regain control of the party, to block new adventures in social reform, and if possible to roll back the New Deal, to contain the President's power, and to bruise him at the polls.

For several years the Republicans in Congress had pursued their objectives in domestic policy with the co-operation of most Southern Democrats. Those Democrats, loyal to their party in election years, in varying degree disliked Roosevelt, many of his advisers, and some of his programs, especially those involving heavy spending. Since 1938 the bipartisan coalition had blocked the recommendations of the administration except on war-related or international matters. On those questions the coalition broke apart, with the Southerners and some Eastern Republicans following the President. After Pearl Harbor, the disenchanted Southern congressmen, further alienated by the FEPC, allied themselves with their Republican colleagues to protect white supremacy, reduce federal spending except on the war, eradicate various New Deal agencies, and challenge the authority of the White House. Aware of the uncertainty of effective support on Capitol Hill, Roosevelt accepted it as if by default. He was preoccupied by the war and prepared to subordinate other issues to his quest for victory. Disinclined therefore to engage in causes he considered peripheral, divisive, and probably futile, he ordinarily gave only rhetorical attention to questions of social or economic justice, including those that the progressive minority within his own party were eager to advance.

As Roosevelt also perceived, his conservative opponents profited from the mood of the electorate. During 1941 and 1942 industrial workers earned wages, often supplemented by overtime pay, far in excess of what many of them had received only several years earlier, whether from their employers or on relief projects. Farmers, so recently dependent on federal subsidies, enjoyed their highest incomes in a generation. Professional men and business executives, no longer in need of federal mortgage assistance to retain or improve their homes, chafed over increased taxes and shortages of steak, golf balls, and heating oil. Though most Americans worried about a postwar depression, their harsh memories of the past tended temporarily to be submerged by their desire to savor their new prosperity more than wartime circumstances and federal regulations permitted. What the Democrats and the President had done for them during the 1930's paled for the while in contrast to the irritations they felt over what the government appeared to have failed to do for them lately.

That mood encouraged those who interpreted all government planning as socialism or worse. The early miseries of the War Production Board and of the Office of Price Administration, two of the most unpopular federal agencies, reflected the difficulties of bringing a great industrial nation to full mobilization, difficulties common to every major belligerent, democratic, fascist, or communist. In the

United States in 1942, the right-wing critics of federal fumbling attributed it, with illogical convenience, at once to Roosevelt's inability as an executive and to the inherent impossibility of managing the economy in war or peace.

A few weeks after Pearl Harbor, the *Wall Street Journal* took characteristic alarm over the formation of the American Public Power Association "to unite in a national organization some 3,000 municipal power plants and several hundred rural cooperatives." The administration, the *Journal* claimed, was "definitely sympathetic to the idea," which forebode an "aggressive campaign for public ownership and operation" of electric power. That campaign in turn effectively declared "a type of internal war" against private enterprise. The *Journal* exaggerated, though Roosevelt did favor the creation of agencies for the Arkansas and Missouri river basins to provide low-cost electricity, important for wartime production, and regional development on the example of the Tennessee Valley Authority. That kind of planning frightened the *Journal* and its clientele, who distrusted existing federal agencies whatever their purpose. The president of the New York Stock Exchange spoke for those of like mind. Of all the plans about which he had heard, he said, he liked best "no blueprint." Private enterprise could be trusted on its own to handle the economic affairs of the nation.

In contrast, Alvin H. Hansen, the distinguished Harvard economist and foremost American Keynesian, prepared for the National Resources Planning Board a pamphlet entitled *After the War—Full Employment*. He proposed, among other things, a broadening of the base for individual income taxes, a decreased reliance on the corporate income tax, forward planning by private enterprise, forward planning by the federal government of public improvement projects for national resources, highways, intraurban transport, and housing, and an expansion of federal aid for education, public health, and old-age pensions. During the war, Hansen also argued, higher personal and corporate income taxes and higher excise taxes were necessary to combat inflation.

Hansen and the NRPB, the *Wall Street Journal* judged, had simply "wasted effort," for during the war "private business can and will do the job of production" so long as government expenditures sustained demand. That process would assure full employment and prevent depression. For the postwar period, Hansen was proposing a "halfway house to socialism." There would be no postwar depression "unless we plan for . . . one by preparing these quack remedies

. . . in advance." As the *Journal* went on a month later, only superficially in jest, the War Production Board had already forced mother to lose her fly swatter, father his cuspidor, teen-agers their jukeboxes and pinball machines, and babies their safety pins. Yet the NRPB was making plans. That exercise was absurd, for unfilled consumer wants would create a demand so large that it would dwarf any projects the NRPB could invent for the government to finance.

Though Roosevelt disagreed with the *Journal* and its fans in Congress, he would not let an ideological dispute dilute his concentration on the imperatives of the war. He may also have felt that any such debate would hurt the Democrats in the fall elections. In July 1942 the NRPB proposed to create a small planning committee to devise a program for demobilization. The State Department for its part had already asked the President to approve a "full-fledged" study of postwar international and economic problems. He declined both requests, though off the record he allowed the NRPB to let a couple of men study demobilization in their spare time. "I finally decided," he officially informed that agency, "that this is no time for a public interest in or discussion of postwar problems—on the broad ground that there will not be any postwar problems if we lose this war. This includes the danger of diverting people's attention from the winning of the war. I am inclined to think, therefore, that any publicity given at this time to the future demobilization of men . . . would be a mistake."

The debate nevertheless continued in the tone the *Wall Street Journal* had sounded, just as party rivalries continued although Roosevelt had implausibly told the press that "politics is out." While politics went on as usual, in 1942 much of politics was as petty as usual. The Republicans found one broad target in the Office of Civilian Defense, which had been created to boost morale, invigorate drives for scrap metal and food conservation, and organize volunteers against the unlikely chance that enemy airplanes would bomb American cities. Fiorello La Guardia served as the first head of the agency, with Eleanor Roosevelt as the first director of its Voluntary Participation Committee, which, among other duties, ran physical-fitness clinics. Mrs. Roosevelt appointed to her committee a modern dancer, Mayris Chaney, and, as a recruiter of volunteers, Melvyn Douglas, a talented, outspokenly liberal Jewish actor and film star. The Republicans gleefully sprang to the attack. "This is just another sample of the way the New Deal operates," John Taber told the House of Representatives. "Under the direction of Mrs. Roosevelt, and certainly with the approval of the President himself, they have taken the old WPA crowd, plus some 'red' recruits and promoted them into the OCD office." The Republican press mocked Miss Chaney and identified Douglas as a fellow traveler, a typical protégé of the President's wife. Roosevelt ducked; at his quiet instigation, La Guardia, Mrs. Roosevelt, and her appointees resigned.

Their opponents seized upon another issue, one that arose from the spontaneous partisanship of the Democratic National chairman, Edward J. Flynn, the boss of the Bronx. Republican criticism, he announced, benefited only Tokyo, Rome, and Berlin. Consequently the election of a Republican majority to the House of Representatives in 1942 would strike the worst blow the nation could suffer except a military disaster. Infuriated, the Republicans charged that Flynn had impugned their patriotism, which of course he deliberately had. But Roosevelt dodged again. Woodrow Wilson's call for a Democratic Congress in 1918, he believed, had backfired. Taking no chances, the President repudiated Flynn's statement. The voters, he said, should cast their ballots for candidates, regardless of party, who would "back up" the government in wartime.

Democratic candidates nevertheless continued to identify as unpatriotic those of their opponents who had, before Pearl Harbor, voted against policies like lend-lease. Furthermore, the President could not really leave politics alone. In 1942 he tried but failed to influence the nomination of a Democratic gubernatorial candidate in New York who would give Republican Thomas E. Dewey a close race. Dewey had established himself as a strong contender for the Republican Presidential nomination in 1944, a position he enhanced by his easy victory in New York. Wendell Willkie, who seemed in 1942 another likely Republican nominee, was providing Dewey with only token support in New York, but Willkie was continually critical of Roosevelt's management of the war, of the colonialism of the President's friend Winston Churchill, and of Roosevelt's equivocations about the grievances of blacks. The President cut off that criticism temporarily by sending Willkie on a trip around the world that coincided with the crucial weeks of the congressional campaigns. Otherwise Roosevelt was unusually subdued, partly because he was too busy for intense partisan activity, partly perhaps because his interventions in congressional races in 1938 had yielded only mixed results. He did urge the defeat of his Dutchess County Republican congressman, the conservative and isolationist Hamilton Fish. He also spoke out vigorously for Republican Senator George Norris, of

Nebraska, a splendid old progressive. "One of the major prophets of America," as the President put it, Norris "transcends State and party lines." Fish won re-election and Norris lost, two victories for the right wing.

Yet partisan maneuvering affected the behavior of American voters less than did the impact on their morale and their daily lives of the gloom of the first year of war. From the Pacific, the only front on which American troops were yet engaged, the news was all bad. The valorous defense of Wake Island and of Bataan and Corregidor prevented any feeling of humiliation about those Japanese victories, but Americans were unaccustomed to the idea of losing battles, especially to an Asian enemy. They were shocked by the inadequacy of the nation's air forces, the damage to the nation's Navy suffered at Pearl Harbor and in the southwest Pacific, the Japanese seizure of Kiska and Attu in the Aleutian Islands, American soil, and the vulnerability even of the legendary marines, who in August 1942 landed on Guadalcanal to begin their dreadful ordeal there. For the United States, the course of war had apparently yet to turn, and the most visible heroes had emerged not from triumph but in defeat. Those conditions bred frustration and doubt among civilians and military trainees who could not be privy to the importance of the great naval victory at Midway or the developing plans for the counterattack in North Africa.

Conditions on the home front affected morale more immediately. Through 1942 the War Production Board operated without the organizing benefit of the Controlled Materials Plan. In the resulting absence of an effective system for priorities, industrial contractors experienced constant difficulties in obtaining the basic metals they needed. For those who went to Washington to ferret out supplies, the government seemed a labyrinth with thousands of paths either totally blocked or opening out upon bureaucratic ignorance or delay. During 1942 the crisis in rubber was exposed but not yet subjected to adequate study, much less to remedy. On strictly political grounds Roosevelt deferred ordering nationwide gasoline rationing, one indispensable program to conserve rubber, until after the elections. So also in 1942, for most farmers the most prosperous year in memory, efforts to produce more food and fiber crops seemed threatened by the decreasing availability of farm labor, especially at harvest time. Labor enjoyed record wages in 1942 but struggled for still more as inflation eroded much of the advance in real income achieved during 1940 and 1941. Working men and women, like the farmers, understood inflation largely as it affected them, not as their demands contributed to it. Accordingly, few Americans were happy with the administration's necessary recourse to selective rationing and to higher taxes.

During 1942 the Office of Price Administration instituted ten major rationing programs, while shortages in still exempt goods drove their prices up, encouraged hoarding, and forebode further rationing to come. The rationing system supplied consumers with coupons necessary for the purchase of designated goods—meat, shoes, fats, coffee, among others—at prices the OPA had frozen. Dealers could not restock their supplies without surrendering the coupons from sales they had made. In every community, local rationing boards set quotas for the coupons for each family, with the members of some boards more lenient toward their neighbors than were others. That unavoidable inequity occasioned fewer complaints than those submitted by producers and vendors, who almost invariably believed the OPA set prices too low. Their ire stimulated accusations, in Congress and out, that OPA, like all New Deal agencies, was an authoritarian bureaucracy, staffed by stupid, radical, and arrogant men. Congressmen of both parties, but particularly Democrats, also deplored the sensible decision of Leon Henderson, the first head of OPA, to insulate his agency from patronage seekers. Beset by critics as well as by other federal agencies, whose objectives conflicted with his, Henderson—a veteran New Dealer, and a tough, learned, and entertaining economist—lampooned his enemies in public statements that suggested he enjoyed controversy even to the point of puckishly promoting it. His manner and his stance exposed him to growing hostility, which he happily ignored or reciprocated.

A thoroughly competent man, Henderson never really had a chance to succeed. He had no control over the most unstable elements in the economy, wages and farm prices, which remained undisciplined until after he left office. Worse, he had no way of attacking the basic cause of inflation. Government spending was pumping money to consumers through industrial wages; but war production constantly diminished the supply of consumer goods. During 1942 the resulting difference between spendable income and available goods, the excess of demand over supply, or the annual "inflationary gap," in the phrase of the day, approached $17 billion. Increased taxes could have narrowed that gap, but Congess hesitated to impose them.

Pressed by Roosevelt, Congress in January 1942 did authorize

OPA to set maximum prices. After allowing for some incentive increases, Henderson in April gave up the practice of selective price controls and issued the General Maximum Price Regulation, "General Max," which ordered a freeze on all retail prices at the highest level reached during March. The ruling proved difficult to enforce. The quality of many items deteriorated though their prices held firm, and for some goods black markets flourished. Still, the order set the stage for the Price Control Act that Congress passed in October at the President's insistence. That new law broadened the field of price control and provided for the control of wages. For the duration of the war the price control program restricted inflation, but during most of 1942 General Max, still less than fully tested, tended to irritate merchants and, in the absence of new taxes, to afford too light a shield against inflation.

In that year, resistance to tax increases emanated from three imcomparably influential sources: individual taxpayers, corporate taxpayers, and (because they understood their constituents) a majority of congressmen. Consequently the administration's recommendations had small chance to prevail. Further, Roosevelt's advisers disagreed among themselves. Their first difference arose from the Treasury's determination to keep war-bond sales on a voluntary basis. Leon Henderson, Budget Director Harold Smith, Federal Reserve Chairman Marriner Eccles, and Vice President Henry Wallace recommended a compulsory savings program to sop up rising personal income. Secretary of the Treasury Henry Morgenthau, Jr., argued that his voluntary program would provide an equivalent brake on inflation, permit the continued use of bond drives to boost civilian morale, and push Congress toward higher taxes. Roosevelt supported that view.

The revenue program that the Treasury, with the President's approval, took to the Hill in March 1942 stressed the need for more money to finance the war. Obviously anti-inflationary, the program was also designedly progressive. Morgenthau proposed increases in surtaxes on personal income and in the corporate income tax, a tightening of the terms of the excess profits tax in order to prevent the possibilities for avoidance that industry was exploiting, increases in estate and gift taxes, and higher excise taxes on luxuries and goods that were scarce. With the dual purpose of expanding and improving postwar social-security coverage and of reducing wartime pay checks, he advocated an immediate increase in social-security taxes. He also proposed a system, then innovative, for collecting income taxes at the source by regular payroll deductions instead of awaiting the end of the year and relying upon taxpayers then to calculate their liabilities and file their returns.

Those recommendations met immediate opposition. Key congressional leaders, convinced that the Treasury was less intent on financing the war than on social reform, which they opposed, considered a sales tax the most effective device to raise revenue and reduce inflation. Henderson, Smith, Eccles, and Wallace also urged the President to support a sales tax. Distressed, Morgenthau wrote Roosevelt in April 1942 that "it would be a mistake to yield to the clamor for a sales tax. . . . It . . . would . . . have shifted the source of revenue to the lower income groups." As Morgenthau saw it, his opponents within the administration and on the Hill had ceased to care about the poorest Americans, the "people in the lower one-third." Ruefully he mused: "I can get all my New Dealers in the bathtub now."

Again the President backed the Treasury, but the House Ways and Means Committee, eager for a sales tax, sat on the revenue bill through April and May. The delay persuaded Morgenthau that the collection of income taxes from payrolls, if Congress adopted that plan, would begin dangerously close to Election Day. Though a postponement would reduce revenues for 1942, the Secretary saw no alternative. "I am more interested," he said privately, "that we get a Democratic Congress than I am in inflation or anything else, as far as domestic . . . economic, political situations go. . . . If you . . . hit . . . people before election . . . it may be very harmful to Democratic congressmen."

The Democrats in the House, who harbored exactly that anxiety, passed a revenue bill in June that contained few of the Treasury's recommendations and promised to yield only $6.25 billion in additional revenue during the ensuing year. "Those fellows," Morgenthau complained, "just don't know there's a war on." Awaiting action by the Senate, Roosevelt was more contained. "Keep on sittin'," he replied, "and no sweatin' and no talkin'. . . . Just stay put."

Anxiety about the pace of inflation provoked Morgenthau to reject that advice. The Congress continued to lean toward a sales tax to combat inflation, Roosevelt to further controls, but the Treasury to a new scheme, a spendings tax. At a steeply graduated rate, that tax would apply to all that an individual or a family spent in a year, less stipulated allowances for necessities. The tax, as the Treasury contrived it, would encourage saving, reduce demand, increase federal revenue, impinge more heavily on the rich than the poor, and obviate the need for a sales tax. But the Senate Finance Committee quickly and unequivocally rejected the proposal. Roosevelt, who had not endorsed it, told Morgenthau that "I never make any recommendations to Congress while a bill is pending." "Well, you know, Henry," the President added, now with candor, "I always have to have a couple of whipping boys." For 1942, if Henderson was perhaps one of them, Morgenthau was surely the other.

In the end the public suffered the whipping, for the Revenue Act of 1942, which Congress passed in October, failed sufficiently to take into account the social inequities that had concerned the Treasury. Though it did increase personal and corporate income taxes, the new act also levied a flat 5-per-cent gross income tax on all annual incomes in excess of $624, a device the Treasury had endorsed only as preferable to a sales tax. The gross income tax bore regressively on low-income families, especially industrial workers, most of whom would now pay an income tax for the first time. The impact of the new taxes and of General Max, and the wage controls Roosevelt imposed in October, together kept inflation in hand. Yet at a time when the war was going badly, the taxes and regulations, though essential for the economy, intensified the irritations that most Americans were already feeling.

In the autumn of 1942, as public-opinion polls revealed, farmers fretted over labor shortages, suburbanites over the price of meat; and industrial workers, the backbone of the New Deal coalition, grumbled about the good life they had begun to lead. Depending upon their place of residence and employment—depending, too, upon the make-up of their families and their age—these grumblings differed, but the general mood could be reflected, more or less, in the gripes of a typical Italian-American machinist from Hartford, Connecticut. Married, in his forties, native-born and, so, a citizen, he resented the status of his parents as enemy aliens. Trained in his skill by the Works Progress Administration, he nevertheless attributed his new affluence, as did most Americans, to his personal virtue and assiduity. Again like most Americans, he had tended to blame himself for his failure to find employment ten years earlier, so he naturally took credit for his recent success. So did his wife take pride in her first paid job, a relief from housework and mothering and a source of an exhilarating income of her own. They had moved in 1941 to Hartford, with its growing aircraft industry, but in Hartford, besides good jobs, they found housing scarce and expensive. They resented the rent they had to pay and the disrepair of the two-family frame house in which they lived. Even more, though they had never known a black, they disliked having a residential district of blacks only a few blocks away and rather feared the increasing number of blacks at the factory. In partial compensation, the machinist had money in his pocket, money in his war bonds, and money in the bank, but he was annoyed that the federal income tax was about to reach him, and he resented his inability to spend his money for the things he wanted, many of them rationed or wholly unavailable. He had bought his first automobile, a used car, but now he had trouble buying gasoline and tires. Rationing in New England even kept him from taking his family for a Sunday drive.

His intelligence told him that the fact of war accounted for his deprivations, such as they were, but his emotions, on which Republican ward politicians played, tempted him to look at the administrators in Washington as incompetents, perhaps "pinkos" who wanted to take away what he had gained. His wife agreed. From Hartford, the military looked incompetent to both of them, and they could not understand why Roosevelt sent the Russians arms that Americans could have used at Guadalcanal against the Japanese. Furthermore, the wife's kid brother wrote from training camp that the lieutenants there were snotty college boys. The machinist had never had it so good, but he wanted it a lot better, and under the circumstances, even though he still thought the Republicans had somehow caused the Depression a decade ago, neither he nor his wife saw any reason to take the trouble to register to vote against the GOP in 1942.

Hadley Cantril, who sampled political attitudes for the President, warned Roosevelt that the public mood was unpromising. Voters, particularly those on whom the Democrats depended, were indifferent. The election bore out that assessment. With Democratic turnout far below normal, even for an off year, and Republican participation close to average, the Grand Old Party gained forty-four seats in the House of Representatives; in all, 209 to the Democrats' 222. The Republicans also added seven seats in the Senate and won several important governorships. When the new Congress convened in January 1943, the coalition of Republicans and Southern Democrats would enjoy a comfortable control. All analysts agreed that the low turnout explained the results. That turnout reflected the absence from the polls of young men in uniform, a majority of them Democrats, and of war workers, also mostly Democrats, who had moved to new communities where they had yet to meet the residence requirements for voting, or who had failed to register. It reflected even more the political indifference produced by wartime irritations, and the political lassitude of farmers and workers who had often supported the Democratic party, a lassitude that contrasted with the ideological and partisan motivations that brought out the Republicans.

Roosevelt's critics exulted in the returns. Raymond Moley, for one, observed gleefully that the belt running from central Ohio through Colorado was solidly Republican again. The new Republican governors there and elsewhere, he judged, were blunt, tough, honest men. "There is nothing visionary about those people," Moley wrote, whereas it had been a bad November for "extremists and prophets of 'new days.' . . . The American people . . . have reminded the 'morale builders' in Washington that they don't want to be told what to think or how to feel. They will take care of their own morale." The *Wall Street Journal* urged the reinvigorated GOP in Congress to "make a program of domestic aims its first business." *Fortune* agreed. "The victorious candidates," it wrote, "rode an anti-Roosevelt and an anti-Washington wave. They were almost entirely normalcy men, quiet, churchgoing, family men, not quite prohibitionists, men whose outlook was limited to their states and their regions. They may be relied upon to investigate Washington thoroughly. Many of them think they have a mandate to repeal all New Deal reforms."

The New Dealers were dispirited. Harold Ickes, in public the most pungent advocate in Roosevelt's Cabinet, had in November 1942 temporarily lost his taste for contest. "You suggested that I call together some of the liberals for a political powwow," he wrote the President. "Upon careful consideration . . . I doubt whether I could accomplish the results that are so necessary if we are to be in any position in 1944 to put up a real fight on the political front." Speaking for the Democratic National Committee, Oscar Ewing, another New Dealer, attributed the losses of 1942 to the "innumerable irritations caused by Government restrictions and the arrogant way in which some of those matters were handled. The Republicans capitalized on all of those irritations and were able to make many voters blame the Administration and completely forget . . . Hitler." Ewing prescribed improved public relations, a better job of "selling" the war.

Probably the shrewdest and most self-interested postelection analysis that Roosevelt received, a strictly partisan document, came from Edwin W. Pauley, secretary of the Democratic National Committee. A successful oil man from California, Pauley had identified himself with the hard-nose Democratic professionals. He was scornful of the party's liberals and intellectuals, who learned to count him a major enemy. His biases overwhelmed theirs in his report to Roosevelt, which on this occasion nonetheless stated accurately what Pauley had found. He had polled a disgruntled, sometimes bitter group, the "Democratic Senatorial and Congressional Candidates, whether they were victorious or not." After sorting their replies, he explained the party's defeat in terms of the politics of resentment, which he itemized and weighted. Five factors in particular had contributed to the Democratic losses:

Resentment of Bureaucracy . . .
   A few of the letters complained . . . that the increase of the power of the Executive Branch . . . at the expense of the Legislative Branch indicated a trend away from our constitutional theory . . . of government. By far the greater complaint was directed at the methods of bureaucrats charging arrogance, etc. . . .

Resentment of the Conduct of the War . . .
   . . . The well-established conviction that America could "lick Japan with one hand tied" but wasn't doing it, was cited. Many letters stated that if . . . the African campaign had preceded rather than followed the Elections, the results would have been different.

Resentment of O.P.A. Particularly of Mr. Henderson . . .
   This was the most universal and serious complaint of all. . . . It appears from the letters that the complaint is directed rather at Mr. Henderson and his attitude and methods than at the abstract question of . . . rationing and price control. . . .

Resentment of Labor Policy . . .
   . . . The Government's "coddling of labor" received a substantial complaint. . . . The refusal to freeze wages and the freezing of farm prices worked a double hardship on the farmers. . . .

Resentment of Farmers . . .
   . . . The freezing of prices as to farmers and the refusal to curtail the profits of others such as industrial laborers. . . .

Roosevelt got the message. However unfair the complaints, the electorate was unhappy about the running of the war. The voters had stayed home or cast their ballots against what they resented. Victory in battle would help in time to renew loyalties to the Democratic party, as could other, more immediate, steps. Pauley had suggested one. The complaints about Leon Henderson, he wrote, were "correctable." The President got rid of Henderson. Pauley had reported no political advantages in pressing for social reform, which the Republicans in Congress could in any event block. Roosevelt during 1943 sponsored no social reforms. The Republicans, their confidence rising on the wave of irritation from which they had profited at the polls, began, with the assistance of their Southern allies, to tear the New Deal apart. They were not merely vindictive. They intended to wipe away enough of the recent past so as to reassert the authority of Congress, to reduce the power of the Presidency, and thereby to prepare to impress upon postwar public policy an unequivocally conservative stamp.

2. The New Deal at Bay

The Republicans elected to the Congress that convened in 1943 brought to Washington a variety of backgrounds, interests, and convictions, but a composite profile of the group revealed a dominant type of formidably conservative leanings. The representative of that type was a middle-aged, white, Protestant male, college educated, a lawyer, the father of two children, a member of the Methodist or Presbyterian church, and of the Elks, Odd Fellows, or Masons (32nd degree). He had served in the Army during World War I, joined the American Legion, invested in a local retail business or newspaper, and held sundry local offices. He hated the New Deal with the venom of Charles A. Halleck, of Indiana, who fit the type exactly.

"The social experimentation and reckless extravagance of the New Deal," Halleck had predicted in 1936, "are on the way out because the common sense of the people is reasserting itself." The war agencies, he later asserted, were "merely an extension and development of what was already in progress . . . under the New Deal." The American people had been deprived of their liberty, he maintained: "We must be free of annoyance . . . of restrictions which cramp . . . our lives. . . . We must be allowed to work, to invest, and to save without making out a bureaucratic blank for every move we make." To that end the GOP, the party of Americanism, challenged the Democrats, "a party riddled . . . with foreign, left-wing, un-Americanism." For Halleck and those of his mind, one first task was to rid the country of the New Deal relief agencies, which they had always considered ineffective.

Before the 1942 elections, friends of those agencies could

defend them only weakly. However useful, even indispensable, work relief had been before 1940, it lost its function after the surge of employment created labor shortages. So, too, even as stand-by offices against the event of a postwar depression, the federal agencies of the 1930's attracted the enthusiasm largely of their own staffs. The most popular of them, the Civilian Conservation Corps, had enrolled unemployed young men to perform worthy tasks in the conservation of woodlands and water resources. But those young men, now in uniform, would be eager after the war to escape forever from camps run by military officers, as the CCC camps had been. There would be a postwar need for federal expenditures for conservation, but the CCC was by no means an ideal institution to preserve for that task. By 1942 public opinion opposed continuation of the corps, which could not any longer describe a persuasive mandate for its existence. Roosevelt, who had a special fondness for the CCC, suggested it might help strengthen and discipline boys below draft age by employing them "for only nominal purposes" in national parks, forests, and historic sites, but Congress declined to appropriate funds except for the orderly liquidation of what had been the first of the New Deal's programs for work relief.

As with the CCC, so with the Works Progress Administration, once Harry Hopkins's agency, the vortex of all kinds of work relief; the staff could not offer a convincing case against dissolution, which the Republicans and Southern Democrats demanded. Roosevelt, interpreting the political climate as too hostile even for a "wartime furlough" for WPA, in December 1942 gave the agency an "honorable discharge." Early in 1943 it made its last relief payment.

The President fought little harder in 1943 to preserve the National Youth Administration, though its director, Aubrey Williams, tried to enlist him, as he had the previous year. During the Depression the NYA had provided funds for the training of needy young men and women and for their part-time employment while they attended school or college. In 1941 Williams, whom anti–New Dealers had long since designated as radical, reorganized the agency's program to focus it exclusively on the training of young people in skills needed by defense industries. Those trainees were hired as rapidly as they became available. Businessmen—Owen D. Young, of General Electric, for one illustrious example—praised Williams for his devotion to the development of American youth and his contributions to the war effort. An investigation of NYA that Roosevelt had commissioned reported in 1942 in the same vein and with a recommendation, similar to one Young had made, to preserve the agency for its potential role during "the period of reconstruction and reorientation that will follow the cessation of hostilities." Congress then appropriated enough to keep NYA going until 1943.

In a letter thanking the President for his support, Williams wrote that "the most sinister thing we face is . . . the school crowd. . . . They would give opportunity to those who already enjoy advantages and . . . those who are now receiving training would . . . be forced out through lack of means." As Williams knew, fierce opposition to NYA programs was building within the National Education Association and the American Vocational Association, the former the premier organization of public-school administrators, the latter a similar body for vocational schools. Jealous of their roles, both groups condemned NYA training centers, demanded the preservation of state and local authority over secondary education, and predicted that federal funding even of vocational training would lead to federal control over the whole educational process.

Those arguments only superficially disguised the primary motive of Southern critics of NYA, who objected particularly to the agency's successful efforts in recruiting and training black workers for war industry. The combination of Southern and Republican hostility to NYA with the influence of the "school crowd" put Williams in a perilous position in 1943. The President, though impressed by his subordinate's purpose and achievements, ducked combat with the NEA, one of the most powerful lobbies in Washington. As one NYA officer admitted, moreover, the necessity for increasing production, with its related requirement of unskilled labor, had reduced the importance of pre-employment training. Even Eleanor Roosevelt could not persuade her husband to give Williams more than rhetorical assistance.

Left to his own resources, Williams presented Congress with a budget that excluded all frills. NYA, he proposed, should offer training in industrial skills to any person, employed or unemployed, between the ages of sixteen and twenty-five. Trainees who lived at home would receive $25 a month, those living in NYA centers, $10.80 a month. Wages and subsistence together would cost only $66 a month. Frugal though that request was, the agency's enemies rejected it. Senator Harry Byrd, of Virginia, a fervid and penurious anti-New Dealer, complained to Roosevelt about Williams's extravagance over the years. In contrast, Paul V. McNutt, then Federal Security Administrator, praised the NYA program and urged its continuance to assure an adequate supply of skilled workers, of whom the vocational schools trained only a small fraction. The president of the United

Transport Service Employers also defended the agency. It had "been responsible," he wrote, "for countless thousands of Negro young men and women being given an opportunity to serve their country on the production and battle fronts because of the skills which they received." But Byrd's view prevailed in Congress, which in June 1943 liquidated NYA.

Hostile as they were to any federal planning, congressional Republicans especially opposed planning for postwar social and economic policy, which they hoped themselves to dominate. Planning, as the *Wall Street Journal* put it, could not be left "to theorists or advocates of collectivism or professional officeholders. Master plans, prepared and administered by agencies of political origin, mean control of industry by such agencies." The *Journal* and the Republican leadership agreed that the National Resources Planning Board sheltered both theorists and collectivists. The NRPB therefore became a prime target for extinction in 1943.

The NRPB served a potentially significant function in which it had not performed consistently well. Its head, Frederick Delano, an uncle of Roosevelt's, presided benignly and without compensation over a small, unpaid board that hired consultants, often college professors, to report on large national problems. The reports over the years had had a learned, empirical, gradualistic tone, though those produced between 1940 and 1943 were not systematic enough to satisfy Harold Smith, the Director of the Budget, who believed his own office could do a better job. At once liberal and practical, Smith, an able manager, was also one of Washington's foremost exponents of Keynesian economics, which congressional conservatives still interpreted as a heresy unsurpassed except by Marxism.

That heresy had imbued the controversial NRPB pamphlet of 1942 *After the War—Full Employment*. It appeared again in 1943 in another pamphlet the NRPB published, and Roosevelt sent to the Hill, *After the War—Toward Security*. The "totalitarian plan" that the President had forwarded, the *Wall Street Journal* editorialized, threatened the use of wartime emergency powers "to make permanent changes in American social and economic institutions." Government, the NRPB pamphlet had noted, was taking part in the management of key war industries which had expanded because of federal expenditures. Consequently government had "a direct responsibility and should participate in the decisions as to . . . what concerns should continue to operate these industries." The industries in question, the *Journal* countered, were basic industries. Federal control of them would involve control of all industry, an unthinkable eventuality.

Wartime experience with controls, the NRPB had suggested, should lead to the development of "such regulation as may stimulate the effective function of competitive business in normal times." To that end, the pamphlet recommended that the government retain control over patents seized from enemy aliens and license their use so as to encourage the development of more private enterprises. The *Journal,* as ever, opposed adventures in antitrust as much as it opposed government controls.

The NRPB had called, too, for a "shelf" or "reservoir" of desirable public construction projects, projects particularly for regional development, for which federal funds would be spent if the postwar economy needed stimulation to reach full employment. The *Journal* for its part insisted that private industry should and would provide jobs for demobilized troops and dismissed war workers. The NRPB approach, in the view of the *Journal* and of anti–New Dealers in Congress, would result only in the waste of billions of dollars. Just as ominous were NRPB proposals for increases in social-security coverage and payments. "To have a decent standard of life," the *Journal* responded for itself and its admirers, "to say nothing of security, the American people in the post-war days will have to work and work hard. They will have to save and save hard. Anyone who tells them that a government can concoct a scheme which will save them from the rigors before them tells them a cruel lie. Unemployment and old age pensions are a national policy. There may be occasions when they can be extended. That occasion is not now when no man can see the end of expenditures necessary for national self-preservation."

For the *Journal,* "now" meant "ever," as it also did to many newspaper editors. According to one typical comment, "there are no lengths to which these planners will not go. They even propose that in time to come every American man and woman who so chooses will be a college graduate at public expense! . . . Everyone is going to have to pay through the nose if those dreams come true." The president of the National Association of Manufacturers summed it all up in one word: "Socialism."

Charles Halleck and his friends and allies in the House of Representatives agreed. In February 1943 the House voted to cut off all further funds for the NRPB. "Dear Uncle Fred," Roosevelt wrote Delano, "I am much upset over the Congressional attitude toward the National Resources Planning Board. The need for such studying and planning as your Board is doing is so obvious that I am very hopeful that the Senate will restore the appropriation." The President also, rather too late, made the rescue of the NRPB an administration measure. He praised the board in public and in two letters to his "old friend," Senator Carter Glass, the aged, influential, but reactionary Virginia Democrat who was chairman, now constantly *in absentia,* of the Appropriations Committee. Union labor, the Conference of Governors, the council for State Governments, the American Municipal Association, and the Committee for Economic Development—founded by business and industry—all spoke out for the NRPB. The Senate subcommittee handling the matter voted 6 to 4 for a little more than $500,000—less than half Delano's request to keep the board alive. The full Appropriations Committee, less generous, cut the sum to $200,000, to which the Senate consented, with twelve Republicans in the affirmative even though Robert Taft led the fight against the board. But in conference committee the adamant spokesmen for the House prevailed. The Independent Office Appropriations Act of 1943 required discontinuance of the work of the NRPB by August 21 and deposit of all its records in the National Archives, where Halleck apparently hoped they would get lost or crumble away. His friends rejoiced in his victory. "It is well that Congress has denied funds to the NRPB," the *Wall Street Journal* jested. "It might be rewriting the Ten Commandments next. Of course, it has already repealed the law of supply and demand."

Congress went on writing laws to its taste. The Farm Security Administration, "the very symbol of the New Deal for small farmers," turned its energies after the war began to help those farmers increase production and to protect agricultural laborers, many of them Mexicans in the Southwest, from exploitation by their employers. Considered socialistic by those employers and their friends in Congress, the FSA survived in name only after budget cuts Congress voted in 1943. The Rural Electrification Administration, which private power companies detested, suffered an identical blow, as did the domestic branch of the Office of War Information. The Senate also strangled a bill for federal aid to education, most of it for teachers' salaries in impoverished areas. The Republicans, aided by short-sighted advocates of civil rights, amended the measure to prevent aid to schools that practiced racial segregation. That provision, as it was intended to, galvanized the opposition of Southern Democrats to the proposal. So, too, Congress rejected the Wagner-Murray-Dingell bill to expand social security.

Anti–New Dealers used the session of 1943 to punish their ancient foes. So it was that the Senate withheld approval of Roosevelt's nomination of Edward J. Flynn as ambassador to Australia. Flynn's obvious lack of qualifications for the post hurt him less than did his deserved reputation as one of Roosevelt's ablest political operatives. Earlier the House threatened to hold up OPA appropriations until Leon Henderson was forced to resign. Democratic Senators Byrd and Kenneth McKellar, of Tennessee, long suspicious of executive extravagance and jealous of congressional prerogatives, had since 1941 orchestrated the campaign for retrenchment and against the New Deal through the operation of a joint congressional committee on unnecessary expenditures, which Byrd chaired. In 1943 that committee harassed war agencies for moving personnel from one office to another without specific congressional approval. McKellar, a pinched, vindictive, and venomous man, introduced a bill to require the Senate's consent for all appointments to positions earning $4,500 a year or more. The measure, which would have hamstrung the President and provided pocketfuls of patronage for congressmen, failed only after Roosevelt's vigorous intercession. McKellar also allied himself with Jesse Jones, the sanctimonious Secretary of Commerce and head of the Federal Loan Agency, who had often been openly hostile to New Deal proposals. McKellar and Jones attacked Vice President Henry A. Wallace, the foremost spokesman of liberal Democrats and the chairman of the Board of Economic Warfare. Like Jones, McKellar considered the policies of that board wasteful and radical.* Hearings over which McKellar presided opened a forum for Jones's myopic views. Partly because of those hearings, the antagonism between Jones and Wallace flared up in 1943, embarrassing the President, who abolished Wallace's board without disciplining Jones, a decision that stunned dedicated New Dealers and the liberal press.

In 1943 the House Un-American Activities Committee, an instrument for the witch hunts of its chairman, Martin Dies, of Texas, identified some forty administrators who it alleged had radical affiliations, ranging from Communism to nudism. The gullible House authorized the committee to conduct hearings on the loyalty of federal employees, three of whom Dies and his fellows found guilty of subversion, wholly on the basis of hearsay. Nevertheless persuaded, the House attached a rider to a deficiency appropriations bill that required the dismissal of the three men unless they were reappointed by the President and confirmed by the Senate. Unable to detach the rider from the bill, the Senate accepted it and Roosevelt signed it. Three years thereafter, too late to protect the three victims, the Supreme Court predictably ruled that Congress had punished them

* For a fuller discussion of the Board of Economic Warfare, see Chapter 8, Part 3.

without a trial and violated the Constitutional prohibition against bills of attainder.

Had it dared, Congress would doubtless have passed a bill of attainder against John L. Lewis, the pugnacious head of the United Mine Workers. Relentless in his recourse to the strike as a weapon to force owners to increase the pay and improve conditions of safety in the mines, Lewis offended American public opinion, even journals like *The New Republic,* by calling the miners out in 1943, as he had in 1941. His obduracy in the pursuit of needed change gave antilabor congressmen an excuse they welcomed for regulating all labor unions. Two Southern Democrats, Howard Smith in the House and Tom Connally in the Senate, introduced separate bills that Congress reconciled and passed as the Smith-Connally Act. It enlarged the President's power to seize plants useful to the war effort, made it a crime to encourage strikes in those plants, provided for a mandatory thirty-day cooling-off period before union leaders could call strikes in other industries, and, as the Republicans had long wanted to, forbade union contributions to political campaigns. Secretaries Stimson and Knox applauded the measure, because both men characteristically looked upon strikers in wartime as no better than deserters. But New Dealers like Harold Ickes and Frances Perkins doubted both the wisdom and the efficacy of the proposal, and Roosevelt, though eager for more authority over strikes, was loath to offend labor or discourage its political contributions. In June, while asking Congress to authorize the conscription of strikers (a prospect at which friends of civil liberties could only shudder), the President submitted a veto of the Smith-Connally bill, which Congress quickly overrode. Labor in 1943 had few advocates at either end of Pennsylvania Avenue.

On the issue of taxation the President's New Deal advisers met even more jarring defeats. The first loss was difficult for the ordinary citizen to understand. Before 1943, individuals were always a year behind in their tax payments. In 1942 they paid their taxes on incomes from 1941. In 1943 they were to pay their taxes on incomes for 1942. There was as yet no system for keeping taxpayers current by withholding taxes through monthly payroll deductions or by anticipatory quarterly payments. In 1943 the Treasury recommended the adoption of that system. Tax rates for 1943 were higher than they had previously been, as were personal incomes, and the Treasury wanted to collect currently in order to reduce take-home pay and thereby combat inflation. In addition, the Revenue Act of 1942 had multiplied the number of Americans who owed income taxes. Most new taxpayers, moreover, had little understanding of income tax forms and no experience in preparing them. Payroll deductions would substantially ease the collection of taxes, especially from industrial workers.

Millions of Americans, however, were reluctant to pay 1943 income taxes currently while they also had to pay the taxes due on their 1942 incomes. In some cases the resulting burden was bound to create hardships. In most instances taxpayers could have managed, but they were alarmed, and even, as it seemed to Congress, on the verge of revolt. Indeed, the degree to which World War II was a foreign war for Americans could be measured by their lack of a sense of emergency and obligation about taxes, which they paid at a far lower rate than did the citizens of any other major belligerent. In its dismay about taxes, the American public embraced a plan proposed by Beardsley Ruml to bring taxpayers up to date immediately by canceling their tax debt for 1942.

The chairman of the New York Federal Reserve Bank, an accomplished economist, as well as the treasurer of R. H. Macy and Company, Ruml solicited support for his plan on the radio and in the press. The forgiveness of 1942 tax obligations, he argued, would permit taxpayers to pay their 1943 taxes currently without strain, even though tax rates were higher for the latter year. Moreover, taxpayers would remain current through the last year of their lives. His plan, Ruml said, would simply move "the tax clock forward, and cost the Treasury nothing until Judgment Day."

The Treasury emphatically disagreed. As its staff observed, forgiveness of 1942 taxes, for whatever purpose, would deprive the government of substantial revenue and grant an appreciable windfall to taxpayers, especially those business executives whose earnings from war-related contracts had been extraordinary in 1942. Ruml admitted himself that his proposal particularly benefited the wealthy. The Treasury also believed that Americans would pay both 1942 and 1943 taxes if the Congress had the courage so to demand. Some congressmen did. Robert Doughton, chairman of the House Ways and Means Committee, called the Ruml plan "the biggest outrage ever attempted." Roosevelt agreed. "I cannot acquiesce," he wrote Doughton, "in the elimination of a whole year's tax burden on the upper income groups during a war period when I must call for an increase in taxes . . . from the mass of our people." But a majority of Congress contentedly adopted the Ruml plan with slight modifications. The Current Tax Payment Act of 1943 forgave 75 per cent of 1942 liabilities while it also made taxpayers current in their payments as of July 1, 1943. As the Treasury and the President had feared, privileged individuals who had enjoyed bonanza incomes in 1942 would never have to pay taxes on them, even when Ruml's tax clock reached eternity.

There remained another $12 billion of revenue to be raised by new taxes. That figure, the Treasury calculated, was essential both to defray half the ongoing cost of the war, the department's target, and to check still ebullient inflationary forces. But Roosevelt, gun shy, in August 1943 instructed Morgenthau to reduce his request for new taxes to $10.5 billion and to leave it entirely to Congress to work out the plans for reaching that goal. Once more the Treasury tried to enlist the President's support for immediate taxes to support the postwar expansion of social-security coverage and services, including medical insurance. Both Roosevelt and his congressional advisers balked. The people, the President told Morgenthau, were "unprepared." So was the White House. "We can't go up against the State Medical Societies," Roosevelt said, "we just can't do it."

Even the proposal for $10.5 billion in new taxes struck most congressmen as an "unbearable burden." The House Ways and Means Committee wrote a bill that would provide only $2 billion, a bill that also froze social-security taxes, which had been scheduled to yield an additional $1.4 billion. Furthermore, the measure granted lush tax favors to business, especially to owners of mines and timber. The Senate added further "private industry grabs," which the House gladly retained in the final draft of the bill Congress passed.

The measure promised so little new revenue and contained so many inequities that Roosevelt's advisers almost unanimously urged him to veto it. On February 22, 1944, he did so in a message that condemned the bill as "wholly ineffective" for meeting national budgetary and economic needs, and as "dangerous" because of its "undefensible privileges for special groups." In that respect, the President wrote, "it is not a tax bill but a tax relief bill providing relief not for the needy but for the greedy."

That assessment, applicable to the whole record of Congress in 1943, stirred a hot retort from the Hill. A majority of the Ways and Means Committee declared the $10.5 billion goal "oppressive to tax payers." More important, Alben Barkley, of Kentucky, the Democratic majority leader in the Senate, long one of Roosevelt's loyal lieutenants, now rebelled. Resigning his post on February 23, Barkley called the President's veto message a "calculated and deliberate assault upon the legislative integrity of every member of Congress. . . . If the Congress . . . has any self-respect left, it will override the veto." It did—the House the next day, the Senate the day after that. Roosevelt had watched Barkley's performance imperturbably, apparently on the accurate assumption that the histrionics would subside once the veto was overridden. In a soothing letter to "Dear Alben," the President urged the Democratic senators to re-elect their leader, as they forthwith did.

The "Dear Alben" episode aside, Roosevelt had suffered a double defeat. In spite of his concessions to conservatives on the Hill, Congress had written a revenue act that defied both the administration and common sense. And for the first time in history Congress had overridden a veto of a revenue bill. Probably more than ever before, Roosevelt's opponents had bruised him.

As *Fortune* had predicted, the Congress during 1943 and early 1944 interpreted the election returns of 1942 as a mandate against the New Deal. Roosevelt had read those results about the same way. He dumped Leon Henderson. Except in the case of the NRPB, and then too late, he did not much resist successful attacks on old New Deal agencies. His passivity, his punishment of Henry Wallace, his only half-hearted approach to tax reform, like his suspension of anti-trust prosecutions and his flirtations abroad with Darlan and Badoglio, grievously disappointed his liberal friends. He had let them down, though from his point of view with reason. He had not wanted to divert his energy and attention from the necessities of warmaking. He understood the restlessness of his constituents with social reform. He recognized the strength of his opposition on the Hill. In one sense, the "Dear Alben" episode proved that Roosevelt could not have put across liberal domestic measures in 1943 even if he had tried. In another sense, that episode was something of a self-fulfilling prophecy. Because Roosevelt had not tried, he had built up no reservoir of liberal public and congressional support on which he could call when he needed to.

Only in December 1943, at least a year after he had begun to drift with the wave in Congress rather than to resist it, did Roosevelt admit to his course, as if it were wholly new. Perhaps he was merely trying to make the best that he could of a dismal record. When he explained that "Dr. New Deal," a specialist in internal medicine, had been succeeded by an orthopedic surgeon, "Dr. Win-the-War," the reporters asked whether the President meant again to take up the social program. New needs might require a new program, he replied, apparently with the postwar period in mind. Privately he also knew he had been hurt and would have to recover some of his losses before

November 1944. To that end, he had already begun quietly to plan a
limited counterattack.

3. Two Bills of Rights

The confidence he exuded, the raillery he enjoyed, the inspiration that
his oratory carried, the slyness of his manipulative manner, the con-
tradictions between his professions and his actions as between the
idealism that he preached and the evasiveness he practiced, all this
made Franklin Roosevelt at times infuriating alike to his admirers
and to his opponents. All this made him in some degree inscrutable
even to those most devoted to his person and his career. He was now
bold, now cautious, now exuberant, now moody, now generous, now
cruel. He never let his right hand know, he once told a friend, what
his left hand was doing. He knew no man, a loving relative suspected,
and no man knew him.

Those qualities apart, during the autumn of 1943 and the winter
of 1944 the President bore the burdens of his days with discretion, to
be sure, but without mystery in the slightest. Those were months of
continual anxiety. The postponement of the cross-channel invasion,
on which the Soviet Union had counted, had strained the Grand
Alliance beyond his ability wholly to repair it when first he met with
Stalin as well as Churchill at Teheran. Nor could he satisfy the exaggerated demands of Chiang Kai-shek, whose hopes for a campaign in
Burma had to yield to the more exigent requirements of the war in
Europe. There and elsewhere the Allies, particularly the Russians,
were making impressive gains, and the outcome of the war was no
longer seriously in doubt; but the pace of victory was slower than
Americans had expected it to be, and the costs in lives were rising
with the rough campaigns in Italy and in the South Pacific. The
President knew that in 1944 those costs would rise still more with the
invasions of France and of the islands that protected the approaches
to Japan. His responsibilities as commander in chief, heavy for two
years, would not soon diminish, and the weight of diplomacy would
increase. Yet he would not delegate to any subordinate the authority
he exercised in order, as he saw it, to hold the alliance together
through victory and beyond.

As ever, Roosevelt remained on a twenty-four-hour basis, flexible, reluctant to commit himself to long-range plans, but nonetheless with some obvious, general goals in international matters. He envisaged a postwar international organization in which the victorious great powers would dominate proceedings, an organization that would operate to prevent future wars and gradually to remove past inequities, including colonialism. The success of that union of nations, he also believed, depended upon preserving and strengthening the ties among the United States, Great Britain, and the Soviet Union—an objective that called for unprecedented reciprocal understanding and tact, but one that he counted on reaching in large part through his friendship and persuasiveness with Churchill and Stalin. Concurrently he would have to sustain the tender *amour-propre* of Chiang Kai-shek and of Charles de Gaulle, both of whom—especially de Gaulle—he had carelessly offended. The foreign relations of the war, the strategy as well as the diplomacy, and the incipient shaping of the postwar settlement, constituted a sufficient burden, wholly apart from domestic problems, for a President at his physical and temperamental best.

Roosevelt in 1944 was not at his best. An old sixty, he was troubled by chronic illnesses that often left him weary. Increasingly it was difficult for him to walk with the heavy braces that he had had to use for twenty years. In late December 1943 he contracted influenza, with its usual symptoms and debilitation, aggravated in his case because of his life-long vulnerability to infections of the sinuses. The discomfort and fatigue that accompanied the disease kept him from attending to major questions awaiting his decisions, among them Anglo-American disagreements about the problem of Jewish refugees, and about currency reserves and lend-lease assistance.

So slow was Roosevelt's recovery from the flu that his personal physician, Surgeon General of the Navy Ross McIntire, persuaded him in March 1944 to undergo a general physical examination at the Bethesda Naval Hospital. That examination disclosed for the first time that the President suffered not only from acute bronchitis but also, to McIntire's surprise, from "hypertension, hypertensive heart disease, cardiac failure (left ventricular)," a not unusual condition for a sedentary man of Roosevelt's age and weight. Following the doctors' recommendations as closely as the demands of his office permitted, the President recuperated rapidly. By early April he looked well and his spirits were excellent. He continued to lose weight, as had been prescribed, ordinarily to sleep well, and rarely to show weariness during the strenuous months from March through November. The hypertension remained, and with it the probability of intermittent periods of discomfort, irritability, and ennui. But during 1944 the President responded, as always, with fresh zeal to the issues arising from the politics of an election year.

In 1941 Roosevelt had not intended to run for a fourth term. Before 1944 the members of his family told him they did not want him to run again, and the state of his health also argued against a race. At times he yearned for privacy and serenity. Had the war been over, he probably would have retired. But victory remained at a distance, as did the strategic decisions to achieve it and to fabricate the peace, and he wanted to control the great events himself. In addition, like many of his countrymen, he doubted the wisdom of a change in leadership in the midst of a war. He would run, he decided, if he was nominated. There was never a chance that his party would not select its champion again. The President liked to think of himself as a good soldier, responding to duty. That was a satisfying and plausible metaphor. Campaigning, moreover, even the prospect of campaigning, exhilarated Franklin Roosevelt. He had even begun in 1943 to prepare his platform and to strengthen his party for the coming election year.

In order to win in 1944, the Democratic party would have to get out the vote, especially the vote of those groups on whom the party could rely for a substantial majority. On that account, labor and youth, particularly servicemen, needed special attention. Yet it was also necessary to preserve the Roosevelt coalition, which included independent liberals, the South, and the large northern cities, with their Democratic machines and increasing black population. Labor and the liberals had to believe that Roosevelt was still the best available custodian of their aspirations. The soldiers had to be enfranchised, the South appeased, the city machines rewarded, and the various ethnic constituencies recognized. Beyond all that, the President, like any other candidate, could not win unless he stood close to the consensus of his countrymen, who were more concerned about preserving their personal well-being than about any precise foreign or domestic issue. Those circumstances led him to his familiar and comfortable role as a broker among many interest groups. Like a good Father Christmas, he needed to arrange to have something for everyone, not too much for anyone, and plenty of glitter to make every gift seem shinier than it really was.

The President's forward shuffle on the issue of social security revealed his election-year techniques. During 1943 he displayed little of the enthusiasm of his New Deal supporters for the postwar proposals for Great Britain put forth in Sir William Beveridge's long report on "Social Insurance and Allied Services." Beveridge recommended, in the phrase of that day, cradle-to-the-grave security—national unemployment, old age, accident, and burial insurance, and comprehensive medical care. The National Resources Planning Board offered less daring suggestions, which the President did not directly endorse, whereas Churchill, no social dreamer, partially endorsed Beveridge's plan. Further, the President offered no encouragement either to Senator Wagner or to Secretary Morgenthau in their separate and unsuccessful efforts to persuade Congress to improve the social-security system.

Where congressional approval was unnecessary, Roosevelt took a limited initiative in domestic postwar planning. In October 1943, in keeping with an earlier suggestion of the NRPB, he instructed Harold Smith to have the Bureau of the Budget correlate "realistic long-range programs" of public-works projects developed by other agencies for use in case of a postwar depression. More important, as he put it, he was "vitally interested in the problem of the returning soldier . . . his retraining . . . his reemployment." That was a genuine personal as well as a political interest, one most of official Washington fully shared. Indeed, the NRPB, shortly before it was dissolved, issued a report on "Demobilization and Readjustment," and at about the same time, in June 1943, the Armed Forces Committee on Postwar Educational Opportunities for Service Personnel sent the President its first preliminary report. Following recommendations in those documents, Roosevelt in the autumn of 1943 asked Congress to provide liberal unemployment, social-security, and educational benefits to veterans.

It was a shrewd request. The President knew that Congress, during the coming months and with an election pending, could not resist promising to men in uniform the postwar benefits that they and their families most wanted. For their part, the beneficiaries would identify their good fortune with Roosevelt, who had spoken for them, and just as much with his "Dr. New Deal" as with his "Dr. Win-the-War." The President could, moreover, ask for and get for the GI's the social benefits that liberals applauded, but that Congress was sure to withhold for the entire society. Broad social reform had no chance to succeed in 1944, and Roosevelt no time or intention vigorously to sponsor it, but he had the zest as well as the opportunity to push through for the "soldier boys" a program for which other Americans would have to wait. The program, distinctly progressive in its objectives, was so confined in its coverage as to suit even a conservative

Congress, whose members, after all, felt no less indebted than did their constituents to the nation's men in arms.

The GI Bill had yet to begin to move through congressional committees when Roosevelt outlined the broader national goals that he did not expect soon to achieve but wanted immediately to establish as a part of his platform for re-election. In his State of the Union message in January 1944 he called for "a *second Bill of Rights* under which a new basis of security and prosperity can be established for all." That economic bill of rights, a projection of the social programs of the New Deal, was to include rights to "useful and remunerative" employment; to earnings sufficient for adequate food, clothing, and recreation; to freedom for businessmen from monopolies at home or abroad; to decent housing for every family; to protection from the economic fears of old age, sickness, accident, and unemployment; to a good education. "*All* of these rights spell security," the President said. "And after this war is won we must be *prepared* to move *forward*, in the implementation of these rights, to new goals of happiness and well-being."

Those goals, as broad as the proposals of the Beveridge plan, stirred the spirits of liberal Democrats, whom the President had intended to reassure. After the war, so he seemed now to be saying, he would turn to building a new America in which his countrymen would enjoy the security and the rich life to which they aspired. Roosevelt spoke out of conviction but also, as was his way, for the record, for history perhaps, and immediately for the mustering of the forces on the left flank of his party.

It was one of the President's preferred roles to raise the expectations of Americans, one he enjoyed, one his speech accomplished. But he made no convincing effort to give substance to his oratory. It had to remain campaign talk, not policy, while the winning of the war absorbed him, and the mood of Congress was mirrored in the Revenue Act of 1944. During that year only the GI could look forward to the postwar advantages that other Americans wanted as much.

Still, the GI Bill served as an objective correlative for the good society Roosevelt outlined for the postwar period. The measure itself contained the kind of provisions Roosevelt had described, but in a form proposed early in 1944 by the American Legion, the strongest veterans' organization, noted for its conservativism, chauvinism, and influence in Congress. The Senate in March voted unanimously for the bill. At the insistence of John Rankin, chairman of the committee handling the measure, the House of Representatives adopted some minor modifications that reduced benefits for unemployment and re-stricted eligibility for education. Even so, the act, which Roosevelt signed in June, guaranteed veterans both the protection and the op-portunities that public-opinion polls had identified with the hopes of the American people. "Freedom from want," Roosevelt's phrase, as most Americans interpreted it, entailed employment at the least and prosperity at the optimum, both antitheses of the experience of the Great Depression, both boons of the wartime economy. The GI Bill did not promise veterans jobs, but it gave them priorities for many jobs and it assured them of occupational guidance and of substantial monthly allowances while they looked for satisfactory work. It also encouraged veterans to establish small businesses of their own, a goal inherent in the national folklore. The Veterans Administration would guarantee half the amount of loans required for the founding or acquisition of small enterprises, or for the purchase of farms, or for buying a home, the objective to which so many Americans gave first priority in planning their postwar lives. The war had demonstrated, alike in the armed services and in private industry, the advantages of training and education, which Americans regularly listed as an oppor-tunity they cherished for themselves and their children. The GI Bill provided full tuition and supplementary allowances to support vet-erans enrolled in educational programs at any level from the elemen-tary through the secondary, collegiate, and professional. As expanded by later legislation, the measure provided for comprehensive medical care for the disabled and for the construction of more veterans' hospi-tals. It also established the administrative apparatus for advising veterans about their benefits and facilitating their use. All in all, the bill did about as much as legislation could to help veterans to attain and to enjoy a secure, bourgeois status in employment, education, and residence—exactly the hope dominant both on the affluent home front and in the nostalgic trenches. Willie and Joe, their parents and the girls they left behind them, were bound to be pleased.

But Willie and Joe were not certain they would be able to vote. The polls showed a strong preference for Roosevelt among the 9 million servicemen, and the Republicans had no desire to ease access to the ballot for the 5 million men overseas or for others within the United States but stationed away from their legal residences. The Southern Democrats resisted any plan for federal intervention in the voting process. They feared it would set a precedent for future use and might lead to the enfranchisement of blacks. Further, as one of their admirers suggested, however loyal to the party Southerners were, many of them were doubtful about the President: they "cannot help but see that if Mr. Roosevelt is elected the radical groups will . . . control the party. . . . Roosevelt's defeat would break the radical popular front and destroy the grip of the job-holders on the party machinery. It would also leave the old-time southern Democrats in possession of the state party machinery. Obviously . . . thousands of Southerners will vote one way and hope another way."

That evaluation overestimated Roosevelt's radicalism, but not Southern antagonism to his proposal for absentee voting. Instead of having to request an absentee ballot from his state, a soldier, according to the terms of the administration's bill, would automatically receive a federal ballot, which he could mark and return at a convenient time before the election. A bipartisan commission would oversee the operation. As Samuel Rosenman said, only such a bill, one that "asserts Federal supremacy over State law," could assure the GI's the vote. That assertion and its consequences were just what Roosevelt's opponents feared. John Rankin characteristically blamed the provisions of the bill on radicals and Jews. Robert Taft flushed as he warned that Roosevelt would march the soldiers to the polls just as he had earlier mustered, so Taft believed, the votes of workers on federal relief. Three redoubtable Dixie Senators, Eastland of Mississippi, McClellan of Arkansas, and McKellar of Tennessee, put forward a "state control" bill which merely recommended that the states pass laws enabling soldiers to cast absentee ballots. During the ensuing debate, Byrd of Virginia talked of a Southern secession from the Democratic party, a prospect that old Joe Guffey, of Pennsylvania, a bellicose and indiscreet reformer of impeccable Democratic credentials, quickly associated with the treason of 1860–61.

The Senate, unmoved, passed the Eastland-McClellan-McKellar bill, to Roosevelt's dismay and that of all Americans who believed those fighting the war were entitled also to vote. The House modified the measure to permit use of a federal ballot if a state so requested, or if a serviceman requested a state ballot, was denied one, and then specifically asked for the federal substitute. That compromise failed to satisfy either the President or informed soldiers, but Roosevelt had no better option than to sign it. As he had often done before, Bill Mauldin put into Willie's mouth a common GI view of the Congress and its voting bill. "That's okay, Joe," Willie said, "at least we can make bets."

Once again it appeared that Congress had defeated the President. As it worked out, however, though only twenty states used the federal ballot, enough of the others mailed out ballots of their own so that more than 4 million soldiers and sailors voted in 1944. That figure was comparable to the civilian turnout and sufficient in size and state-by-state distribution to meet the reasonable expectations of Roosevelt Democrats.

Those expectations also rested upon a better turnout of union labor than had occurred in 1942. In 1944 the unions had more members than ever before, but there had been little change in the level of irritation among working men and women about rationing, taxation, and the ethnic issues that had bred apathy two years earlier. Philip Murray and Sidney Hillman, of the CIO, encouraged by the President and resentful of the Smith-Connally Act, in 1943 organized a Political Action Committee to work for Democrats friendly to labor, for northern Democrats in general, and for Roosevelt in particular. They also helped to found the National Citizens Political Action Committee, which recruited civil-rights advocates and the liberal intellectuals and professional men and women who had traditionally supported the President and the New Deal. The CIO-PAC, committed to the principles of the economic bill of rights, as Roosevelt had realized in framing that bill, exerted a liberal force within the Democratic party that helped to balance the organized strength of the South and of the northern urban political machines. More important in 1944, the PAC raised funds, as unions could not, alerted its constituency to the importance of voting, and operated systematically and successfully, especially in urban areas, to get out the vote.

Yet Roosevelt had no intention of permitting himself to be identified exclusively with the liberal-labor Democrats. He could not afford to, for he needed the support of the regulars in the South and the cities and he needed the votes of much of the middle class. The Republicans exploited middle-class suspicions by attempting to associate the PAC, and through it the President, with Communism. But that tactic converted few besides those already so persuaded, partly because the PAC was not really radical, partly because Roosevelt continued, as he had for several years, to make convincing overtures to the moderates among Democratic and independent voters.

Late in 1943, the President had satisfied the business community by appointing Bernard Baruch, a matchless symbol of prudential public policy, and his associate of many years, John Hancock, to report to Congress about reconversion. To the gratification of most of Congress and all conservatives, their recommendations of May 1944 about the timing and conditions for termination of war contracts emphasized the interests of private industry. But to the distress of New Dealers in the Senate, the Baruch-Hancock report ignored what they called the "human side of reconversion," including the question of transitional unemployment. To the further pleasure of industrial interests and their friends in Congress, Roosevelt was also during the spring of 1944 speeding the resignations of both Donald Nelson and the Treasury official most responsible for the rejected version of revenue legislation.

Throughout the year the President found a convenient coincidence between the exigencies of war and the desiderata of politics as those conditions touched public policy. That coincidence marked the ongoing debate about manpower, during which Roosevelt changed sides. The expansion of the armed services during 1943 and 1944 reduced the labor pool, but the demand for agricultural and industrial production continued to grow. Sporadic strikes, some of them in vital industries, lowered output and irritated both soldiers and civilians whose patriotic ardor often exceeded their understanding of the issues.

For Henry Stimson and his associates in the War Department, the answer was simple: work or fight. The unions vigorously resisted national-service legislation, which, as Stimson recommended, would have subjected all men and women to federal assignment to jobs or uniforms. That constituted "involuntary servitude" in the view alike of the CIO and the AFL.

National service would provide the government with the threat of the draft as a weapon against strikes, and give the government the authority to move a worker from one job to another. Since the advocates of national service would not agree to enroll workers in unions at plants to which they were assigned, labor leaders predicted employers would exploit the operation to break unions. Labor had every right to argue further that industry, profiting from cost-plus contracts, cheap government loans, and the suspension of antitrust actions, had suffered from no kind of conscription at all. On some of those accounts, business leaders, too, opposed national service as a threat to their autonomy in hiring, a lever for increased government regulation of industry, and an excuse for limitations on profits.

Assisted by its friends on the Hill, organized labor was the most vocal and powerful enemy of the plan. "Labor," Stimson lamented while the national-service bill was caught in committees, "has indoctrinated everybody with the idea that national service is . . . slavery." It could be. National service did work democratically in Great Britain, but as it was practiced in Germany it was akin to slavery, and it might have become so in the United States if the War Department had controlled it with the high-handed singleness of purpose that had crippled the War Production Board. Stimson himself had rarely displayed much patience with or sympathy for labor. Yet at the time he complained, it was he, not labor, who had convinced Franklin Roosevelt.

Through 1943 the President, tacitly assisting the unions, had kept out of the debate and let the national-service bill languish. Then in his State of the Union message of January 1944, perhaps as a counterweight to the economic bill of rights, he came out unexpectedly for national-service legislation that would "extend the principles of democracy and justice more evenly throughout our population," provide a "unifying moral force," "prevent strikes," and "make available for war production or for any other essential services every able-bodied adult."

For political purposes Roosevelt's timing was typically superb, though that timing was not necessarily contrived. He had recently visited American troops in the Mediterranean zone, where the fighting was tough. He had also recently worried about a national railway strike, which was barely avoided. He was annoyed by the selfish preoccupations of civilians and by the unabating hostility to the White House in Congress. On the last account, he linked his endorsement of national service to the enactment of a satisfactory tax law, a reduction in profits on war contracts, and the authorization of further price ceilings—conditions Congress was determined to reject, as the President knew. So his advocacy, while open, was less than ardent.

More important for the careful politics of 1944, he included his support for national service, a form of hard discipline for immediate war-related reasons, in the very message that announced the bill of rights that suggested a later Valhalla, presumably easy to achieve once victory was won. During the rest of 1944 the President exerted no significant influence nor did he employ extensive rhetoric in behalf of either the close or the distant goal.

In a campaign year, the favors that liberals and labor received from Roosevelt's left hand his right hand effectively took away. The PAC in the late spring of 1944 had little to show for all its efforts except promises, whereas the captains of industry and finance had little about which to complain in the President's actual performance. It would not have been entirely inaccurate, though it would have been indiscreet, for the President to have said, as he had in 1936, "We planned it that way." If he had not deliberately been planning, he had at least been balancing the interests that Democratic strategy had to take into account in constructing both the platform and the ticket for 1944. In that process, not yet completed, no man knew just what Roosevelt's intentions were; no man, in fact, wholly knew him.

8 / Referendum on Roosevelt

1. Mise en Scène

The message that reached the President directly merely affirmed what discerning politicians in both parties sensed about conditions for victory in 1944. Samuel Rosenman conveyed that message formally to Roosevelt in a memo of November 24, 1943, to which he attached a report prepared by Hadley Cantril, the head of the Office of Public Opinion Research at Princeton. "People are almost twice as much interested in domestic affairs as international affairs," Cantril had written. Two-thirds of the people thought "we should not give aid to foreign countries after the war" if so doing "would lower our own standard of living," and almost half the people believed it would. The administration, Cantril recommended, "solely as a method of obtaining public support for its international plans . . . should carefully avoid giving the impression to the nation that foreign affairs will be carried on at the expense of domestic programs."

That advice also applied to the politics of the election year. As Cantril reported elsewhere, for the poorer and less educated Americans on whose votes Democratic victory depended, domestic issues were especially salient. In an informed and thoughtful book, *Mandate from the People*, Jerome S. Bruner, Associate Director of Cantril's office, in 1944 stated similar conclusions: "To the average American, the domestic and international are far from equivalent in either personal significance or interest. In spite of the years of war, the events and problems which beset the world beyond our boundaries are of secondary interest to the man in the street. To him, the payoff is what happens right here at home and what is likely to happen. . . . A good job is still the bench-mark for the best of possible worlds."

During 1943 and 1944, Democrats and Republicans alike developed strategies to reach that average American. In that time both parties moved to patch over their internal divisions, to achieve at least an operating unity during the election campaign, and, to that end, to nominate national tickets that would at once reinforce partisan loyalties and attract as many independent or wavering voters as possible. Those objectives, like the dominant interests of the man in the street, led to the submerging of potentially divisive international questions, and that evasion appeared to postpone important decisions about the appropriate role of the United States in the postwar world. Yet the range of disagreement about those decisions was revealed by the ongoing debates within the parties, and the outcome of those debates was registered in the platforms and tickets on which the parties separately agreed. Behind the apparent partisan evasions about international issues, there emerged during the election year a bipartisan stance that conformed to the preferences Cantril identified a year before the voters cast their ballots.

Wholly apart from their concern about party unity, Roosevelt and the Republican leadership during the preliminary phases of the campaign both maneuvered to avoid taking firm positions about foreign policy. The President realized that there was no more effective way for him to run than again to make the race against the memory of Herbert Hoover. Roosevelt's declaration of his economic bill of rights and his handling of veterans' legislation prepared the way for that campaign. Roosevelt also remained loath openly to address at home any of the many questions that would provoke dissension within the Grand Alliance and possibly interfere with the prosecution of the war. Though he permitted the State Department to work on postwar plans, he did not much trust that department; he never had. He still believed, as he had put it earlier, that winning the war was "the single and supreme objective of the United Nations." Others within and without the administration were speculating about the broad objectives and detailed structure of the forthcoming peace. As late as May 1944, Roosevelt was still concentrating on two principal war aims: "to defeat our common enemies as effectively as possible, and . . . to do everything . . . possible to prevent Germany, Japan, or any other aggressor nation from plunging us into another war in the next couple of generations."

Those were the primary objectives, too, of most Americans who, in Bruner's careful analysis, were "not after an ideology" but "after the prevention of war . . . the end of war." Because they closely associated both the prosecution of the war and the management of the Grand Alliance with the President, Roosevelt profited politically from what was in any case his preferred wartime role as commander in chief. The Big Three meeting at Teheran in November 1943, the Anglo-American cross-channel invasion in June 1944, the American landings in the Philippines in October 1944, all built public awareness of and confidence in the Commander in Chief. In his role as a candidate, Roosevelt could suffer only if fellow Democrats fought publicly about foreign policy, as it was or as it should be, or if his subordinates put forth schemes for the postwar period that were at once identifiable with him and offensive to any considerable number of voters. With the recollection of Woodrow Wilson so often in his mind, Roosevelt was equally eager not to alienate the Republicans in the Senate. His sense of the need for circumspection in managing the war and making the peace was identical with his sense of the need for circumspection in discussing foreign policy during the campaign. When discussion was necessary, he preferred to say little, that largely in platitudes, and those phrased to stress victory and security, objectives as politically safe as blueberry pie.

So it was at a press conference of May 26, 1944. The United States, the President then announced, had invited the other United Nations to a conference on international monetary and financial matters to meet that July. That meeting, like an earlier one about food and agriculture, proved "we had got a good deal further ahead in the discussion of things" than during World War I. He did not need to say that among those "things," money and finance related to American prosperity. A reporter asked about other postwar planning. Roosevelt hedged. He had talked with Churchill and Stalin, he said. He was not yet ready to provide specifics. But his purpose was clear: "I am trying to eliminate a third World War." In mid-June the President inched further. He had his press secretary give out a statement describing the State Department's outline for a postwar international organization. It was to maintain "peace and security," the statement read, but "we are not thinking of a superstate with its own police forces."

Such infrequent and calculated banalities about foreign policy served two ends. They helped to quiet, though not to settle, disagreements among Democrats. They also helped to obscure the direction that the President's own foreign policy had been taking. That direction, perhaps less obvious in 1944 than it became in retrospect, was less the product of any self-conscious or overarching scheme for world order than it was the function of the President's spontaneous, sometimes untidy, and ordinarily unexpressed leanings. In the abstract, he believed in his own neo-Wilsonian rhetoric. But as a world politician he operated to reconstruct a pattern of power characteristic of the decade between the end of World War I and the onset of the Great Depression—the status quo antecedent to the growth of Axis influence, but one to be tipped for the future to the advantage of the United States.

Of that purpose, inadvertently but conveniently in accord with the sentiments of most Americans, the British and the Russians had abundant evidence. The "special relationship" with Great Britain, always a troubled marriage, was continually strained but never broken by the wartime rivalry of the two partners over the extent of each other's postwar political and economic control. The President continually interceded to support those of his subordinates who were trying to hold down British gold and dollar balances by confining lend-lease aid, and trying, too, to open American avenues for postwar trade both in the British Dominions and in areas of the world Great Britain had long dominated commercially. Yet the President accepted much of Churchill's views about appropriate spheres of British political influence and about an Anglo-American monopoly of the still secret atomic bomb, at least for the while. Opposed to colonialism in Churchill's pronouncements, Roosevelt nevertheless expected England and the European colonial powers to rule in their prewar overseas domains during the years in which their subject peoples were gradually to be prepared for self-government. Unlike the State Department, the President also privately had little fault to find with the division of middle Europe into areas of influence on which Churchill and Stalin secretly agreed.

So, too, Roosevelt accepted the map of western Europe as of 1932 as proper for the future. He intended, to be sure, to weaken Germany, partly by dividing that country into several states, but he expected the other western European nations then to function safely and traditionally under the governments the Nazis had deposed—the governments-in-exile, mostly royal governments. His fear that political strife within a major European nation might disrupt the continental order accounted in part for his policies toward France and Italy. He had dealt with the Vichy regime in France, a puppet of the Germans, largely in order to try to neutralize it, but also because he doubted that the French people really supported Charles de Gaulle, as in 1940 they certainly did not. Until after France was liberated, Roosevelt remained skeptical about de Gaulle's popularity and ability to govern, and he continued to be uneasy about France's postwar role in Europe. In Italy he had embraced Badoglio largely to gain a military advantage that in the end eluded him, but also because Badoglio and King Victor Emmanuel represented authority and the prospect of order. Roosevelt saw western Europe in terms of states and heads of state more than as a home of sundry peoples, of whom many were poor and miserable. In his view the area was not to become a site for political upheaval, and most emphatically not for social revolution.

The Soviet Union had no great quarrel with that deeply conservative vision. Twice invaded from the west within three decades, the Russians intended to build a safe belt between themselves and Germany, to fragment German power, and to dominate the policies of the countries of middle and eastern Europe. But during the war years at least, Stalin viewed with equanimity the restoration of the status quo in western Europe and its colonial areas. In those years the great powers were not at odds about Asia, either. There Roosevelt, like the European allies against the Axis, looked forward to stripping Japan of areas she had gained by conquest or treaty since the turn of the century, a prospect that would enhance the safety of Soviet Asia as well as of colonial areas dominated by the British, the Dutch, and the French. As for China, where civil war and revolution were already under way, only Roosevelt of the senior statesmen of the United Nations expressed a guarded but genuine confidence in the future of Chiang Kai-shek. It was a sentimental fantasy the others tolerated. In contrast, they accepted completely the President's assumption that the United States would continue to exert an unchallenged political and economic influence in Latin America, where Washington's wartime policies sustained friendly but usually reactionary regimes.

Roosevelt did not plan it all that way. He was not a part of conspiracy either to stem social change or to expand American commerce, much less to perpetuate a global *ancien régime*. Rather, he did not seriously question the general desirability of the political state of the world as it had existed in his early manhood. He did not doubt that a gradual amelioration of social conditions by more or less representative governments would suffice to bring a decent life to most people. He believed that the prevention of war depended upon a mutuality of interests among the great powers, a continuing balancing of their needs and responsibilities, and their own continuing and self-interested co-operation in preserving order. Those were his essential assumptions, confirmed as he saw it by a lifetime of observation and experience, never subjected to his own systematic inquiry or exposed to systematic challenge from other men. Those assumptions had become so comfortable that he did not express them except obliquely, when some friend or critic proposed a course that contradicted them.

Then he responded not as a debater but, depending on the occasion, with charm or sarcasm or recourse to the power of his office. That power brought Churchill to yield to his American friend when their differences could not otherwise be resolved. That power, used disarmingly, kept the Democratic party within the lines of march that Roosevelt devised for 1944. It also during 1943 and 1944 kept those Democrats who disagreed with his governing assumptions about international affairs, with his resulting decisions about American policy, or with the priorities he allowed for domestic matters from achieving or exercising an authority that could effectively counter his own.

Determined though they were to unseat the President, the Republicans, as they tempered their factional differences, came more to replicate than to alter the contours of his program. Rivals for office, they were also rivals for the support he was soliciting, politicians as eager as he was to resonate to the culture of the electorate. Accordingly the developing consensus within the Grand Old Party departed only marginally from the President's own positions, domestic and international, though at the fringes some Republicans, like some Democrats, considered Roosevelt either too timid or too bold. As members of a party out of power, the Republicans particularly felt the need for unity in 1944, though in their eleven lean years they had lost the advantage of a visible and acknowledged leader with the prestige to impose that unity. The party's factions, uneasy with each other, did share, besides a thirst for office, an aversion to Roosevelt. But it was not enough to hate that man. They had to have a convincing platform from which to attack him and an attractive man of their own to outrace him.

For those purposes, on several counts the troglodytes who led the party in the Congress would not do. Their long record of opposition to New Deal programs, their success in pushing those programs back in 1943 and 1944, their addiction to the rhetoric of the *Wall Street Journal,* marked the incompatibility between them and the mood of the country. To be sure, there was no wave of reform sweeping through the precincts. But the solid burghers of the Middle West who controlled the GOP on the Hill were still battling the reforms of the 1930's that the bulk of the American people had accepted as beneficial parts of a permanent national social and political structure. The Republican leaders in the House and Senate seemed to battle, too, the dominant popular concern with postwar jobs and income. Whatever the personal integrity of Robert Taft or the parliamentary skill of Charles Halleck, indeed, whatever the merit of their goals, neither man, nor any of their like-minded fellows, could command enough votes to be elected President, or therefore enough support for the nomination.

Most of the Republicans in Congress had also to live down a prewar record of opposition to military preparedness and to aid to the victims of Axis aggression, now allies of the United States. If the electorate was not ready to brook any deprivations for the sake of postwar assistance to those allies, it was also, as the public-opinion polls showed, deeply persuaded of the need for some kind of American involvement in a postwar settlement that would prevent another war. Roosevelt and the Democrats could play convincingly to that temper; the Republican leaders in Congress could not. Taft, the ablest of them, aware that he was in trouble because of his record on foreign policy and uncertain about his party's chances, deferred his Presidential ambitions in 1944 and, instead, supported his fellow Ohioan Governor John Bricker, a vapid conservative. As it worked out, Taft had that year the hardest struggle of his life for re-election to the Senate. It was not a year for the Republican right.

Yet that faction did represent the mass of Republicans in the solid party belt between central Ohio and western Colorado. Consequently the forces of the party's right had the strength to influence, even if they could not control, decisions about the national platform and nominations. They exerted that influence with a practical respect for the mood of the electorate but an unvarying determination to confine to the minimum any concessions to that mood. They continued to identify the New Deal as radical, Roosevelt as an authoritarian, the CIO as socialistic, public spending as rash, intellectuals in public life as irresponsible, and all of those targets as somehow un-American. They also continued to defend American nationalism against any postwar internationalism that might involve some infringement on American sovereignty or some obligation to use American troops or money other than on the basis of a unilateral American decision. They were wary of Great Britain only less than they were afraid of the Soviet Union, and they warned constantly against Roosevelt's alleged proclivity for selling out to either or both of those powers.

No more than the Democrats, however, were the Republicans of a single mind. In both parties there were eloquent advocates of courses less cautious than Roosevelt adopted, far less than Taft or Halleck contemplated. So it was with Democrats like Senators Robert Wagner, of New York, and Claude Pepper, of Florida, who continued to champion immediate social reforms, and with Republican Governor Thomas E. Dewey, of New York, who presided over an increase in the social services rendered by the government of his state. So it was, too, with Republican Senators Joseph H. Ball, of Minnesota, and Harold H. Burton, of Ohio, and Democratic Senators Lister Hill, of Alabama, and Carl A. Hatch, of New Mexico—the four who had sponsored, unsuccessfully, a resolution calling on the United States to ask the United Nations to create a permanent international organization with a broad mandate for relief and rehabilitation, and with a police arm to "suppress by immediate use of . . . force any future attempt at military aggression by any nation."

To that resolution, as to the proposals of the National Resources Planning Board and the program of the National Association for the Advancement of Colored People, there rallied politicians in both parties who believed, however naïvely in the view of their colleagues, that after the war, at home and abroad, there could be at least a better and perhaps even a brave new world. Those idealists of a kind stirred the often latent aspirations of a sufficient part of the American people to create a countercurrent against the conservative drift of national politics.

From 1942 forward, the proponents of liberalism at home and internationalism abroad had two special champions, each a candidate for national office, each a threat to the establishment within his own party. By 1944 those candidates had raised large questions, the answers to which would determine much of the direction of American policy in later years. Each, as he knew, was a general in search of an army rather than an accepted tribune of the people. Each was eager to assume the latter role, a prospect alarming to the leaders of his party. Accordingly, senior Republicans were continually involved in opposing the campaign of Wendell Willkie for the Presidential nomination, while their Democratic counterparts resisted the efforts of Henry A. Wallace to infuse his party with his own reforming zeal without giving up his office as Vice President.

2. Willkie: Republican Heretic

Wendell Willkie during his Presidential campaign in 1940, his hair tousled, his language folksy, his voice hoarse, wanted the voters to think of him as a barefoot boy from Indiana now grown up, toughened, and ready, because the nation needed him, to take over its management. His successful career as a head of Commonwealth and Southern, the giant utilities holding company, made him at the same time a paragon of industrial virtues. The message was obvious: hard work and high character, in the best American tradition, had brought Willkie to the top. But the images he projected made him seem, even potentially, an unlikely champion of liberal internationalism. By 1943 he had achieved a metamorphosis.

"We used not to think much of Wendell Willkie," the editors of *The Commonweal* then wrote. "In 1940 he seemed a symbol of big business and Republican reaction—a somewhat appealing symbol. . . . Our sympathies were aroused, but we were against him. In 1942 we liked him even less. Listening one night to a . . . broadcast, we suddenly heard that eminent gravel-voice referring to . . . 'the day when American democracy will rule the world.' Quite frankly, we thought to ourselves, 'This man is a dope—and a dangerous dope at that. . . .' But Mr. Willkie . . . has now come back strong. . . . He has written a very remarkable little book, 'One World' . . . a moving appeal for self-government among the peoples of the Far and Middle East, for the end of white imperialism everywhere, and for the immediate creation of international machinery that will learn to keep the peace by helping to win the war."

That little book and its reception marked the third and final peak in Willkie's extraordinary career. The first two were notable in themselves. The grandson of German immigrants, the son of a leading family of Elwood, Indiana, Willkie suffered no disadvantages during his youth, which included graduation from the college and law school of Indiana University and service as an artillery lieutenant during World War I. After a decade in a lucrative law practice in Akron, Ohio, he moved in 1929 to New York City as the junior partner in the firm that was counsel to the Commonwealth and Southern Corporation, which then had a hundred sixty-five subsidiaries. In 1933 he became president of that company. In that role he emerged as the most effective and celebrated foe of the Tennessee Valley Authority. In particular his negotiation of a $78 million TVA payment for C. and S. properties made him something of a hero on Wall Street and with the conservative press, which was also impressed by his thoughtful testimony during his occasional appearance before congressional committees, by his easy humor and expansive presence.

Success in business opened a new forum for Willkie, who began in 1937 to publish magazine articles attacking the New Deal for concentrating too much power in the federal government and for excessive federal spending. Those tendencies, he argued, endangered American freedom, a conclusion wholly in accord with the view of the Republican party, which he joined in 1939. Previously a Democrat, he had voted for Roosevelt in 1932 but against him in 1936. In July 1939 *Time* devoted a cover story to Willkie as "the only businessman in the United States who is ever mentioned as a Presidential possibility for 1940." That possibility grew after Willkie's stunning performance on a popular radio quiz show, "Information, Please," and especially after Oren Root, Jr., a young New York Republican, undertook on his own initiative in April 1940 to organize a Willkie movement.

Managed at first by amateurs like Root, but always generously financed, the Willkie campaign caught the fancy of the Luce magazines, the Cowles family newspapers, the New York *Herald Tribune,* and those many Republicans, a significant minority of the party, who were gravely worried at once about the Nazi victories in Europe and the isolationist convictions of the leading party candidates, Robert Taft and Thomas Dewey. In contrast to them, Willkie called for all possible aid to the Allies short of war, and for an unstinting program of American military preparation. His was, too, a fresh face. He seemed to stand above politics; he had the personal vibrancy Taft and Dewey lacked; and he appealed to independent voters. By June 1940 almost 30 per cent of all Republicans, with that percentage growing daily, had decided that Willkie could provide the best leadership for their party. With the galleries shouting "We want Willkie" throughout six exciting ballots, the Republican National Convention, in a momentous upset, reached the same conclusion. Willkie had climbed a second peak.

It was downhill for the next five months. The Willkie organization, notoriously inefficient, alienated the Republican professionals by ignoring them. Willkie disappointed his eager followers by the loose repetitiveness of his speeches, which became increasingly conventional in their anti-New Deal rhetoric as Roosevelt's campaign gathered speed. Before November Willkie had so emphasized his commitment to avoiding war that he seemed to have moved close to the position of Taft and Dewey, though they continued to regard him, quite accurately, as an adventurous outsider. After the election he dismissed his closing speeches as campaign oratory, an honest evaluation that further offended the party's regulars. On Election Day he ran better than Hoover or Landon had, better probably than any other Republican could have, but particularly in the cities Roosevelt prevailed, in all by a comfortable margin both in the popular vote and in the electoral college.

More than his success in business, his defeat in politics shaped Willkie's mature character. He had come close enough to the Presidency to feel its great burdens. Although he remained ambitious, he came more and more to eschew the expediencies of politics. Increasingly he spoke and acted out of convictions, changing as he learned and grew, that were at once selfless, essentially patriotic, and yet unconventional and often politically poisonous. For one example, in 1941, in contrast to Hoover and Landon, Taft and Dewey, he supported the Lend-Lease bill. During that year and the next, he visited Great Britain, admired the courage of the English people, irritated Churchill with anticolonial pronouncements, and scolded Roosevelt for the government's sluggishness in advancing civil rights for blacks.

In July 1942, moved partly by suggestions of several American newspapermen in the Soviet Union, Willkie wrote the President to propose a trip "to the Middle East, into Russia and perhaps China." Roosevelt leaped at the idea for reasons that Willkie later reported: "I went for three chief purposes: (1) The first was to demonstrate to our Allies and a good many neutral countries that there is unity in the United States. . . . That was my idea. (2) The second purpose of my trip was to accomplish certain things for the President. . . . (3) The third job I set out to do was to find out as much as I could, both for myself and for the American people, about the war and how it can be won—won quickly . . . and won securely." Accompanied by Gardner Cowles, publisher of *Look*, and Joseph Barnes, formerly a foreign correspondent of the New York *Herald Tribune,* both of whom were then on the staff of the Office of War Information, Willkie left New York on August 26 aboard a converted army bomber, and during the next forty-nine days flew 31,000 miles, with several long stops. His account of his trip, published in 1943 under the title *One World,* incidentally boosted Willkie's campaign for the 1944 Republican nomination and immediately carried his observations and ardent recommendations to his potential constituency, with special impact on converts as sympathetic as were the frank editors of *The Commonweal.*

As in so much else that Willkie said or wrote, there was a gee-whiz tone to *One World,* a quality of open-mouthed wonder, of delighted surprise. As so often, too, Willkie punctuated that apparent innocence with criticisms carefully aimed at the targets of his dislikes. Some of those barbs were unfair, just as some of Willkie's naïveté was invincible rather than contrived. But those qualities were insignificant compared to the central meaning of his book, the essential understanding and foresightednes that it contained.

Willkie began on what was still a fresh note: "There are no distant points in the world any longer." Accordingly, in the future what concerned the "myriad millions" of the Far East and Middle East would also have to concern Americans. Throughout the Middle East, as he saw it, there was "some sort of yeast at work." Arabs and Jews alike were skeptical about the political ideas of the West, ideas that had failed to prevent war, racial and religious prejudice, and colonialism. They were also eager to employ Western technology to overcome the "backwardness and squalor" of their own societies. Indeed, they had begun to do so where they enjoyed a measure of self-government, as in the cases of the Zionists in Palestine and the Arabs in Bagdad. The people of the Middle East, Willkie wrote, needed "more education . . . more public-health . . . more industry . . . more of the social dignity and self-confidence that comes with freedom and self-rule." Especially among the young, the recognition of those needs was stirring up "intense, almost fanatical, aspirations," which were both revolutionary and fervidly nationalistic. Yet when, while in Egypt, he had tried to "draw out . . . experienced and able administrators of the British Empire, on what they saw in the future" all he got "was Rudyard Kipling, untainted even with the liberalism of Cecil Rhodes." British officials both in London and in the stations of empire "had no idea that the world was changing." Unhappily, Willkie concluded, "brilliant victories," like Montgomery's at El Alamein (Willkie liked Monty and his "passionate addiction to work"), would not win the war; "only new men and new ideas . . . can win the victory without which any peace will be only another armistice."

Those deliberate slaps provoked Winston Churchill to remark that "I did not become the King's First Minister to preside over the liquidation of the British Empire." But it was Churchill, not Willkie, whose sense of the future was wrong. More obliquely, Willkie also attacked Roosevelt. After talking with French soldiers and sailors in North Africa, Willkie wrote, he "never accepted without discount stories of the probable losses we would have sustained at the hands of the French if we had gone in directly as Americans without dealing with Darlan." Unlike Churchill, Roosevelt made no public reply, but he persisted, as did the State Department, in relying on French fascists to impose order rather than freedom in North Africa. As Willkie implied, that kind of policy threatened to liquidate, as ultimately it did, the "reservoir of good will" that the United States then still enjoyed among colonial peoples.

For those peoples, with their resolve for independence and development, Willkie found a model, though he did not call it that, in the Soviet Union. Since 1917 it had become what they now wanted to become. Russia was ruled, Willkie discovered to his admitted surprise, almost entirely by men and women whose parents had had neither property nor education. Furthermore, "there is hardly a resident of Russia today whose lot is not as good or better than his parents' lot was prior to the revolution." Naturally, then, the Russians tended to overlook "the ruthless means" by which the improvements had been accomplished. There was much more to be done: agriculture and industry were still less efficient than in the United States; even the intellectuals, trained since childhood "in a system of absolutes," thought too much "in blacks and whites." But much impressed Willkie: the importance of hard work which brought those who practiced it material rewards and social mobility; the wide employment of women; the apparent absence of racial exclusiveness; perhaps above all the courage of the Russians in the face of the devastation that the war had brought them. "The Russian people," Willkie wrote, "—not just their leaders—the Russian people . . . had chosen victory or death. They talked only of victory."

One World put forward three categorical conclusions about the Soviet Union: "First, Russia is an effective society. It works. It has survival value. . . . Second, Russia is our ally in this war. The Russians, more sorely tested by Hitler's might even than the British, have met the test magnificently. . . . Third, we must work with Russia after the war. . . . There can be no continued peace unless we learn to do so." Those points provided reassurance for those Americans who were in 1943 still dubious about Russia. Though the doubt lingered, Willkie's observations tempered it for a time.

So did his report about Joseph Stalin, whom he called "a hard man, perhaps even a cruel man, but a very able one." Willkie found him also unaffected, robust in humor, tenacious in thought, direct in conversation. That directness exploded in private conversations during which Stalin, as well as some of his associates, were openly abusive about Great Britain and the United States, particularly about the diversion to the United Kingdom of certain airplanes designated originally for the Soviet Union and about the delays in opening what Willkie called "a real second front in Europe." In *One World* Willkie did not mention the abusiveness, but, at a press conference in Moscow, he had called for a second front "at the earliest possible moment our military leaders will approve. . . . Next summer might be too late."

Doubtless Willkie's concern about public reactions in the United States accounted for the omission in *One World*. His Moscow statement reflected in part his sympathy for the extent of Russian sacrifices, in part his misunderstanding about the limits of Anglo-American strength, which Roosevelt had failed to make clear to him, and in part his sure sense of the state of mind of Soviet officials. The Moscow statement, if it embarrassed Anglo-American leaders at all, did so only briefly. The continued postponement of the second front, which to be sure was not a military possibility in 1942 and only a risky one in 1943, roiled Anglo-American relations with the Soviet Union for two years.

Then and later Willkie's detractors dismissed his report about the Soviet Union as naïve. Certainly he wanted to believe in the Soviet Union and its intentions, and he was taken in by the special conditions arranged for his inspections in Siberia. But he was not on those accounts wholly naïve. On the contrary, he saw how much the Soviet Union was bearing the brunt of the war in Europe, he realized that the Russian people supported their government and its policies, he knew Stalin was at once a powerful and a merciless man but by no means insane, he stated the primacy of the issue of the second front, and he warned that continuing peace would depend upon continuing Soviet-American co-operation. "Russia," Willkie wrote, "is a dynamic country, a vital new society, a force that cannot be bypassed in any future world." That circumstance did not alarm him. "Russia is neither going to eat us or seduce us," he judged. "That is . . . unless our democratic institutions and our free economy become so frail . . . as to make us soft and vulnerable. The best answer to Communism is a living, vibrant, fearless democracy—economic, social, and political."

That confidence remained, though Willkie's insights faltered, when he went on to China, where he fell into every trap that caught most of his countrymen who visited wartime Chungking. Like them and like Roosevelt, he was entranced by Chiang Kai-shek, the Soong family, and their friends, whose popularity with their people and effectuality in their responsibilities he vastly overrated. In addition, he paid too much heed to General Clair Chennault, an admirer of Chiang and a romantic about the potentialities of air power in the war in China, and too little attention to General Joseph Stilwell, who did not succeed in correcting what he knew to be Willkie's false impressions of the other. One of those misreadings evaluated Chou En-lai as an impatient reformer rather than the resolute revolutionary that he really was. If all Chinese Communists were like Chou, Willkie concluded, their movement was "more a national and agrarian awakening then an international or proletarian conspiracy." In that judgment he was less wrong than his critics claimed, for the Chinese Communists were not conspirators but open advocates of systematic changes that they intended to make over the Chinese economy entirely, and to strengthen the Chinese nation substantially. They might have become less hostile than they did if the United States had not persisted in its futile support of the Kuomintang. Still, Willkie did not see China clearly. Rather, like Roosevelt, he viewed it through the prisms of sentimental expectations. Yet, like Roosevelt again, Willkie, whatever his errors, did recognize that the millions of Chinese were determined, and destined, to play a major and an independent role in postwar Asia. And he saw that a new and genuine freedom for the peoples of Asia—as for the peoples of the Middle East—had to be just as important a war aim as the liberation of the conquered peoples of Europe.

As Willkie concluded *One World,* he criticized Stalin's silence about the aspirations of the eastern Europeans just as he criticized Churchill's imperialism, indeed the whole European colonial system, and Roosevelt's expediencies as well as the domestic American imperialism of white over black. "If hopeful billions of human beings are not to be disappointed," he wrote, with a prescience that outweighed any naïveté, "the United Nations must become a common council, not only for the winning of the war but for the future welfare of mankind. . . . We must formulate now the principles which will govern our actions as we move . . . to the freeing of the conquered countries. And we must set up a joint machinery to deal with the multiple problems that will accompany every forward step of our victorious armies. Otherwise we will find ourselves moving from one expediency to another, sowing the seeds of future discontents—racial, religious, political—not alone among the peoples we seek to free, but even among the United Nations themselves. It is such discontents that have wrecked the hopes of men of good will throughout the ages."

It was such discontents that Roosevelt's unexpressed assumptions about wartime and postwar policies were inviting; such discontents that the State Department was blindly cultivating in its unchanging daily operations; such discontents that a bipartisan majority of the Congress preferred to those unvoted alternative programs that might have troubled their constituents; such discontents that most of the American people could barely imagine—or if they did imagine them, could blithely ignore. As Willkie's fellow Republicans knew, it was far easier for the American people to read and applaud his book than to take his advice.

Read and applaud it they did, as did professional critics. The review in *Foreign Affairs,* the journal of the Foreign Policy Association, the voice of the establishment, called the "little volume . . . one of the hardest blows ever struck against the intellectual and moral isolationism of the American people." Willkie, according to Clifton Fadiman, then chief book reviewer for *The New Yorker,* had "simply, by an inexplicable miracle, burst all the bonds which might have confined him within the narrow limits of big-business thinking and become . . . one of those 'new men' . . . for whom he pleads in his book." *The Atlantic,* which spoke, like *The New Yorker,* to the concerned and urbane upper middle class, praised Willkie's "courage and candor." *The Nation* and *The New Republic,* addressed to the explicitly liberal, also praised Willkie, though more critically. In the former journal, Reinhold Niebuhr called Willkie a "proponent of a wise internationalism" but regretted his failure expressly to disavow his onetime attacks against a "managed" economy at home. Malcolm Cowley in *The New Republic* also remembered Willkie's intemperate oratory of 1940, but he commended Willkie's "naturally sound instincts," condemned his critics within the State Department, and concluded that he was "exactly 128 years ahead" of that department.

Cowley and Niebuhr, their readers, and subscribers to *The New Yorker* and *The Atlantic* constituted only a few precincts that the Grand Old Party had not carried regularly for several decades and did not plan seriously to canvass in 1944. The acclaim for *One World* helped Willkie's campaign among the disfranchised as the Republican leadership saw them. The content of *One World* annoyed, even angered, those leaders—the stalwarts of the party in Congress and their comrades in so many of the statehouses of the country. Willkie, so those men had long believed, sounded too much like Roosevelt even when he criticized the President. Willkie's proposals went beyond those even of the party's intelligentsia on foreign policy. Willkie's hopes flew out of sight of the narrow vision of Senators like Vandenberg and Taft.

Worse, Willkie knew it and seemed not to care. In the months after the publication of *One World* he incorporated its themes in his speeches, stressed more than he had before the urgency of civil rights for blacks—an objective the Republicans in Congress opposed as much as did the Southern Democrats—and went so far as to attack big business, a Republican shibboleth, as well as big government, Roosevelt, and the New Deal. At times he sounded like an anti-truster. Once, in St. Louis in October 1943, he let himself go entirely. He was introduced on that occasion by an industrialist who, irritated by Willkie's recent behavior, described him as "America's leading ingrate." "I don't know whether you are going to support me or not," Willkie told his hostile audience, "and I don't give a damn. You're a bunch of political liabilities, anyway." His support, he believed, would come from the people. He expected to display it in the Republican Presidential primaries in 1944.

For their part, the strong factions of Republican regulars were moving toward their own accommodation on foreign policy. One identifiable though not wholly cohesive group adhered to the position of the dominant Republican bloc in Congress and looked for leadership to Senators Taft and Vandenberg. The other, rather more flexible, attracted most Republican regulars in the states east of Ohio and west of the Sierras. Its obvious, though for many months undeclared, candidate was Governor Dewey of New York. As Taft, Vandenberg, and Dewey knew, by 1944 the question could no longer be whether the United States would have a postwar involvement in international affairs, but what form that involvement would take. To that question Taft and Vandenberg formulated reconcilable but not identical answers.

Exposed though he was to an eastern education, to Yale College and the Harvard Law School, where he excelled academically, Robert Taft retained throughout his life the provincial attitude toward Europe of the good burghers of Ohio River cities like Cincinnati, where he resided. Like Herbert Hoover, under whom he had served in the administration of food relief in Europe after World War I, Taft felt a kind of disdain for Europeans. The good Europeans, Hoover thought, had emigrated to the United States; those who remained were a lesser breed. Taft agreed and, like Hoover again, had little interest in the culture of Europe and England, and he attributed to eminent European rivalries the onset of both world wars and the economic dislocations of the interwar period, disasters that infected the United States. He was at once an intelligent and an unsophisticated man, an informed but a dull speaker, an ambitious politician but an inept political organizer, and an astute lawyer but, like his father, a narrow legalist. Those qualities, along with his personal shyness, made him seem to his critics a rigid reactionary, though he had always a patrician sense of responsibility for the helpless and a zealous capacity for study and learning. Even among his most bitter opponents, there were those who recognized his large talents and listened to him carefully. Even among his admirers, few expected his thoughtful remarks, written or spoken, to be stirring.

Characteristic provincialism, legalism, and ponderosity tinctured Taft's approach to international issues. The "war crowd," he warned, was eager to commit the United States prematurely to an international organization. Agreeable to undertaking "some obligations," Taft opposed policing Europe or becoming involved "in every little boundary dispute that there may be among the bitterly prejudiced and badly mixed races of Central Europe." He wanted to postpone territorial and economic decisions until after victory, only then to join a world league, and one so built as to rely primarily on regional councils. Meanwhile he counted on the big three powers to prevent disorder. He hoped the United States would permanently keep out of an alliance with Great Britain, which would be costly and imperialistic, and a world superstate, which would fail in its oversized mission. It was a measure of Roosevelt's conservatism that, for reasons of his own, he had similar preferences.

But whereas Roosevelt looked to a continuing great-power domination through the agency of a world league, Taft turned to the law as the only appropriate long-term basis for international relations. He called on the United Nations to develop and promulgate the principles of international law, and to establish a world court to interpret and apply them. Until world opinion accepted principles of that kind, which collective sanctions could then enforce, no league, in his judgment, could work. In that conclusion Taft stood, as was so often the case, almost precisely with the two former Presidents who had most influenced his thinking—his father, the late William Howard Taft, and Herbert Hoover, who had (with Hugh Gibson) published in 1942 his answers to *The Problems of a Lasting Peace*. The hope that world opinion would support the development, adjudication, and enforcement of international law was no less naïve, though it was less inspirational, than Willkie's belief in one world, which Taft so derided. Taft's alternative, however, allowed him to appear something of an internationalist but also hold out the prospect that he and Hoover had continually as their objective—a minimal postwar engagement of American force and money in world affairs. As much as Taft's belief in the law as a higher good, that purpose, constant for twenty-five years, molded his proposal.

It presented no obstacle to political accommodation between him and his colleague Arthur Vandenberg, the senior Republican on the Senate Foreign Relations Committee. Vandenberg, at one time a newspaper editor, was a large, pleasant, pompous man with somewhat Edwardian manners, a trite mind, and a garrulous tongue. As much opposed to the New Deal and to Roosevelt's prewar foreign policy as was Taft, he later won a reputation for enlightened bipartisanship by supporting various postwar ventures which he was willing to bless only after extracting concessions, substantive and rhetorical, to his own sharply nationalistic and anti-Russian biases. He also made concessions, because his outsized vanity fed on praise alike from Democrats who flattered him, Republicans who heeded him, and reporters who overrated him. Though he never made a vigorous run for the Presidential nomination, the press considered him a possible candidate in 1936, 1940, and 1944, and in any of those years he probably would have been happy, though he was not eager, to accept. Satisfied with his role and prestige in the Senate and in the party, Vandenberg, as he contemplated the campaign of 1944, cared primarily about placing his personal stamp on the Republican foreign-policy plank.

It could not be his without putting American sovereignty first. "We shall vigorously cooperate with the forces of international progress," Vandenberg explained to the press, "but we shall remain irrevocably the independent and sovereign United States of America." In the Senate he sponsored a resolution that called for "the participation of the United States in postwar cooperation between sovereign nations" but "with faithful recognition of American responsibilities and American interests." That wording, he confided to his diary, would repel all "World Staters" and guarantee "the continuance of the American Flag over the Capitol." With Vandenberg's kind of internationalism, Robert Taft was comfortable.

Both Senators were uncertain, as were Republicans in general, about the policy of Governor Thomas E. Dewey, of New York, in 1943 the still unannounced but leading candidate for the nomination. In 1940, when he came close to victory at the convention, Dewey had been in or near the isolationist wing of the party. But he was a man who moved with the political winds, kept his own council, and, as Taft put it in a private letter, was "very arrogant and bossy." Dewey believed avidly in the Gallup poll. "It has never been wrong," he once said, "and I very much doubt whether it will be, so long as George Gallup runs it." Like the other major national polls, Gallup's during 1943 conveyed two general messages that registered with Dewey: the primacy of domestic issues in voter interest, especially the issue of personal economic security; and yet the large saliency for voters of national security as an issue, of the avoidance of future wars through some kind of international organization. The latter pushed Dewey toward a cautious internationalism, by no means the "one world" of Willkie, whom he was determined to defeat.

For advice about foreign affairs Dewey had turned to John Foster Dulles, a New York lawyer, a specialist in international practice with a lucrative corporate clientele, a spokesman for the Protestant churches, and a grandson of one former Secretary of State and the nephew of another. Dulles had been on the staff of the American delegation to Paris in 1919, a role in which he made a characteristic contribution by drafting the clause of the Versailles Treaty that attributed guilt for the war to Germany. Dulles's absolute conviction of his own righteousness and his related habit of defining public issues in moral terms, less pronounced in 1943 than it was to become, even then tended to obfuscate his thinking, otherwise the product of a good and serious mind. The Atlantic Charter, he had pointed out, broadcast "lofty ends, loftily expressed," that accorded with proper spiritual and civic values, but through them all ran the "unifying conception . . . that the postwar should reproduce and stabilize the political organization of the prewar world." That observation, both shrewd and sanctimonious, preceded Dulles's assertion that "the old politico-economic order has failed." Consequently he advocated "some form of federation for continental Europe" and some kind of international body. "Not greatly concerned" with the immediate powers and initial jurisdiction of that body, Dulles believed that once it existed it would "increasingly become the repository of duties and responsibilities of international import."

Those responsibilities, he implied, involved especially a police function to prevent wars, which the United States, he argued, would be unwilling long to assume, even in alliance with Great Britain. (He did not so much as contemplate an alliance with the Soviet Union.) Equally essential, the international body was to study and determine national economic needs, particularly "the domestic need for price and wage stability" and the need of all people for "free and equal access to . . . colonial resources and trade" and for the use of the world's oceans and skies. The intentional vagueness of Dulles's proposal for a league of nations that he feared the United States might again reject contrasted with the calculated precision of his demand for arrangements affecting international trade and commerce, from which American corporations would benefit. His "essentially ethical" approach, his "moral initiative," changed the old politico-economic order chiefly by enhancing American mercantile advantages within it. Republicans like Taft and Vandenberg could accept that prospect although they cringed at Dulles's Wilsonian language.

To resolve their differences, the party's principal foreign-policy spokesmen, with Willkie omitted by their design, met in September 1943 at the Grand Hotel on Mackinac Island in what *Newsweek* called "the most important Republican deliberative assembly since the national convention of 1940." Vandenberg intended to put through a resolution that would not mention an international league as such but would emphasize the protection of American sovereignty and interests. Dewey, however, revealed both his passion for publicity and his disregard for the different schemes of Vandenberg, Taft, and even Dulles. To the press on the eve of the conference, the New York Governor unexpectedly came out for a permanent Anglo-American alliance, though he also suggested he was willing to include the Soviet Union and China in a larger security agreement. After the conference began, Governor Raymond Baldwin, of Connecticut, in 1940 a Willkie supporter, and Governor Earl Warren, of California, led the opposition to Vandenberg's proposal, and indirectly also to Dewey's. In the end the two governors, along with Vandenberg and others, drafted the resolution that the conference then unanimously adopted. The Republican party, it said, approved "responsible participation by the United States in a postwar cooperative organization among sovereign nations to prevent military aggression and to attain permanent peace with organized justice."

The governors got the word "organization" into the statement, Taft and Vandenberg the word "sovereign" and the phrase "peace with . . . justice" as an alternative to "collective security." Dewey got the publicity he wanted and a declaration so cast that, if nominated, he could run on it or around it. The phrasing, cautious and ambiguous though it was, also put the party recognizably on the side of internationalism. As Willkie commented, the declaration was "a very distinct step in the right direction."

Yet the conference marked a distinct Republican step away from Willkie. Just before the Mackinac meeting, he had announced that he would run again on a "liberal, progressive" platform. The party, he held, had to repudiate "the ultra-nationalists, the economically selfish . . . the religious and racial bigots." His own "five indispensable planks," as he explained them later in September 1943, indicated that he, too, recognized the primacy of domestic issues, and, further, that on those issues he was at war with the Republican leadership in Congress. One of his planks did call for an international organization and for planning at once for the future of subject peoples and new nations. That was expected. Another, a necessary dig at Roosevelt, promised "efficient . . . administration." But Willkie also demanded "protection of minorities," and "absolute guarantees . . . against unemployment—and against want because of old age, injury and incapacity." That language resembled the President's new bill of rights, which the majority of Republicans bitterly opposed.

Perhaps even worse, Willkie challenged the party's business creed. "Some of the talk we hear," he said, "about 'free enterprise' or 'private enterprise' is just propaganda on the part of powerful groups who have not practiced real enterprise in a generation. . . . A corporation may be privately owned and still be the worst enemy of free enterprise." Workers, he went on, were also enterprisers, or should be, and therefore deserved "a fair share of the profits derived from their efforts." "A free enterprise system," Willkie concluded, "does not belong to a few at the top. That is a vested interest system. A free enterprise system belongs to everybody in it." By Republican standards, that statement rang of socialism, the ultimate heresy, an apogee of the wild preachings of a consistently heretical man. Yet he was, at least until the 1944 convention nominated someone else, the titular head of the party.

Accordingly the stop-Willkie movement had gathered strength by the time Willkie declared. Every Republican Presidential candidate or near-candidate since 1928—Herbert Hoover, Alfred M. Landon, Robert Taft, and Thomas E. Dewey—opposed Willkie's renomination and directly or indirectly assisted efforts to prevent it. Furthermore, Willkie had sailed past the consensus of the party's rank and file. From November 1943 on, public-opinion polls showed him slipping in popularity among Republican voters, who had preferred Dewey since the previous March. The January 1944 Gallup poll credited Willkie with the support of only 23 per cent of Republicans, with 42 per cent for Dewey, who still denied he was a candidate. Willkie had either to prove the polls wrong or to give up the race.

He selected a perilous battlefield on which to fight, the Wisconsin Presidential primary. It was a quixotic choice. Wisconsin Republicans had been consistently isolationist before the war. Voters could cross party lines in the primary, but Wisconsin Democrats were either satisfied with Roosevelt or from ethnic backgrounds, Polish and German particularly, that tended to make them as averse to Willkie's foreign-policy statements as to Roosevelt's. The La Follettes, who still controlled the Republican organization, opposed Willkie. While he campaigned hard in Wisconsin during March, his rivals remained far away and let their friends handle their affairs. Willkie found himself without visible targets. Dewey officially disapproved the filing of his name in the primary, though his sponsors used it nevertheless. Governor Harold Stassen, of Minnesota, in 1944 stricken with the Presidential bug that plagued him the rest of his life, was in the Navy in the Pacific. He had the reputation of a liberal and an internationalist, but contrary to Willkie's expectation of his support, Stassen said that while he did not seek nomination, he would accept it. General Douglas MacArthur, the egoistic hero of the Republican right wing, also indicated that he was available. As he crisscrossed the state, Willkie swung away at all of them, and at Roosevelt, too, while he repeated and elaborated the points in his personal platform and best-selling book. But not many voters were listening.

The Wisconsin Republicans in the primary of April 4, 1944, selected seventeen delegates to the national convention pledged to Dewey, four to Stassen, three to MacArthur, and none to Willkie. In the popular vote, Dewey's delegates won 40 per cent, MacArthur's 24, Stassen's 20, and Willkie's 16. The results were unequivocal. As even Willkie's friends realized, Gardner Cowles for one, the vote reflected national as well as local attitudes. He was "amazed," Cowles wrote Willkie, at the conservatism of Republicans all over the country. They had "shifted to the right substantially in the last twelve months." They favored Dewey, Cowles continued, "because they think he is 'safe' and you are not." The Chicago *Tribune,* the trumpet of Republican conservatism, gloated: "Mr. Willkie can be dismissed as a minor nuisance." A few days later Willkie terminated his candidacy, though he also expressed the unrealistic hope that the Republicans would adopt the principles for which he had fought. He had just attacked Roosevelt for failing to do so. In a mischievous spirit, but partly to lure Willkie, the President wrote him to suggest that they might someday form a new, unequivocally liberal party. Willkie did not swallow the bait. On October 8, 1944, before endorsing either candidate, he died of a coronary thrombosis. That part of his spirit that infused *One World* had perished politically in Wisconsin.

Dewey emerged from the Wisconsin primary with a clear road to the nomination. Stassen never had a chance, nor did MacArthur, who had an insignificant following and an urgent military responsibility. Governor John W. Bricker, of Ohio, appealed to conservatives but was too fatuous to pit against Roosevelt. Besides, Dewey had many assets. He had barely lost the nomination in 1940, when he had been known largely for his racket-busting success as district attorney in New York City. By 1942 he had won election as governor of New York, the first Republican to do so in two decades, and put through the legislature successful programs to subsidize low-income housing, improve workmen's compensation, restructure legislative apportionment, and modernize the state's fiscal system. He had also discarded his earlier isolationism and built a smooth political organization.

The easy victor on the first ballot at the Republican Convention in July, Dewey nevertheless also had liabilities. His youth might have helped him in other years, but at forty-two he seemed young to assume the enormous wartime responsibilities of the Presidency. He was stiff, egocentric, and dull, in the words of his critics, "the Boy Orator of the Platitude," afflicted with "intellectual halitosis," a man "who could strut sitting down." As Taft wrote during the campaign: "The whole headquarters is run by New Yorkers, and Tom Dewey takes no advice from anyone." Dewey was the favorite of the Eastern, internationalist wing of the party, and Middle America did not wholly trust him even though he took Bricker as his running mate. His platform, derived from the Mackinac declaration, contained "rubber words" about internationalism that appeased the several Republican factions without satisfying them. Dewey knew that the incumbent President, the Commander in Chief, would in any event have the advantage in developing issues of military and foreign policy. But prepared though he was to emphasize domestic matters, those the voters cared about, Dewey started off on an ineffective tack. It was not just, as Taft thought, that Dewey listened to Eastern advisers. Rather, he dared not offend an electorate concerned about postwar employment and personal security. Therefore he avoided a direct assault on Roosevelt's proposed economic and social program. Instead, he charged that the Roosevelt administration was "old," "tired," and "quarrelsome."

Whatever the truth in that contention, Roosevelt could handle it. Indeed, Dewey's weaknesses were enough to encourage the Democratic regulars to believe that they could win with just about any unifying platform, and any unifying ticket so long as Roosevelt led it. Yet Dewey's assets suggested that the fissures within the Democratic party, unless they were closed, might open the way to defeat. Those considerations, predictable for months before the summer of 1944, had been pushing the President and his counselors toward the very politics of calculated blandness that Willkie, for his part, had been unable to prevent.

3. Wallace: Democratic Pariah

The near Democratic equivalent to Wendell Willkie was Vice President Henry A. Wallace. During 1943 and 1944 Wallace mounted no campaign to retain his office, disliked most of its duties and limitations, and yet desired renomination and resented those opposing it. Roosevelt created much of that ambiguity. At his idealistic best, the President engaged Wallace's transcendental faith in progress and brotherhood. But privately Wallace recognized his own accepting vulnerability to the guile as well as the greatness that characterized his chief. Publicly he had to yield to those decisions of Roosevelt that bore adversely on policies to which he attached large importance. The resulting relationship between the two men, both what it was and what it entailed, deeply affected American politics during the war years.

Henry A. Wallace was a child of the middle border. Involved from youth with the good land of Iowa, alike with farming and with nature, he learned from his distinguished father and grandfather, from their friends, and from his own assiduous study about the scientific, economic, and political problems of American agriculture. In his young manhood, during the 1920's, he became an expert on plant genetics, statistical correlations, and commodity prices. He edited the influential family newspaper, the Des Moines *Wallaces' Farmer,* and he founded the profitable Pioneer Hi-Bred Corn Company. During and after his father's tenure as Secretary of Agriculture he observed the ablest intelligences in that department at work. He also committed himself to the scientific management of agriculture, the economics of abundance in both production and distribution, and therefore to low tariffs and the vigorous enforcement of the antitrust laws.

Yet Wallace kept out of politics. A shy man, sometimes blunt and often preoccupied with his own thoughts, he disliked small talk, crowds, and back-room caucuses. For their part, politicians drew back from him. They were especially puzzled by Wallace's spirituality, by his quest for God in nature and in religion, by his attendant asceticism and mysticism. He was an austere moralist, a witness for the Lord, ultimately a devout Episcopalian, and always a social gospeler who brought tenderness and charity to his good works. To the secular, urbane men of Washington politics, Wallace's faith seemed quaint, his devotion incomprehensible, his mysticism distracting, his witnessing eccentric. In Washington, to which Roosevelt brought him in 1933 as his Secretary of Agriculture, Wallace, spiritually and culturally uncomfortable, felt often out of place.

Still, Wallace excelled in his post as Secretary of Agriculture. He proved to be a good manager, quick to delegate to able subordinates whom he selected, as adaptable as he was learned and therefore capable of learning more. He defended first the destruction and later the limitation of crops only as necessary, emergency resorts to bring commodity prices into balance with industrial prices. For the future, he contemplated open international trade, which would encourage the United States to produce at maximum and to export agricultural as well as other surpluses, and to import goods more efficiently produced elsewhere. Increasingly he argued the case for the interrelationship of free trade, economic growth, and full employment. Before the war, moreover, and instructed by his own enthusiasms and the social conscience of his staff, Wallace became an outstanding champion of soil conservation; mortgage and other assistance to the poorest farmers—tenants and migrants, many of them black; federal programs to relieve and in time erase urban poverty; the control of business power and the dissolution of industrial monopolies; and the economic theories of J. M. Keynes, with their implications for federal countercyclical spending and for full employment.

Wallace's expanding sympathies, his liberal credentials, and his popularity with Western farmers recommended him to Roosevelt as a Vice Presidential candidate in 1940. So did Wallace's views about the war, views close to most Americans' at the time. Although he opposed American entry, Wallace openly attacked Nazism, warned against the dangers of a German victory, and preached hemispheric solidarity and national defense. He believed that mobilization need not entail a surrender to generals and financiers, and that a good neighbor should sponsor democratization and economic development in Latin America.

Those credentials appealed to the President but not to the

Democratic chieftains. Wallace had always disliked the powerful captains of the great Democratic city machines. His zeal for civil rights for blacks and for relieving the poverty of sharecroppers offended most of the senior Southern Democrats in Congress. His aloofness cost him the confidence of Roosevelt's circle of immediate advisers, whom Wallace never quite trusted. In spite of the hostility he knew he provoked, he was a receptive though not an active candidate. As he realized, Roosevelt would select his own running mate. He also understood the President's foibles. "In this administration," he wrote, ". . . there is certain to be, from the White House down, a certain amount of . . . intrigue." As he also observed, "the predominant element" in Roosevelt's make-up was his "desire . . . to demonstrate on all occasions his superiority. He changes his standards of superiority many times during the day. But having set for himself a particular standard . . . he then glories in being the dominating figure along that particular line. In that way he fills out his artistic sense of the fitness of things."

That sense in 1940 brought Roosevelt to dictate Wallace's name to a reluctant convention. The President even contemplated withdrawing himself if the convention should reject his selection. It almost did, but Roosevelt's adamancy, the energetic politicking of his agents, and the timely appearance of Eleanor Roosevelt as her husband's ambassador for Wallace, brought the unhappy delegates around. Roosevelt had made Wallace Vice President in 1940, and it was up to him to decide whether to repeat that performance; but during 1942 and much of 1943 his intentions remained unclear, and Wallace's identification with the causes of social justice at home and abroad became increasingly firm.

After the attack on Pearl Harbor, Roosevelt assigned Wallace large responsibilities as chairman of the Board of Economic Warfare. He also approved Wallace's self-imposed role as an adventurous spokesman to test the responses of national and international audiences to bolder proposals for wartime and postwar policy than the President could politically afford to venture himself. The two functions were complementary, for Wallace intended the Board of Economic Warfare immediately to initiate programs of the kind he hoped the United States would expand after victory.

Of the several duties of the Board of Economic Warfare, Wallace concentrated on the procurement of strategic materials, particularly in Latin America. He defined policy with the influential assistance of Milo Perkins, a close friend formerly in the Department of Agriculture and now his senior subordinate. Perkins also handled daily administrative and technical matters. Together they endeavored to develop sources of essential materials of war, among them rubber and quinine, that the United States had previously obtained from areas the Japanese had conquered. The procurement of adequate supplies, Wallace believed, depended upon increasing the productivity of Latin-American workers, whose physical strength and morale suffered from malnutrition, disease, miserable sanitation and housing, and skimpy wages. Efficiency demanded social reform, as did the first step toward a decent future for the laborers.

The BEW attempted to take that step by writing into model procurement contracts obligations on the part of Latin-American governments or entrepreneurs "to furnish adequate shelter, water, safety appliances, etc.," to consult with the BEW "as to whether the wage scale is such as to maximize production," and to co-operate "in a plan to improve conditions of health and sanitation," a plan for which the United States would pay half the costs. As Wallace saw it, the proliferation and execution of such contracts would break the hold of conservative Latin-American governments on their impoverished peoples, would bring Latin-American countries along the route of economic development, which postwar loans would further, and would consequently prepare the nations of Latin America for participation as equals in postwar regional and global organizations. He was trying to foster in Latin America, so long an area of commercial exploitation by American businesses, the kind of social and economic development that he, like Willkie, prescribed for all colonial and disadvantaged peoples.

That policy aroused the opposition of two of the most powerful conservatives within the administration, both influential on the Hill, Secretary of State Cordell Hull and Secretary of Commerce Jesse H. Jones, who was also the head of the federal lending agencies. Both men always defended any apparent invasion of what they jealously regarded as their personal domains. Both also disagreed with Wallace's social and political objectives. There evolved a bureaucratic struggle that entailed substantive issues, issues of fundamental principle.

Hull and his associates in the State Department endorsed some of the BEW's conditions for procurement contracts in Latin America, but Hull also complained that the contracts as a whole constituted interference in the domestic affairs of foreign nations, a course the department claimed to eschew. Noninterference, as practiced by the State Department, had special connotations. The doctrine served for several years as Hull's excuse for protecting the pro-Nazi but officially neutral government of Argentina from the disciplinary measures of economic warfare recommended continually by Army Intelligence and the Treasury Department. The State Department also helped to arrange shipments of lend-lease arms to Latin-American governments, nonfighting allies against the Axis, that were openly repressive toward workers and peasants. Hull knew that Wallace welcomed social change in Latin America. The degree of that change boded too much revolution for Hull, and therefore, by his standard, too much interference.

Like Wallace, Hull was an ardent proponent of freeing international trade from artificial restraints, but he associated that goal with the spread of American institutions. He expected Latin-American trading partners to become capitalist republics on the model, political and economic, of the United States. Wallace did not. He sought postwar trade with any nation, whatever its system of government or of property ownership, and he regarded the nature of the system as less important than the willingness and capacity of a government to deliver social and economic equity to its citizens. Instead of attacking Wallace's purpose directly, Hull insisted upon State Department supervision over American nonmilitary activities in foreign countries. Through that supervision, his agents could block the negotiations, and thus the intentions, of those sent to the field by the BEW.

Jesse Jones employed a similar tactic. The interminable delays and invariable parsimony of his lending agencies retarded the negotiation and limited the scope of BEW procurement contracts. Jones claimed those contracts were needlessly costly. Preoccupied with prices and interest rates, he never grasped the greater importance of productivity, one of Wallace's goals. He did understand and reject Wallace's long-range social concerns, which he scoffed at as an international WPA. With malicious perseverance he contested every effort by Wallace to obtain independent funds for BEW contracts.

Only Roosevelt could resolve the differences among his subordinates. The President had little confidence in Hull and he actively disliked "Jesus H. Jones." Yet he was wary of their popularity with conservative Democrats, whom he was loath to offend, and he revealed no clear comprehension of what Wallace was trying to accomplish. Under continual pressure from both Wallace and his antagonists, Roosevelt during 1942 gave the BEW more the semblance than the sinew of the authority it sought. He also admonished his competing subordinates to stop their bickering.

Prodded by Milo Perkins, Wallace nevertheless tried again in the spring of 1943 to break Jones's hold on BEW loans, while Jones, for his part, enlisted his sympathetic friends in the Senate in an offensive against Wallace. Senator Kenneth McKellar arranged hearings that provided Jones with a platform from which to attack the BEW, its policies, and the limited concessions that Roosevelt had made to it. Jones's inaccuracies and vitriol infuriated Perkins, who struck back in salty language. So, though less colorfully, did Wallace. The now open warfare delighted the Republicans but embarrassed the administration, a condition Roosevelt would not tolerate. In June 1943 he abolished the BEW and transferred its authority to a newly created superagency, the Office of Economic Warfare, soon the Foreign Economic Administration. In that new office neither Perkins nor Wallace had a place, but Jones retained all the power he had held so long.

The outcome, like so much of Roosevelt's politics in 1943, dismayed many liberal Democrats. Wallace just swallowed his hurt. In a meeting with some of his staff, he expressed his feelings about the President characteristically by quoting the Bible: "Though He slay me, yet will I trust him." He also wrote Roosevelt that Milo Perkins had said of him: "Well, he is still the greatest President this country has ever had." Roosevelt replied that, although circumstances had forced him to abolish the BEW, "It is needless for me to tell you that the incident has not lessened my personal affection for you."

Probably it had not, but it did leave Wallace without a base from which to carry out his program, and it did signify, in the view of most commentators, that the President would accept a different running mate in 1944. The Republican gains in 1942 and the conservatism of Congress in 1943 also suggested the political advantages Roosevelt might garner from abandoning the most insistent liberal in his administration. But he had not yet openly done so, nor had he discouraged Wallace from continuing his advocacy of controversial domestic and international policies. After June 1943, as before, the Vice President attempted, partly because the President did not, to spell out a vision of the future that would lift the spirits and galvanize the wills of men everywhere.

In May 1942, in a speech entitled "The Price of Free World Victory," Wallace had taken his stand. He spoke partly in response to Henry Luce's celebrated *Life* editorial of 1941, "The American Century." In it Luce called upon the American people "to assume the leadership of the world," to accept "wholeheartedly our duty and responsibility as the most powerful and vital nation in the world." Succeeding editorials made it clear that Luce, like earlier proponents of American manifest destiny, contemplated a political, economic, and religious imperialism indistinguishable, except by nationality, from the doctrines of Kipling and Churchill. "Some have spoken of the 'American Century,' " Wallace replied in his address. "I say that the century on which we are entering—the century which will come out of this war—can and must be the century of the common man." In his testamental prose Wallace continued: "The people's revolution is on the march, and the devil and all his angels cannot prevail against it. They cannot prevail for on the side of the people is the Lord."

That was a Lord whom Luce did not worship. Neither did his attractive and spirited wife, the talented playwright and incumbent Republican congresswoman from Connecticut, Clare Boothe Luce. A lady with an acid tongue, she characterized Wallace as an advocate of "globaloney." Another critic complained, accurately enough, that Wallace was trying "to extend the New Deal throughout the earth." Wallace had suggested, "half in fun," as he said, that one object of the war was to make sure "that everybody in the world has the privilege of drinking a quart of milk a day." The metaphor was half serious, but appropriate as a symbol of his intentions. His detractors picked it up. Wallace, one of them said, wanted to provide milk for the Hottentots.

The *Wall Street Journal* expressed the reservations that public-opinion polls attributed, more diffusely, to much of the electorate. "High ideals and good intentions," the *Journal* wrote, "are in themselves wholly admirable, but delivered to a world in its present condition of chaos and misery, from a country which that world regards as inexhaustively rich . . . their expression is extremely likely to sound . . . like promises." No one had "stated those ideas and intentions more eloquently and appealingly than . . . Wallace," and the war did portend a "change in our relations with the world." But "we are still very far from appraising the nature and extent of that change and are even farther from committing ourselves to any . . . departure from our tradition. . . . We have had enough of eloquence . . . at least from men whom other people regard as speaking for this nation. . . . Whatever commitments this country will make will be made cautiously, one step at a time, when the necessity for that step is clearly seen."

There was a vague and mystical quality to Wallace's oratory that invited those comments. Still, his address received worldwide attention. It anticipated the themes that penetrated his later wartime speeches, which he included in a collection, *Democracy Reborn*, published in 1944. Less anecdotal than Willkie's *One World*, Wallace's addresses communicated his purpose only in the context of actual issues, like the Board of Economic Warfare. Nevertheless, the total impression he made, the gist of his international and domestic proposals, was amply clear to his contemporaries, including the President and the various factions of the Democratic party.

For Wallace, the establishment and preservation of peace demanded a world community of nations and peoples linked economically and politically through the agency of a United Nations. He expected that agency, after the war, to adopt his views about trade and to pursue his theories about the development of underdeveloped areas. Wallace realized that especially in the preliminary work of rehabilitation, a United Nations would have to draw primarily upon American resources, but he believed that American wealth should not give the United States a proportionate influence. Rather, a UN would have to bend to multinational direction in order to serve multinational interests.

The internationalizing of responsibility for providing relief, nourishment, and development throughout the world depended upon political internationalism. It could evolve, Wallace argued, only with the end of imperialism and the abandonment of balance-of-power politics. He was especially critical of the British, particularly Winston Churchill. Continued British dominion over India violated the whole purpose of the war as Wallace interpreted it. So did Churchill's impulse for empire, his unabashed belief in Anglo-Saxon superiority, his disdain for China and his distrust of Russia, his preference for secret negotiations, and his obvious intention to hold the reins of world leadership, whatever the semblance of world government, in British, American, and—unavoidably—Soviet hands.

Many of those Churchillian intentions Roosevelt either shared or accepted with only limited reservations, though the President wrapped his own power politics in the American flag. Furthermore, Roosevelt's preoccupation with victory resulted in decisions, like those in North Africa and Italy, that struck Wallace as ominous for the future. The United States, Wallace believed, had to align itself unequivocally with the forces of democracy everywhere. Unlike Roosevelt, he also advocated wartime planning for a United Nations that would exercise a broadly multinational responsibility for postwar peace-keeping and disarmament. He contemplated, as did Willkie, a degree of surrender of national sovereignty larger by far than was acceptable to any but an insignificant few in high office in any government of the major partners in the war against the Axis. The State Department looked upon his proposals as fanciful. So did the senior Senators of both parties. So probably did Roosevelt. He approved the texts of many of Wallace's addresses, but, as always, that approval signified not his agreement with the content but his permission for Wallace to speak.

In domestic as well as international policy, Wallace by 1944 had advanced beyond the consensus of the American people and their representatives. The gap reflected both their conservatism and Wallace's continuing dedication to the general principles of twentieth-century American reformers. Indeed, Wallace's concept of the century of the common man was quintessentially American. The political democracy he championed meant representative government, universal suffrage, and the civil liberties guaranteed by the Constitution of the United States. Even in the United States those conditions did not wholly obtain during the war, and Wallace worried about distortions of representation arising out of the disfranchisement of blacks, the power of Democratic city machines, and the influence that wealthy individuals and corporations exerted on Congress and on some wartime agencies. The economic program Wallace proposed grew out of his concern for promoting and distributing abundance. And because the experience of war verified the theories of the new economics, Wallace concluded, with Roosevelt, that the federal government had the ability as well as the duty to establish and preserve postwar conditions that would provide 60 million jobs, a figure that seemed outrageously high in 1944 to adherents of conventional economics. Many of them were no less outraged by Wallace's attacks on cartels and his advocacy of stringent enforcement of the antitrust laws. Yet he was not antibusiness, but antibigness; he was not an opponent of capitalism but a proponent of competition.

Wallace also supported union efforts to increase labor's share in the growing industrial profits of the war years, federal taxes that redistributed income and wealth, and proposals for federal aid to education and small business and for enlarging social-security coverage and benefits. As he put it to Roosevelt in 1943, "it would be sound economics and good politics for you to ask Congress for a series of authorizations . . . lend-lease on the domestic front. These authorizations would deal generously with soldiers, workers, farmers and businessmen." Through such measures, his economic democracy would help the common man.

What Wallace called genetic democracy, a concept close to that of Christian brotherhood, brought him to urge equal employment regardless of race, religion, or sex; to encourage the Zionists; and to defend the wartime demands of American blacks. His belief in the possibility of brotherhood and in the inherent virtue of husbandmen sparked his spontaneous admiration for the common people of China and Soviet Asia, many of whom he met on his tour of the north Pacific triangle in 1944. "All of them," he wrote about those he saw in Siberia, ". . . were people of plain living and robust minds, not unlike our farming people in the United States. Much that is interpreted as 'Russian distrust' can be written off to the natural cautiousness of farm-bred people. . . . Beneath the . . . new urban culture, one catches glimpses of the sound, wary, rural mind." Such men and women, he was persuaded, if they were given the chance, would temper and contain international rivalries. He hoped that after the war there would arise in the area he had traversed a new frontier, which the collective efforts of American and Asian peoples would structure into a thriving modern society.

The century of the common man resembled Willkie's one world and aroused similar alarm. Like Willkie, Wallace asked more from the American people than most of them were willing to give up. Wallace also confirmed the antipathies of his political enemies. He offended the Democratic South, insulted the Democratic bosses, and threatened powerful business interests and their friends in government.

The major oil companies, for example, had a stake in the synthetic rubber industry and therefore in tariff protection for American-made rubber. That course conflicted with Wallace's plans for the expansion of natural rubber production in Brazil and with his proposed policy for postwar tariff reduction. The same companies had ambitions in the Middle East equally in conflict with Wallace's hopes for an independent Jewish state and for the development of that region through the instruments of the United Nations. Within Democratic councils, besides Jesse Jones, Edwin Pauley, the treasurer of the national committee, himself a California oil man, spoke for the objectives of the industry and against the renomination of Wallace. So also, Robert E. Hannegan, a St. Louis Democrat and in 1944 the chairman of the national committee, opposed Wallace, as did his onetime predecessor in that office, Ed Flynn, the boss of the Bronx. So did the senior members of the President's personal staff, Steve Early and Pa Watson, who were ever alert to protect their chief's political interests. As they argued, Southern Senators, their restlessness with Roosevelt overt during the debate on the soldiers' vote, were determined to stop Wallace. There was a chance that if they failed they would throw the electoral votes of their states to Harry Byrd, or at most bolt the party.

The combined strength of those forces whetted the ambitions of aspirants to Wallace's office, all of whom cautioned Roosevelt that he could win only with a running mate who would conciliate disenchanted Democrats. Even though he had ill-concealed hopes of his own, Secretary of the Interior Harold Ickes, no peacemaker by temperament, sponsored his friend William O. Douglas, a liberal and a veteran New Dealer whom Roosevelt had elevated to the Supreme Court. James M. Byrnes thought he deserved the Vice Presidential nomination, as did many of his Southern admirers. Alben Barkley, who was less conservative than Byrnes, also had strong friends in the South. Senator Harry S Truman, of Missouri, openly for Byrnes, privately encouraged Wallace and yet enjoyed the important sponsorship of Bob Hannegan. Wallace formed no organization to head off the others, though Senator Joseph Guffey patched one together for him, and Wallace's personal friends kept him informed about the activities of the various contenders and the results of polls that showed him the sentimental favorite of the party's rank and file. He was also the clear choice of the leadership of the CIO and of Eleanor Roosevelt. Those were potential assets, but the Vice President realized that neither they nor his political liabilities mattered much except as Roosevelt weighed them.

For his part, the President encouraged Byrnes, who would not otherwise have been a contender, and indirectly discouraged Wallace. After the BEW episode, Wallace felt that his fortunes were declining. Later, in the weeks immediately preceding the national convention in 1944, Wallace was in Siberia and China on a mission for Roosevelt. He welcomed that assignment as an opportunity for instructive observation even though, like everyone in Washington, he knew that the President used China as a kind of exile for those whom he wanted politely to dismiss from his councils. As Guffey told Wallace, while the Vice President was abroad "some of the people around the White House" were saying: " 'We need a new face.' "

At the President's request, immediately upon Wallace's return to the United States, Harold Ickes and Samuel Rosenman asked him to meet them in Washington for luncheon. Wallace suspected they had been designated to tell him to withdraw. He flew from Seattle to see

them on July 10, 1944, but before luncheon, he made an appointment with Roosevelt for late afternoon. At luncheon he refused to
talk politics until after reporting to the President about China. Later,
after hearing the report at the White House, Roosevelt said: "I am
now talking to the ceiling about political matters." Wallace was his
choice as a running mate, he continued, and he was willing to make a
public statement that: "If I were a delegate . . . I would vote for
Henry Wallace." He "did not want to be pushed down anybody's
throat," Wallace replied, "but . . . did want to know definitely
whether . . . he really wanted me." So reassured, Wallace added
that Roosevelt should pick anyone else if that alternative would add
more strength to the ticket. Given that opening, the President wrig-
gled. He could not, he said, bear the thought of Wallace's "name
being put up before the convention and rejected." Wallace answered
that he was used to hard situations. "But you have your family to
think of," Roosevelt said. "I am not worried about my family,"
Wallace replied. ". . . I am much more worried about the Democratic party and you." Wallace then left to catch up on the political
situation and to return the next day. At that second meeting, of July
11, he produced polls that showed the Democratic ticket strongest
with him on it, and victorious. Roosevelt said many people looked on
Wallace "as a communist or worse," though of course there was "no
one more American . . . no one more of the American soil." With
nothing settled, they arranged to talk again two days later.

On the evening of July 11, without the knowledge of either
Wallace or Byrnes, Roosevelt sat down with his political professionals—Flynn, Hannegan, former Postmaster General Walker, and
others, a randy group—to discuss the nomination for Vice President.
They quickly dismissed Wallace. Byrnes, they advised, would alienate
liberals and blacks, and also Catholics, because he had left the
Church. Barkley was too old. Roosevelt suggested Douglas, but his
guests were dubious. They were enthusiastic only for Truman, like
Hannegan an alumnus of the Pendergast school, a staunch partisan, a
regular supporter of administration measures, a representative of a
border state and of its partly Southern, partly Midwestern culture.
Roosevelt agreed that Truman was the man.

Characteristically, the President would tell neither Byrnes nor
Wallace that he had been ruled out. On the contrary, after Hannegan
and Walker informed Byrnes about the secret meeting, Byrnes called
Roosevelt, who told him he was "not favoring anybody." He had
merely said, the President continued, that he would not object to

Truman. Byrnes stayed in the race. So did Wallace, whom Hannegan urged to withdraw on the morning of July 12. The following noon Roosevelt repeated his earlier promise, which he kept, to write a public letter saying that if he were a delegate, he would vote for Wallace. But the President also reported about his evening meeting with the political dealers. They all thought, Roosevelt said, that Wallace would hurt the ticket. They also thought Truman would help it most. He would not dictate to the convention, the President said, adding later—with a full smile, a hearty handshake—"While I cannot put it that way in public, I hope it will be the same old team." Then, as Wallace departed, a final sentence: "Even though they beat you out at Chicago, we will have a job for you in world economic affairs."

Roosevelt's virtuoso performance had not yet ended. Persevering in spite of the President's calculated ambiguity, Wallace went to Chicago to fight to win. So did Byrnes. So did Hannegan, as Truman's manager. Hannegan had to beat off the other two without letting either run away with the nomination, a difficult task, probably impossible without the President's connivance. While Wallace was using Roosevelt's letter to round up delegates, while Byrnes was using Truman to the same end, Hannegan in Truman's presence called Roosevelt, who said over the telephone: "Well, tell the Senator that if he wants to break up the Democratic party by staying out, he can; but he knows . . . what that might mean at this dangerous time in the world." That statement convinced Truman.

Wallace had not heard it when he seconded Roosevelt's nomination in a bold speech. The future, Wallace said, belonged to those "who go down the line unswervingly for the liberal principles of both political and economic democracy regardless of race, color, or religion . . . regardless of sex." The delegates responded with the noisiest demonstration of the convention. They might, Wallace believed, have renominated him by acclamation had not Hannegan had the session adjourned. The following day Wallace led on the first ballot, but on the second the South and the city machines, as Hannegan and Roosevelt had written the script, combined to deliver the nomination to Truman. Roosevelt had a last scene to complete. In a telegram to Wallace, meant also for Mrs. Wallace, whom he had always particularly liked, Roosevelt said: "You made a grand fight and I am proud of you. Tell Ilo not to plan to leave Washington next January."

There was little in the Democratic platform that resembled the spirit of Wallace's seconding speech, even less that resembled his century of the common man. Roosevelt and the party leadership planned it that way. In politics as in war, victory, not freedom, was their first priority. In discarding Wallace they turned away, just as had the Republicans in rejecting Willkie, from the promise of a brave new world. There remained for the voters a choice between Roosevelt with Truman and Dewey with Bricker, but it was a far different choice from what would have been offered by Roosevelt with Wallace in a race against Willkie.

The nominations involved no sudden departure from past practice. They depended upon no special chicanery, though a good deal of plotting went into them. Rather, the two parties, in their respective preparations for their conventions and in their deliberations at them, continued to move along paths they had followed since the beginning of the war. They continued to trim sail to stay close to the wind of the public mood, in 1944 a prudential and unheroic mood.

4. And Still the Champion

His doubts suppressed, Henry Wallace chose in 1944 not only to support but even to campaign for Franklin Roosevelt. It was, as Eleanor Roosevelt saw it, a symbolic choice, for Wallace, she said, was "the outstanding symbol of liberalism in the United States." For his part, Wallace considered Roosevelt "a symbol of liberalism . . . in the whole world." To be sure, the President had "double-crossed" him at Chicago, but soon even Wallace could laugh about the transparent maneuvering there. "You had better get over to the White House," one of Wallace's friends told Paul Porter, a one-hundred-per-cent Roosevelt man, "and straighten yourself out with Anna Boettiger, the President's daughter. She told me the other day that you are a son-of-a-bitch because of the way you treated Henry Wallace at Chicago." Porter at once replied: "You go tell Anna Boettiger, 'So's your old man.' " Wallace enjoyed repeating the story. More important, he distrusted Dewey and the Republicans, he believed Roosevelt was the most effective champion of decency and enlightenment in the contest, and he declared that a Roosevelt victory would provide a mandate for "a permanent, enforceable peace" and for "the goal of full employment . . . sixty million jobs."

Wallace's loyalty to Roosevelt probably made it easier for other liberal Democrats to make the same choice he did, though they would doubtless have done so anyway. Those Democrats in Congress who had supported wartime proposals for social reform, men largely from Northern, urban districts or states, were solid party men who wanted Roosevelt's name at the top of the ticket on which they were going to run. Even mavericks in New York City who had joined the American Labor party or its anti-Communist rival, the Liberal party, kept FDR as their Presidential candidate. The liberals had no place else to go. The President by their standards had done well enough, long enough, to hold their support while he cultivated the less certain allegiance of more conservative groups. In so doing he succeeded in holding the party together and in capturing for himself that position only slightly to the left of center which he had found to his taste and advantage in the three preceding national elections.

Those who considered themselves liberals in 1944 had been, for the most part, Democrats since 1932. They agreed about the necessity for eradicating Nazism and they accepted Roosevelt's foreign policy as sufficient for victory and postwar political security on the basis of some kind of international organization. The President cultivated those and other cementing attitudes. In the summer of 1944 he signed the GI Bill, the model in miniature designed to satisfy his liberal constituency's domestic concerns. Later that summer delegations from each of the United Nations met at Dumbarton Oaks in Washington for preliminary discussions about a postwar political organization.

In arranging the nomination of Truman, Roosevelt even managed to avoid offending the CIO-PAC, which he needed to supplement the political strength of the South and the city machines. Sidney Hillman and other CIO leaders had preferred Wallace to Truman, but they much preferred Truman to Byrnes, who never forgave the President for his disingenuousness. Unlike Byrnes, Truman had an acceptable record on issues important to labor. Moreover, Roosevelt kept Hillman informed and engaged in the maneuvering at Chicago. Above all, like the liberals, the CIO had no other place to go, certainly not to the openly hostile Republicans. So the Democratic troops to the left of the party's center marched with the President again in 1944, while those to the right were sufficiently placated to stifle their recent temptation to desert. They had cause for satisfaction. They had stopped reform in Congress and stopped Wallace at the convention.

With the Democratic party, the majority party, united behind his candidacy, if united about little else, Roosevelt entered the campaign with a formidable advantage. He could count upon considerable special support. His equivocal record on civil rights looked better than the Republican record, and Wallace's stumping for him gave him an added allure for black voters, a significant factor in Northern cities as they had been in 1940. Though some Italian-Americans had enlisted with the Republicans, Roosevelt was still popular among Jews and most other large, urban, ethnic groups. More important, among Americans in general, the course of the war itself enhanced the role and reputation of the President as commander in chief. During the summer and fall, Allied forces in France and Soviet armies in eastern Europe were advancing with a rapidity that promised an early victory over Germany. In the Pacific, American sea, air, and infantry units were successfully attacking the Mariannas and Philippine Islands, and closing in on Japan. No Republican, as even Robert Taft came to believe, could beat the Commander in Chief.

Yet Roosevelt and the Democrats took nothing for granted, for Dewey embarked on his campaign with determination and confidence. The New York governor used an efficient organization, staffed by experienced politicians and assisted by specialists in techniques of opinion sampling and public relations. His own "modern Republicanism," which accepted the basic New Deal reforms as permanent, minimized the anxieties of those voters eager to keep the benefits they had gained and unwilling to lose any income or prerogatives to those less fortunate than they were. Dewey would have lost support had he followed the hard conservative line of Taft and Bricker. He also dissociated himself from their current and his past reputation as an isolationist. He committed himself to joining the international organization taking shape at Dumbarton Oaks, while he also warned against the coercion of that organization by the great powers. When Hull at once denied that the conference was planning any coercive superstate, Dewey sent John Foster Dulles to consult the Secretary of State, and the two issued a joint statement that eliminated the issue from the campaign. Yet Dewey had managed, as he continued to, to suggest that he was suspicious of the intentions of the Soviet Union and Great Britain, and that he would not permit himself to be deceived by either nation.

Dewey was trying to capture the center of public opinion, the consensus that the polls reported. Furthermore, he was not just an opportunist. His own record as governor lent credence to the positions he struck, and he had the statesmanship to keep silent about

MAGIC, the American compromise of secret Japanese codes. He could have embarrassed Roosevelt, as the Chicago *Tribune* tried to, by revealing that the government had broken the codes before Pearl Harbor, but that revelation would have damaged American intelligence operations in the Pacific. Yet, as Dewey knew, efficient organization and modern Republicanism did not in themselves provide a sufficient basis for his campaign. He needed to take the offensive, to develop positive issues. He settled on two, both of which he approached obliquely: the President's health, and the influence of Communism in Democratic councils.

The familiar Republican charges, Dewey's charges, that the administration relied on one-man government, that it was old and tired, had a new impact in 1944, for the President looked and often acted just that way. In fact, as a Republican whispering campaign suggested, he was much closer to death than he or his doctors realized. The whispering seemed ghoulish to the President, but it was no more deceitful than the efforts of his aides, in which he co-operated, to demonstrate his vigor and stamina by arranging for flattering photographs and for exhausting motorcades, one even in a pouring rain. In fact, Roosevelt was not honest with himself. Soon after the Chicago convention he suffered an episode of "horrible pain," perhaps a heart attack, which he did not report to his doctors. Later, upon returning from a brief trip to the Pacific in July, the President spoke from the deck of a destroyer moored in Bremerton, Washington. The guns visible behind him did not obscure the vague and wandering nature of his address. During "the early part of the speech," according to his physician's later account, Roosevelt "for the first time experienced substernal oppression with radiation to both shoulders"—an attack of angina pectoris. A series of electrocardiograms revealed "no unusual abnormalities," but the attack remained secret for many years.

During the campaign those who were willing to look could see that the President had aged and could sense that he was often weary. The weight he had deliberately lost gave his face a rather haggard appearance. He was a cripple; he had served for almost twelve difficult years; and he was sixty-two. But most observers believed, as he did, that he was strong enough to run and serve four more. Public knowledge about his high blood pressure and about the painful cardiac episodes he had experienced would probably have hurt the Democrats, because Truman, like Dewey, had established no credentials as a wartime leader. But even if the facts about Roosevelt's health had been made public his party would still have been strong, and its marginal adherents, as well as independent voters, dubious though they might have been about Truman, would very likely have voted for Roosevelt nevertheless, just as Americans did twelve years later for a popular incumbent who had had a heart attack.

As it was, eager for victory and after a July full of vigor and a sense of well-being, Roosevelt deeply resented Republican rumors about his health. As his doctor later wrote, during October the President disregarded his prescribed regimen of rest but, in spite of the stress of the campaign, "was very animated . . . really enjoyed going to the 'hustings,' and his blood pressure levels . . . were lower than before." On September 23, 1944, the President delivered an exhilarating speech to a dinner of union officials and, over the radio, to the American people. He was, he admitted, four years older than he had been during his previous campaign. (So much for the health issue.) Indeed, more than eleven years had passed since the Democrats had begun to "clear up the mess that was dumped into our laps in 1933." (As he liked to, Roosevelt was running against Hoover again.) The liberal elements in the Republican party, he continued, had not driven the Old Guard away, and the Old Guard could not "pass itself off as the New Deal." (So much for modern Republicanism.) His audience was with him, cheering and laughing, as he happily answered the "Republican fiction" that he had left his dog, Fala, on an Aleutian Island and sent a destroyer to fetch the beast at the taxpayers' expense. Fala was Scotch, Roosevelt said, and angry about the story. As for himself: "I am accustomed to hearing malicious falsehoods . . . such as that old, worm-eaten chestnut that I have represented myself as indispensable. But I think I have a right to resent . . . libellous statements about my dog."

The speech hit its targets. Its verve reassured those Democrats who had begun to wonder if their champion had lost his old touch. It also reassured voters who heard nothing old and tired in it. And it infuriated Dewey, who considered the speech "snide" but knew that Roosevelt had scored. "The race," Paul Porter wrote the President on September 28, "is between Roosevelt's dog and Dewey's goat."

Dewey retaliated by giving new emphasis to an issue the Republicans had already raised, the danger of Communism in the United States, and by pointing that issue directly at Roosevelt. In the Congress, the Republican Old Guard, at least in their rhetoric, had habitually identified the New Deal with Communism. In 1944 some of them concluded to their own satisfaction that Communists controlled the CIO and were using the CIO-PAC to take over the Democratic party and the country. The confusion had a convenient utility for the campaign. The Opinion Research Corporation, the polling organization on which the Republicans relied, reported that the election hinged upon the vote along the Atlantic seaboard and of ethnic groups residing there and in other urban areas; that Dewey should concentrate "on home-front issues"; and that he needed to split the labor vote by underlining the differences between the American Federation of Labor, whose support he could perhaps win, and the CIO. To that end, the report advised him "to castigate the Hillman group as extremists and radicals." As early as July, *Time* undertook to advance that purpose in a cover story about Hillman that emphasized his Lithuanian birth, his Jewish faith, and the socialism of his friends. Late in August, Governors Warren and Baldwin, presumably modern Republicans, gave radio addresses charging that Communists in the CIO-PAC were leading the Democratic campaign. The right-wing Republican press stressed the same theme. The New York *Daily Mirror* wrote about "the New Deal–Communist–Left Wing axis"; the Chicago *Tribune* warned that "the people who support the New Deal ticket . . . are supporting the Communists and building them up for the day when they plan to bring the Red Terror sweeping down upon America."

In New England, where the large Roman Catholic population was supposedly responsive to what he had to say, John Bricker, Dewey's running mate, kept repeating one sonorous paragraph: "Insidious and ominous are the forces of communism linked with irreligion that are worming their way into national life. . . . First the New Deal took over the Democratic Party and destroyed its very foundation. Now these Communist forces have taken over the New Deal and will destroy the very foundations of the Republic." Dewey let Bricker preach that gospel until after Roosevelt's Fala speech. Two days later Dewey himself declared that FDR was "indispensable to Earl Browder," the head of the Communist party in the United States. In March 1941 Browder had received a harsh sentence for fraud in the use of a passport. In 1942 the President, moved by a concern for national unity and civil liberties, had commuted that sentence. Yet Dewey asserted that Roosevelt had pardoned Browder in order to assure his help during the 1944 election. Thereafter the Communist theme imbued Dewey's addresses. A Communist, he said in Boston, was anyone "who supports the fourth term so our form of government may more easily be changed." Other Republicans, Taft for one, made the same point and frequently added warnings about the possible perfidy of the Soviet Union, warnings pitched to the anxieties of Polish-Americans about the future of their homeland and to the belief of many Catholics in the perils of an expansionist Russian atheism.

The tactic appeared to be effective. The pollster Hadley Cantril warned the President about the "anti-Roosevelt sentiment of the Irish Catholics, Italians, and Germans in New York City." Dewey, he judged, had made gains among low-income voters. Connecticut Democrats reported increasing Dewey sentiment among Italians. Similar reports reached Roosevelt about Poles in the Middle West. The President responded with a statement to a delegation of American Poles that "Poland must be reconstituted a great nation" and with a speech to party workers in which he said he had not "sought" and did not "welcome" Communist support. Yet Rosenman told Roosevelt that Dewey was still "making dangerous progress in his campaign to hang the Communist label" on the President, and Attorney General Biddle wrote in the first week of November that Democratic politicians "seemed to agree that the Republican 'Communism' attack is doing the party harm."

Roosevelt, more angry than worried, assured a Boston audience three days before the election that he and the Democratic party wanted to continue "to live under the Constitution . . . for another hundred and fifty-five years. . . . When any politician . . . stands up and says, solemnly, that there is danger that the Government of the United States . . . could be sold out to the Communists—then I say that that candidate reveals . . . a shocking lack of trust in America." "Never before in my lifetime," the President declared, "has a campaign been filled with such misrepresentation, distortion, and falsehood."

Though he regarded Dewey with "unvarnished contempt," Roosevelt did not abandon his campaign plan for vituperation. As he had intended to for a year, he beguiled Americans with the picture of a rosy economic future that his bill of rights had sketched. In Chicago he repeated the provisions of that bill, promised 60 million jobs, more homes, more highways, more health, more hospitals, cheap cars, new airports, increased foreign trade, a permanent FEPC, the development of the large Western river basins, help for small business, and rewards for skillful entrepreneurs. He promised what wartime advertising had displayed. He promised what the polls said the people wanted. He promised the kind of society to which the GI's wanted to return. Attuned to the culture of his countrymen, he spoke to them not as a frightening old radical that the Republicans had tried to make him seem, but as the avuncular Santa Claus they had known for a dozen years.

Roosevelt won easily. Though his margin was smaller than in any of his previous elections, he ran better than any of the polls had predicted. He did lose ground among farmers and Italian-Americans, but other large ethnic groups remained safely Democratic, as did the cities, labor, the soldiers, the South, and the northeastern states. Dewey ran best in the Middle West, where the Republicans had made their gains in 1942. Roosevelt carried thirty-six states and 432 electoral votes, and drew just short of 55 per cent of the popular vote. The big cities were decisive. In those with populations over 100,000, taken as an aggregate, Roosevelt received almost 61 per cent of the vote. That urban margin stamped the congressional contests. Apart from the South, the Democrats elected to Congress in 1944 came primarily from urban constituencies. The party gained twenty-two seats in the House but lost one in the Senate, a change too small to weaken the coalition of Republicans and Southern Democrats that controlled Congress.

The election did not significantly alter the national political configuration of the previous several years, or the political prospects for the next two. Just as the war itself, with all it entailed, had driven even Taft and Vandenberg to a reluctant and circumscribed internationalism, just as it had invited Ball and Burton, Hatch and Hill to take a Wilsonian stance, so the election removed from office several last-ditch isolationists. But Taft and Vandenberg still led the Republicans in the Senate, and their Democratic counterparts were conditioned to follow the President on foreign policy, not to lead him. So, too, domestic policy remained the captive of the reigning coalition that had yielded little to Roosevelt's sporadic efforts to advance reform. The President could not count upon having his way during a fourth term. Further, campaign oratory aside, there was little evidence that he contemplated altering his previous intentions in foreign policy or embarking on novel programs for social progress at home.

The election was only an episode in the long continuum of the wartime national mood. It left the American people in the state of mind that Cantril and Bruner had described a year earlier. The fourth term opened with a reprise of the themes of the third. A month after the election, following the resignation of Cordell Hull, Roosevelt nominated seven new senior officers for the State Department. Democratic liberals approved of only one, Archibald MacLeish, designated as Assistant Secretary for Public Information. Senators Claude Pepper and Joseph Guffey, onetime Wallace supporters, led a lonely and futile fight against the confirmation of the others. All of them were either closely connected with great wealth, as were Secretary Edward Stettinius, Jr., and Assistant Secretaries Nelson Rockefeller and William L. Clayton, or long since involved in tacit collaboration with French or Italian fascists, as were Assistant Secretaries James C. Dunne and Julius C. Holmes. As Henry Wallace put it to Pepper, he "couldn't figure what the President was up to unless, like a man in a rowboat, he was looking one way and rowing another."

Soon after the fourth inaugural Roosevelt forced Jesse Jones to resign and nominated Wallace to take his place as Secretary of Commerce. But the President, off to the Yalta Conference, played a passive role in the ensuing donnybrook in the Senate. The Southern Democrats and Republicans first stripped the Commerce Department of the lending agencies, which they refused to trust to Wallace, and then almost blocked his confirmation, which passed with a margin of only one vote. In whatever direction the President was rowing, he still faced strong opposing currents.

While resting at Warm Springs, Georgia, on April 11, 1945, Roosevelt dictated a draft of an address for Jefferson Day. The Nazi state, he said, was crumbling; the Japanese were receiving their retribution. The victory he had pursued so ardently was near. "We must go on," he said, ". . . to conquer the doubts and the fears, the ignorance and the greed, which made this horror possible. . . . The only limit to our realization of tomorrow will be our doubts of today. Let us move forward with strong and active faith." The rhetoric was still splendid, but no one could predict what relationship it might have to the substance of what the President planned. It had not had much for many months. Throughout the years of war, the difference between what he professed and what he achieved had persisted at home and abroad. Yet, as the election had demonstrated, he was still the choice of the people, still the champion. On April 12, 1945, minutes after a massive cerebral hemorrhage, Franklin D. Roosevelt died. The magic and the mystery of the man remained a part of history.

9 / Victory

1. Fundamental Issues

The news that the Japanese were about to surrender came as no surprise to the officers who were passengers aboard a Navy transport en route from Saipan to Hawaii. They had already celebrated the dropping of atomic bombs at Hiroshima and Nagasaki, bombs they then understood only as miraculous superweapons that guaranteed them against a return to the Pacific after they got home. They had realized the previous May, when Germany collapsed, that the end of the war was in sight, though it seemed far away. Now in August the war was over. The Armed Forces Radio cut in on a broadcast of one of the national networks on which a reporter, his voice happy with excitement, described the jubilation of American civilians. *Oklahoma* was the boast of Broadway that year, and the reporter spoke from a locale that was the subject of one of the musical's songs. "Everything's up to date in Kansas City," he said, "and no one's ready yet to go to bed."

Victory had come at last, victory over Germany, victory over Japan, and victory, too, achieved much earlier, over the Great Depression. For servicemen returning or about to return, as for their families, victory at first seemed ultimate, unequivocal, even palpable. Victory was what it had all been about—the fear, the discomfort, the nostalgia. Now the near future beckoned, just as memory and the media had asserted for four years that it would beckon. Loving arms, cold beer, hot baths, sweet corn with real butter. For the further future, the shape of American society and of world politics were supposed to preserve the glow. Of course the euphoria would pass, but the designs for tomorrow, as a reward for the war, were to expand and preserve contentment.

Those who were up to date on national politics knew that the transitory euphoria of the GI's made the aims of victory seem closer and easier of achievement than they were. Yet even the wisest of social critics shared the common hopes. In 1944 Walter Lippmann, for several decades the most respected of observers, published a small book, *United States War Aims,* that ended with an "Afterword." As a young man, he wrote, he had been unconcerned with "the hard questions . . . of history," those postwar America would have to answer:

> I did not think about the security of the Republic and how to defend it. I did not think about intercourse with the rest of the world and how to maintain it. I did not think about the internal order of the nation. . . . Yet those very things . . . have since proved to be paramount. . . . The crucial difference . . . between the recent American past and the American future is that we have come to the end of our effortless security and of our limitless opportunity. . . . The security and prosperity and welfare of the United States cannot be left to chance . . . the fundamental issues of national existence have now to be dealt with consciously and positively.

As the war ended, Americans in and out of public life were trying consciously and positively to assure the nation's safety and prosperity. They were proceeding from sets of premises as varied as had been the sets of assumptions that characterized public debate during the war years. They were engaged, as before, in partisan as well as ideological wrangles. Yet the President and the Congress wanted to satisfy the yearnings of the GI's, the two great yearnings of the American people—safety from war and security from depression. Committed to those American wartime dreams, those in Washington who were responsible for the designs for tomorrow were also for the most part decent and earnest men. Still, they had grown up in the recent American past; they stood within the culture that had imbued the war years. That culture, as well as their sense of their constituents' mood, conditioned the way they thought about the hard questions they could no longer leave to chance.

2. Limits of American Internationalism

Among those with a major voice in answering those momentous questions was the new President, Harry S Truman, whose accession to office made less immediate difference in the early postwar months than his detractors thought. To be sure, he commanded none of the love that the electorate had had for Roosevelt, or of its confidence in him, and little of the prestige that Roosevelt had had on the Hill or in international councils. Truman was uncomfortable with the most dedicated New Dealers in the Cabinet, especially Wallace, Ickes, Morgenthau, and Biddle, all of whom distrusted him. Liberals in general, from the first suspicious of Truman, were dismayed by his early appointments and by his growing reliance upon wealthy financiers, courthouse politicians, and professional military men. Now seemingly cocky, now seemingly indecisive, lacking both eloquence and elegance, Truman simply did not fit the picture of a President that Roosevelt had engraved so deeply on the consciousness of Americans.

In contrast to Truman, Roosevelt, had he lived, might have pursued a foreign policy more generous financially to Great Britain and more flexible politically toward the Soviet Union. Perhaps Roosevelt would also have commanded the course of domestic policy as Truman did not. But Roosevelt had not been notably successful in his initiatives on the Hill for some years, nor had he repudiated the kinds of advisers whom liberals blamed Truman for consulting. More important, the momentum of public opinion as it bore upon public policy would have affected the Congress no matter who was President. Roosevelt would have had a better grasp of the issues before the Congress than did Truman. He might, had he wished to, have dramatized those issues, appealed to the country, and possibly swayed some votes on the Hill, whereas Truman for too long let things drift. But the Congress was tired of executive intrusions, the Democratic party was still divided, and Roosevelt in 1944 had bowed to the circumstances that Truman was unwilling to challenge during the next two years. The course of domestic affairs in the early postwar months followed channels that wartime developments had dredged.

At no time during World War II did the majority of the American people or of their representatives contemplate a postwar settlement in which the United States would not circumscribe its commitments, economic, political, and military, to other nations. On the contrary, at all times the majority of Americans seemed resolved that there would be no unconditional postwar giveaways—of American wealth, of American sovereignty, of American weapons. There was no chance that Uncle Sam would play Uncle Sucker. The question, rather, involved the degree to which the United States would depart from its prewar traditions, the lengths to which the country would go toward internationalism. How much, and under what conditions, would the nation contribute to the reconstruction of the countries of devastated Europe and to the development of the countries of Asia, Africa, and Latin America? How willing would the nation be to subject its foreign policies to the supervision or the opposition of the collective voice of the United Nations? How ready was Washington to share its unequaled reserves of manpower, its unsurpassed arsenal of modern weapons, its superior knowledge of the technology for producing them? As each of those three lines of policy was drawn, the center of the resulting triangle would provide a fix, as in navigation, of the immediate postwar internationalism of the United States.

In drawing those lines, the American people and their representatives felt the influence alike of wartime sentiments and of national habits of thought that had prevailed for many decades. They believed, as the Founding Fathers had, in the special virtue of the American experiment and those participating in it. To the particular goodness they saw in their own country they contrasted the supposed corruption of the Old World. The presumption of national innocence carried two self-serving corollaries: one, that American adventures in the affairs of other peoples were exceptional, unique in their selflessness, untainted by imperialism; the other, that the devious spokesmen of Europe—whether representatives of an effete *ancien régime,* as in the case of Great Britain, or of a sinister radical giant, as in the case of the Soviet Union—exploited their characterological advantage over guileless Yankee negotiators. It followed that Americans had to take special care not to be fooled, special care not to subordinate their interests to the influence of others, particularly at a time when the great economic and military strength of the nation, a reflection of its natural virtue, made it unnecessary to yield to any antagonist. That strength, so the theory went, if properly deployed, could assure the safety of the United States for an indefinite future, though the cost of safety would drop if the country's wartime allies were willing to construct collective agencies amenable to American guidance. For that desirable end, limited concessions were appropriate. But few Americans were comfortable with the prospect of large concessions to the enormous needs or the perceived fears of the other United Nations, even the indispensable partners to the Grand Alliance.

Those attitudes had characterized congressional debates about foreign economic policy throughout the war. In 1941, months before Pearl Harbor, opponents of lend-lease continually attacked British imperialism and past British offenses against the Irish, the Indians, and the Boers. Others warned against British acquisitiveness. As they saw it, Parliament had created the system of imperial preference, a system that encouraged commerce among Great Britain and the Dominions, in order to cut Americans out of their legitimate share of world trade. That argument made no allowance for American protectionism or for the extraordinary drain on British wealth during World War I. Still, it appealed to Middle American patriots like Republican Representative Everett Dirksen, then in the dawn of his unceasing grandiloquence. "If the lend-lease bill is enacted into law," he declaimed, "the President should insist that Great Britain secure some of her agricultural . . . commodities from this country. Great Britain has been buying her wheat, flour, cotton and meat from Canada, Argentina, Australia, and other colonies."

Dirksen attached an unusual meaning to the definition of colonies, but he also evoked a familiar suggestion of British conspiracy against the good folk of the American farm belt. So, too, he implied that the President might not adequately protect American interests. That possibility seemed predictable to the isolationists of 1941, especially Robert Taft, who feared that Roosevelt would administer lend-lease in a manner that would lead the United States into war, and that the Lend-Lease Act, entirely apart from the policy it incorporated, placed too much power in the White House. On that account Taft and others sponsored successful amendments that required a regular accounting of the disbursement of lend-lease money, an accounting that would precede necessary new appropriations of further funds. The amendment assured Congress of a continuing influence on foreign economic policy, an influence it retained throughout the war.

The Senate's traditional concern with its own Constitutional authority, along with strong and bipartisan senatorial instincts for parsimony overseas, instincts that represented constituent attitudes, conditioned the development of American policy for the relief of war-torn areas of Europe. In 1943 the Department of State began work on the construction of the United Nations Relief and Rehabilitation Administration. As Roosevelt informed the Democratic and Republican leaders in Congress, he did not plan to submit the UNRRA agreement to the Senate for its approval. He proposed instead to pursue it as an executive agreement among heads of state, though of course Congress would have to appropriate the American funds for relief. Arthur Vandenberg immediately protested, as did his colleagues in both parties on the Senate Foreign Relations Committee, against any precedent of ignoring the Senate in the creation of

United Nations agencies. After a strained session with the committee, Secretary of State Hull got the point. He accepted some changes in the draft for UNRRA. Thereafter he moved forward, as Vandenberg put it, "with total Congressional consultation." Further, Hull incorporated the UNRRA agreement in the resolution sent to Congress to authorize money for the agency. By virtue of approving the expenditures, Congress approved the agreement itself, though a few Republicans, Taft for one, would have preferred a formal treaty requiring a two-thirds vote in the Senate, a vote they might have been able to prevent.

Others rejected the whole idea of international relief and rehabilitation. Jessie Sumner, a Republican from Illinois, hostile to UNRRA, grieved by public expenditures to assist areas overrun by the Nazis, called instead "for a tremendous drive for voluntary contributions, to be distributed by the Red Cross and the Quakers and such organizations." She viewed the problem of international relief rather as Herbert Hoover in 1931 had viewed the problem of domestic relief. Neither understood the size and severity of the issue; neither understood the need for mobilizing resources in relatively rich regions for distribution in impoverished areas. That kind of myopia was bipartisan. Senator Guy Gillette, an Iowa Democrat, opposed appropriating funds "for an organization which we did not create, an organization in which we have . . . only one vote, an organization . . . with power to determine and change . . . its policies under which the money is to be expended and over which we have no control except the negative control over appropriations." Perhaps Gillette wanted a majority vote for the United States in UNRRA. Certainly he did not realize that the necessity for relief and rehabilitation would grow with the continuing reconquest of Europe. But probably, like others of similar mind, he was distressed primarily at the thought of committing the United States to participation in a world organization whose function would involve continuing expenditures abroad largely of American money.

The support of relief, a humanitarian function, appealed to Americans more than did the support of rehabilitation. The latter effort had to be closely defined, the *Wall Street Journal* argued, so as not to imply spending American dollars to restore whole foreign economies. That was also the sense of the Congress, which expressly forbade the use of UNRRA appropriations for reconstruction. Even for the purpose of providing relief, congressmen preferred spending that would bring their constituents some direct share of the benefits.

With its ordinary altruism at work, the Southern and Western farm bloc amended the UNRRA agreement so as to earmark $50 million for the purchase of American agricultural products, particularly raw wool and cotton. In that provision congressional internationalism revealed its special fiber. The United States subscribed to UNRRA but the issues raised during the debate indicated that enthusiasm for world organization wavered in the absence of provisions for the primacy of American influence and self-interest.

Those concerns marked the negotiation of the agreements signed in 1944 at Bretton Woods, New Hampshire, to establish the International Monetary Fund and the International Bank for Reconstruction and Development. Preliminary discussions about those agreements proceeded largely between the United States and Great Britain. Initially each nation submitted plans designed partly to advance its own interests, partly to satisfy the technical preferences of its own experts. As the British knew, the Americans could dominate the ensuing process of accommodation, for the United States would emerge from the war with huge accumulations of gold and credits, and with undamaged industrial facilities of demonstrated productivity. Still, the British had intangible assets in the superior intelligence of their negotiators, J. M. Keynes in particular, and the superior virtue of their proposals. Those looked to a postwar clearinghouse that would serve as a kind of international central bank to manage the relative values of currencies of all subscribing nations, and to do so in a manner that would at once preserve stability in exchange values and equity as well as openness in world trade. The alternative American scheme was contrived to protect the monetary value of gold, of which the United States had so much, and to establish the dollar as the dominant currency in the world, as well as to prevent the kind of competitive devaluation that had characterized international monetary rivalry during the years of the Depression. Secretary of the Treasury Henry Morgenthau, Jr., suspected that the British were maneuvering to acquire authority to kill "anything that they don't like." Dean Acheson, then Assistant Secretary of State, had a more accurate and sympathetic reading of the situation. "I think," he said, "there is a rather pathetic feeling on the part of the British that we really are going to write the ticket, and all they want is a chance to go over it with us, pointing out their views, and to be allowed to come in on the formulation from the start." In their weakened economic state, they would be satisfied with the appearance of participation as equals.

That was all they got. Both before and during the Bretton

Woods conference the Americans made the most of their incomparable financial and economic resources. In order to win the other United Nations to the agreements, the United States had to make some concessions, essentially political and cosmetic, to the *amour-propre* of wartime partners, but on the whole the American plan prevailed. The charter for the International Monetary Fund further aggrandized the role of the dollar among currencies, as the United States Treasury and Congress had intended it to. The United States, the major contributor to the resources of the Fund, obtained a major voice in the weighted voting procedures established to govern Fund policies. So, too, American proposals, American resources, and American influence dominated the plan for the International Bank.

That outcome appeared to subordinate Anglo-American arrangements to multilateral co-operation in postwar economic affairs, but the internationalism associated with the multilateral emphasis was superficial. The American negotiators looked forward to American leadership in the Fund and the Bank, to a consequent diminution of the British voice and share in world trade, and to the continuing position, achieved during the war, of New York as the capital city of world finance. The technical aspects of American monetary diplomacy, which proceeded from and confirmed those objectives, also needlessly constricted the operation of the Fund. Any more flexible or less nationalistic scheme would probably have exceeded the tolerance of congressional opinion, as the United States Treasury constantly maintained.

The difficulty lay not so much with some peculiar American nationalism. All the nations represented at Bretton Woods consistently cultivated their self-interest and prestige—the Soviet Union, China, and France perhaps most noisily. Rather, the United States was partner to a general failure of vision, a general failure to foresee the severity of postwar monetary and economic problems, a general failure to comprehend the huge size of the task of restoring the economy of Europe and developing the economies of the Southern Hemisphere. Like other Western peoples, Americans, though eager for a new and better postwar world, had yet to shake off the grip of the old one.

The right wing in the Senate remained in its full clasp. "The United States," Robert Taft asserted in 1945 during the debate on the Bretton Woods pacts, "should involve itself in no long-range program." The Bank, in his oblique interpretation, "was part of the general New Deal program to create new methods of deficit spending." As for the Fund, joining it "would be like pouring money down a sewer." Arrangements for the Fund to provide debtor nations with assistance from creditor nations were "childishly absurd." That attitude, common among Republican congressmen and small business men, and within the American Bankers Association, had constrained the President and his agents during the negotiations. Now, with senatorial approval of the agreements at issue, the executive branch promised more than the Fund and Bank could deliver. They were capable, as Roosevelt had said, of enhancing prospects for "expanded production, employment, exchange and consumption." But the Treasury and State departments claimed more in their public-relations statements, which held that the Fund and Bank would guarantee permanent stabilization of currency values, rapid growth of American commerce, and even, according to one speech Morgenthau made in Detroit, a standing export market for 1 million American automobiles a year. That kind of perception of American advantage affected the image of the Bretton Woods pacts, which the Senate approved on July 19, 1945, by a vote of 61 to 16.

By that time there was no longer a majority in the Senate for ventures that seemed disadvantageous and expensive, especially appropriations for lend-lease. Congress had been tiring of that wartime program and was suspicious of the President's plans for it. Partly out of deference to congressional sentiment, the Treasury strove continually to limit lend-lease to the United Kingdom in order to keep British gold and dollar balances below a billion dollars. Over the sensible objections of the State Department, the President supported that objective, which allowed much too little for the postwar health of the British economy.

For his part, Secretary of State Hull had used lend-lease to further his obsession with free trade. That purpose allied him, inadvertently and uncomprehendingly, with much of American business, for if free trade were to prevail, with markets open to all sellers and raw materials available to all buyers, then American business, which survived the war intact, would reap great postwar gains in sales and profits. Further, many American businesses were zealous to break into channels of distribution and trade in the Middle East that Great Britain commanded. That issue continually intruded upon wartime discussions of lend-lease. The State Department made Article VII of the Lend-Lease Act of 1941 as binding as it could in each protocol it negotiated. That article, along with lend-lease agreements incorporating its terms, called for the postwar settlement of lend-lease obligations according to conditions that would promote trade, reduce tariffs, and eliminate "all forms of discriminatory treatment in international commerce." But for that purpose the British argued justifiably that they had to accumulate foreign exchange, particularly gold and dollars. Unrelenting on that issue, the Treasury tried to tie the volume and type of lend-lease to the United Kingdom to British relaxation of practices that hampered American trade in the Dominions and colonies. The British, successful in resisting that effort, antagonized Treasury representatives involved in the continual bargaining.

The approach of victory altered the context in which the Lend-Lease Act had originally been proposed. As the title of the legislation said, it was an act to defend America, and the assistance rendered to Great Britain and the Soviet Union did provide in return vital time and help for the United States. Further, many recipients, Great Britain and the Dominions especially, rendered significant amounts of reverse lend-lease in the forms, among others, of facilities, services, and victuals for American forces. Yet the State Department had obscured the basic purpose of the act by attaching to it conditions that related to postwar commerce. And for their part the British expected continuing lend-lease, a wartime aberration, to relieve their admittedly serious postwar economic problems which deserved attention but not as essentials for the defense of America. The United States had persuasive reasons, by no means wholly generous, for providing postwar economic assistance to its wartime allies, but many Americans were not convinced of that, still others wanted postwar economic policy subjected to congressional controls, and few considered lend-lease an appropriate vehicle for the rehabilitation of the world.

As early as 1943 the debate over lend-lease appropriations alerted the dubious to the possibility that the program might be "bent to purposes other than the war itself." As the *Wall Street Journal* then put it, "the Act is a war measure and should be strictly so construed. Whatever other assistance the United States may render other nations when the war is over should be given only under specific authorization of Congress."

Even the proponents of that kind of assistance were careful of congressional sensitivities. In *Fortune* in January 1945, Eugene V. Rostow, then Dean Acheson's assistant on lend-lease affairs, published a cogent plea for American aid for the reconstruction of European economies during the period of transition that would follow the end of the war. In the absence of American credits, for one example he cited, the British would have to confine much of their trade to the sterling bloc; for another, the Italians would be unable to replace the food and medicine they were receiving from the American military. Rostow did not propose to pamper any country, but, rather, to prevent food riots, pestilence, and autarky. The Lend-Lease Act, he suggested, "is the best guide to the future." It looked "to a liberal and multilateral . . . approach to the solution of world economic problems." He envisaged, as did so many others in the State Department, a world economy unfettered by restrictions, and he urged that the United States "provide emergency credits to tide over the acute period of readjustment immediately after the war." The resulting prosperity and the possibilities it would generate for trade would ultimately benefit the United States. In the precedent of lend-lease, but not as an extension of it, Rostow called specifically for "new and special loan agreements" that would require congressional approval.

Senator Taft opposed any such approval. He had led the opposition to the Lend-Lease bill in 1941, and in 1945 he continued to object to the program as a precedent. Taft feared that after victory, unless Congress interceded, the President would divert lend-lease goods and funds to purposes unrelated to the defense of the United States. In 1945 he therefore introduced an amendment to the bill extending lend-lease that prohibited executive action to sell for postwar relief, rehabilitation, or reconstruction abroad "defense articles" originally shipped under lend-lease contracts. Bipartisan support for the amendment produced a tie vote broken only by Vice President Truman's negative. Even so, Congress forbade the use of new funds for the purposes Taft had specified. That provision undercut Roosevelt's private intention temporarily to extend lend-lease to Great Britain after the end of the Pacific war.

While Congress, exultant over victory in Europe, indulged in an economy wave during May and June 1945, President Truman, then adjusting to his new office and still sensitive to the disposition of his former colleagues on the Hill, broke promises Roosevelt had made the previous January about the extent of lend-lease to the United Kingdom during the period between the fall of Germany and the surrender of Japan. Moreover, Congress in June 1945 voted lend-lease appropriations only after Truman renewed his assurances that the program would not be used directly or indirectly for postwar rehabilitation and that he would discontinue it within a month of the end of the Japanese war.

It was then supposed that the end of that war lay six to twelve months away; in fact, it was only six weeks away. Yet without regard for the consequences of precipitate action, Truman in August abruptly terminated lend-lease. The United States first canceled all sailings of ships bearing lend-lease cargoes to Britain. Then, before promised discussions with the British government could begin, Truman on August 21, without qualification, announced the end of lend-lease. That order of course applied also to the Soviet Union, to which Truman in May, without notification, had already cut off all shipments unrelated to the "direct military purposes" of the American war in the Pacific.

The British, who had been the major beneficiaries of lend-lease, recognized how crucial the program had been for them during the war years. Both the British press and the British government commented on the indispensability of the American contribution to the common war effort. Now Winston Churchill, to whom Roosevelt had given assurances on which Truman reneged, called the decision to terminate lend-lease "rough and harsh." Clement Attlee, who had just become Prime Minister, regretted the abandonment of lend-lease without consultation. His statement carried an overtone of anxiety about the degree to which the United States would go it alone in matters vital to the interests of Great Britain and other allies. For his part, Stalin, long since suspicious of American postwar intentions, interpreted the cessation of lend-lease as a deliberately hostile act, though it was at least equally a penurious one.

There were, as the pollsters disclosed, few Americans, in office or out, who believed that reconstruction was a legitimate part of the cost of war, few who believed that equality of sacrifice—so much discussed during the war—applied to the rebuilding of the societies the war had rent. Congress vigorously endorsed Truman's policy, a policy it had provoked him to follow in order to preserve American resources, not to penalize England or to threaten Russia. When Churchill made his critical remark, Senator Robert Wagner, neither an isolationist nor a conservative, said that he could not understand it. Sol Bloom, the internationalist Democratic chairman of the House Foreign Relations Committee, found the British response unreasonable. The British Embassy in Washington had its own, trenchant view. "The dollar sign," the Embassy cabled London, "is back in the Anglo-American equation."

The sign of the bald eagle had already appeared in American negotiations and approval of the charter of the United Nations. As at Bretton Woods, so at Dumbarton Oaks, where preliminary discussions of the charter took place in August 1944, and at San Francisco, where the charter was completed and signed in May and June 1945, no participating country neglected its self-interest. The resulting disagreements among major powers, especially between the United States and the Soviet Union, gave an adversary tone to the proceedings that much contemporary reporting stressed. There were differences that needed to be resolved, but the two mightiest powers had a disguised but basically common view of the key arrangements for the United Nations.

The debates about the UN charter revealed the determination of the partners of the Grand Alliance to avoid the establishment of a world government. The United States and the Soviet Union had no intention of subjecting national policy, especially in the event of some future dispute, to the collective control either of small countries or of other major powers. The major powers—the United States, the Soviet Union, the United Kingdom, China, and France—were to be permanent members of the Security Council, on which six other seats would rotate periodically among other member nations. That council had the authority to employ stand-by military forces, to which all members were to contribute, in order to prevent aggression. All member nations were to have places in the General Assembly. But as Roosevelt had said during a discussion about how many seats the Soviet Union should have in that body: "It is not really of any great importance. It is an investigatory body only." What was important was the voting procedure in the Security Council and the latitude of the council to use force. The Soviet Union insisted upon an absolute veto power for each member of the council. The British, preferring a procedure like that of the League of Nations, proposed that members should abstain from voting in cases in which they were involved. At first Roosevelt had sided with the British, but he soon yielded to Stalin's intransigency about the veto. For their part, the Russians later accepted the American demand that there should be no veto of the discussion of an issue. On the surface the two nations seemed to have traded concessions, but on two accounts Roosevelt and in the end Truman gave away nothing they could have won. In the future, if the UN was to proceed with force against any great power, it might thereby resist aggression, but it would in the process start a major war. In the event of hostility between major powers, the machinery of voting, whatever it was, as the New York *Times* observed, could not in itself "suffice to keep the peace." In the short term, moreover, the absence of a veto for the United States would probably have prevented Senate approval of the charter. The Republicans with few exceptions had consistently demanded the retention of national sovereignty, an objective to which many Democrats also subscribed, as did most of the American people. Indeed, the very men who criticized the veto, Robert Taft for one, were most adamant in opposing the creation of a superstate, and the defeats of Willkie and Wallace in 1944 had presaged the political necessity of the principle of the veto.

So did the Senate's position on the use of force. Expressing the opinion of a majority of his colleagues, Arthur Vandenberg consistently argued that the American delegate to the Security Council could not commit armed forces for a UN action without the explicit consent of Congress, for any such action would constitute an act of war which, under the Constitution, the Congress had to declare. The State Department appeased Vandenberg by suggesting that the Senate could approve the charter on its merits, and later, in a separate vote, decide whether Congress had thereafter to approve each use of force. That ambiguous formulation bothered Senator Joseph Ball. The employment of UN troops, including an American contingent, to stop aggression was not the equivalent of a declaration of war, Ball contended, but an effort to preserve the peace. To delay the use of force while Congress debated would vitiate the whole idea of collective security. "If we make that kind of crippling reservation," Ball said, "so will every other nation . . . and we will have no more certainty of international law enforcement than we had in the twenties and thirties." But the United States, like the Soviet Union, was not prepared to surrender its independence of action for the possibility of the kind of collective security that Ball envisaged.

Ball still stood for a degree of internationalism to which there was commanding bipartisan opposition. That sort of bipartisanship distinguished the American delegation to the San Francisco conference and the tone of the Senate's debate about the charter of the UN which the conference wrote. In the delegation and during the debate the dominating figures were Vandenberg and his fellow Senator Tom Connally, of Texas, the Democratic chairman of the Foreign Relations Committee. As vain as Vandenberg but less informed and intelligent, Connally construed statesmanship to consist of about one-third string tie, one-third volubility, and one-third party loyalty. He wore one, practiced the other two, and nurtured his ego while posing at rest or pontificating in motion. He and Vandenberg got along famously. Addressing the Senate during the debate on the charter, Connally referred to the "splendid . . . harmony within the United States delegation," and Vandenberg reported that at San Francisco there had been not the "faintest hint of partisanship." Connally plucked the strings of sentimentality. "Many representatives of foreign nations," he said, "are still doubtful as to what the vote on the Charter will be here in the Senate. They remember 1919. They know how the League of Nations was slaughtered here on the floor. Can you not still see the blood?" Neither he nor his colleagues wanted to miss their second chance. Vandenberg assured those colleagues that the charter posed no risks. The United Nations would provide "mechanisms for constant and friendly consultations" among countries, but the United States retained "every basic attribute of its sovereignty." More, "the use of force" was not consistent with "the genius of this great experiment." There was, Connally had asserted, practical unanimity among the American people in support of the United Nations. So, too, in the light of Vandenberg's incontrovertible analysis, was there practical unanimity in the Senate, which voted 89 to 2 approve the charter.

That unanimity rested on the blandness, the soft consensus, of senatorial internationalism. Senator William Fulbright, of Arkansas, already an incisive critic of American foreign policy, exposed the ambiguity inherent in the vote. "Practically no measure of real importance," he said, "has ever been accepted with such docility. . . . Can it be that the Senators do not recognize that, if we accepted the charter in good faith, it means a complete departure from our traditional policy in international relations?" Fulbright welcomed that kind of departure but identified the docility that belied it. Yet only he and Senator Claude Pepper questioned the comforting belief that somehow approval of the charter and participation in "constant and friendly consultations" would provide collective security. Fulbright observed that assertive national sovereignty had long impeded the resolution of international problems. It still did. He could at best only hope that the United Nations would offer a balance to national sovereignty "for the lack of which we and the world have narrowly missed domination by . . . ruthless forces."

That hope elicited the scorn of Robert Taft. With unabashed prescience, he saw the United Nations as useful merely for settling arguments between small states, as a kind of sheriff on the payroll of the major powers, a convenience for eliminating nuisances like border squabbles in Latin America. His affirmative vote for the charter contravened the meaning that Connally had attached to unanimity. Fulbright's dismay over the Senate's docility afforded the accurate interpretation of the intention of that body and of a considerable fraction, probably a considerable majority, of its constituents. Concerned as they were with their own national safety, persuaded as they were of their own national virtue and national strength, Americans, with exceptions like Ball and Fulbright, saw in the concept of collective security only the image of their own conceit.

They had no interest, nor did their government, in sharing their military power. During the war years, Americans drew back, as they always had before, from the prospect of committing troops for an extended period to the occupation of enemy territories. Sensitive as he was to public opinion, Franklin Roosevelt assumed that the pressure to bring the boys home would, within a few years after the end of the fighting, overcome any contrary policy, whatever the need. As he had expected, at the time of victory that pressure was strong both from GI's, who had had enough, and from their families. The premise of a quick American evacuation of Europe underlay Roosevelt's planning for the occupation and reconstruction of Germany. He did not contemplate a postwar rivalry with the Soviet Union of the kind or degree that materialized. No more than his countrymen did he imagine that large American forces and billions of American dollars would protect and rebuild a new German state. Rather, fearing German resurgence, he envisaged Germany, denazified after several years of occupation, stripped of its military powei and of its industrial capacity for war. Safe from Germany, postwar Europe, in that view, would have no need for the troops that the American people were reluctant to supply. The presumption that the Yanks were coming home to stay may even on occasion have moved Roosevelt out of necessity to regard the possibilities of postwar Russian friendship through deceptively rosy lenses.

Military planners in Washington were less optimistic. Before the end of 1945 they were beginning to redefine national defense policy according to the modified prewar assumption of an impregnable "fortress America." Though the submarine warfare in the Atlantic and the attack on Pearl Harbor in the Pacific had demonstrated the inadequacy of the oceans as a basis for security (in Lippmann's phrase, of "effortless security"), the development of long-range strategic aircraft and of the atomic bomb seemed at the time to promise an alternative barrier. That was especially the doctrine of Air Force planners, who wanted in any event to contrive a scheme of national security that would command an independent status and unprecedented peacetime appropriations for their branch of the service. They identified the Soviet Union as the likely threat to the national future, and they proposed to counter that threat with a circle of American bases for strategic bombers armed with atomic weapons. In less than a year after Hiroshima the pattern of American occupation of territory formerly under Japanese control and of American development of military outposts formed that strategic circle. From its perimeter the United States, however benign its intentions, had, as other nations saw it, an unequaled capacity not only for counterattack but also for unilateral adventures that the United Nations could not possibly prevent.

But the building of a fortress America had begun during the war, as American policy toward the invention and use of weapons revealed. The admitted suspicions of American officials toward their allies continually reflected the common assumption that only the United States was sufficiently innocent safely to command the whole arsenal of new devices for warmaking and destruction. One relatively minor example of American reservations about the reliability of the Soviet Union involved the electric enciphering machine, the ECM, that scrambled secret messages beyond the ability, in the existing state of technology, of the uninitiated to compromise them. The United States shared the secret of the ECM with the British but not with the Russians, and even the British had no access to various interchangeable units of the machine that were reserved for American communications only.

So, too, at its most ominous level of technology, American wartime policy attempted to exclude the Soviet Union from knowledge about the development of the atomic bomb, and tried also to include the United Kingdom and Canada only partially. Even before Pearl Harbor the Americans and the British had begun to exchange information crucial to the continuing refinement of such innovative equipment as radar, underwater search systems to locate and attack submarines, and fuses to detonate shells and mines without contact with the object to be demolished. As late as February 1943, Churchill expected a full exchange of information about the atomic bomb, then still many months from production. He proposed to Roosevelt that the two nations pool their data, work together on equal terms, and share the results. To that partnership he seemed to believe the President agreed. Churchill erred. On several counts the United States decided seriously to restrict information about the highly classified technology of the bomb. Partly it was a matter of both pride and pique. As Secretary of War Stimson told the President late in 1943, the United States was doing 90 per cent of the work on the manufacture of the bomb and had only negligible need for further British data. The British, who lacked the wealth to produce a bomb on their own, were still dragging their feet about a cross-channel invasion, to Stimson's resentment. More important, the President's scientific advisers had a limiting view about the exchange of technical information. It was justified, they believed, only when it advanced the war effort. But for that purpose they saw no need to enlighten the British about processes for the production of fissionable material that were wholly American in origin and development.

Those two advisers, Vannevar Bush and James B. Conant, were not narrow, secretive, or inflexible men, or superpatriots, Anglophobes, or philistines. Bush, a lean, shrewd Yankee scientist in his middle age, an inventive mathematician and electrical engineer, had been vice-president of the Massachusetts Institute of Technology and then president of the Carnegie Institution before he drew up the plan for the National Defense Research Council, the agency that provided the central liaison between science and the military during the war. In 1941, again on Bush's recommendation, Roosevelt created the Office of Scientific Research and Development, a center to mobilize the country's scientific resources for defense and later war work. Bush took over the new office and Conant replaced him as head of NDRC, which continued to have a large voice in research and development, including the program to build the bomb. Conant, the president of Harvard University, was an eminent chemist, an experienced organizer of science and education, a convinced democrat who had tried at Harvard to replace an aristocracy of family with one of talent, an ardent interventionist, and a wry companion on friendly fishing expeditions. Spare of frame, impatient with ceremony, he was bright, tough, and energetic, and he was determined, as was Bush, that the United States would beat the Germans in the race for the bomb. As time passed and the race began to go their way, Bush and Conant and their immediate associates developed a natural proprietary interest in their unprecedented project. They also became increasingly involved in the prototypical defense establishment of the war years, the unofficial company, brought together by common interests in a common cause, of senior men from the War Department, the armed services, the war industries, and the scientific community. It was therefore easy for Bush and Conant to conclude, when in 1943 the British requested a free exchange of information, that secret data, like the construction and operation of facilities for its use, should remain accessible exclusively to Americans. It was easy for the scientists to believe that theirs had become the only safe hands.

They had other reasons for arriving at the same conclusion about secrecy. Though neither Great Britain nor Canada intended to manufacture fissionable materials during the war, both wanted American information in order to use it for postwar construction of new sources of energy. With characteristic Yankee doubts about English motives, doubts that most Americans probably would have harbored had they known about the issue, Bush and Conant suspected that the British were interested primarily in postwar commercial enterprises. Those suspicions drew apparent confirmation from the association of William A. Akers, the engineer in charge of the British uranium programs, with Imperial Chemical Industries, the great international corporation. As it happened, Great Britain and Canada lack the resources seriously to challenge the United States in any postwar effort to apply atomic energy to industrial use, but that circumstance abated none of the wartime suspicions.

Even more important, Bush and Conant worried about security. They considered it prudent to confine secret technological information strictly to those who needed to know in order to advance the program. The Anglo-Russian agreement of September 1942 for the exchange of new weapons gravely disturbed the scientists, and Stimson, too. Worse, the prospect of any breach of secrecy that might lead to Soviet knowledge about American developments drove General Leslie R. Groves into paroxysms of panic. Groves, the chief administrator of the bomb project, was a xenophobic superpatriot. He brought unusual administrative force to his difficult task, but he had little if any understanding of the ways of science or the life of the mind. While Bush and Conant knew that basic knowledge about atomic science could never be the special province of any nation, although American technology was a different matter, Groves acted as if the entire bomb program existed within a little black box whose magic somehow belonged only to the United States. The worries of Bush and Conant about security played to Groves's fetish. Their recommendation, which the President accepted, permitted only a restricted exchange of information with the British, primarily about scientific research and only negligibly about development and production—the keys to making the weapon. The British accepted those terms because they could obtain none better. As it was, in return for what they received they had to agree with the United States that neither nation would ever use the weapon against the other, or communicate about it with, or employ it against, any third party except after mutual consent.

The American mood that produced that decision anticipated

postwar policy. Neither the President, the Congress, nor the American people were properly prepared to confront the momentous question of how to control the new instrument of terror. During the war, Roosevelt and his advisers concentrated on making the bomb as rapidly as possible in order to beat the Germans to that goal. A failure to do so could have been a disaster. But preoccupied as they were, they gave little consideration to what to do with the bomb once they had it, or to how they might prevent its proliferation and possible use in the future.

Henry Stimson urged the employment of the bomb against Japan in the genuine belief that it would obviate the need for a costly invasion, hasten victory, and thereby save both American and Japanese lives. The vindictiveness generated by a long and ghastly war contributed to that plausible conclusion, which Truman reached partly for the very reasons Stimson advanced. A more dispassionate assessment of Japan's military weakness and political condition would have recommended an alternative course, either the negotiation of an armistice on terms very like those ultimately obtained, terms the Japanese were already contemplating, or at the most a demonstration of the bomb to warn the Japanese of its destructiveness. Most of the leading scientists who had engaged in the construction of the weapon asked the President to pursue one or the other of those options. But even if their counsel had prevailed, the bomb would still have been there, a dark shadow over the postwar world. Further, there were among the President's advisers some, like Groves and Secretary of State James Byrnes, who welcomed the use of the bomb as a warning to the Soviet Union that the United States held a superior counter for postwar bargaining. Whatever the weight of their influence on the President, Truman's decision to order the drops, along with the antecedent policy of secrecy, whetted Soviet suspicions about American intentions, and would continue to even after the Russians were able to produce a bomb of their own.

That was bound to be only a matter of time, because Soviet science was sophisticated, the Soviet Union had the resources to make the weapon, the American success proved it could be done, and in those circumstances the relevant technology could not long remain a mystery. As it happened, moreover, Soviet espionage had punctured the veil of secrecy, a penetration that accelerated the Russian program. Still, in the immediate postwar period, the preservation of a monopoly over the bomb motivated American policy, domestic and international. The motives remained mixed, functions partly of an

ingenuous conviction in American benignity, partly of a growing anxiety about the Soviet Union, partly of a naïve quest for military
impregnability.

The military obtained some of what they wanted in the legislation governing the control of atomic energy within the United States.
The measure first drafted for that purpose and introduced in the
House in October 1945, the May-Johnson bill, established clear mili-
tary control. It had the support of the War Department, Groves,
Bush, and Conant. Opponents of that measure, particularly officers in
the Bureau of the Budget and liberal Democrats on the Hill, persuaded President Truman in June 1946 to use his influence for the
McMahon bill, an alternative proposal to create a civilian agency, the
Atomic Energy Commission, with clear control over atomic policy.
But Senator Vandenberg, who had worked with Senator Brien Mc-
Mahon in the drafting of the bill, mobilized his usual following behind a disabling amendment. McMahon's draft placed upon a
military board on atomic energy the responsibility for consulting with
the civilian commission. Vandenberg's amendment directed the commission to consult with the military board and gave the board a right
to appeal to the President about questions it considered relevant to
national defense. "To call this civilian control," the official historians
of the Atomic Energy Commission later judged, "seemed a
mockery."

So it did at the time to McMahon; to Henry Wallace, then still
Secretary of Commerce, who warned that the Vandenberg amendment might deliver the United States into the hands of "military
fascism"; and to both the Federation of Atomic Scientists and the
Committee on Atomic Information, which included most of the best-
informed scientists in the country. But the administration saw considerable merit in the Vandenberg amendment, as did a clear majority
of the Senate, and, according to the polls, a large majority of the
American people. The compromise that evolved in the face of those
sentiments gave the President rather than the commission control
over the stockpile of atomic weapons; made the commission responsible for keeping the military board informed; and routed appeals by
the military against policies of the commission through the service
secretaries, who could then decide whether or not to involve the
President. Those changes, large and small, put the military back into
the center of atomic policy, where they remained after the amended
bill was enacted in July 1946.

The military spokesmen in Truman's Cabinet—Secretary of the

Navy James Forrestal and Secretary of War Robert Patterson, who succeeded Stimson in 1946—opposed exchanging even basic scientific information with other nations, especially the Soviet Union, though Stimson had favored that kind of exchange, as did Wallace. Furthermore, the American proposal for international control of atomic energy, the plan that Bernard Baruch took to the United Nations, contained what Henry Wallace properly labeled a "fatal defect." The United States would release scientific information about atomic energy in a series of stages provided that on matters of atomic energy there would be no veto. A new international atomic authority of the UN would control, inspect, and license all uses of atomic energy and, in the final stage, outlaw the bomb, which until then the United States would retain. For their part, other nations, the Soviet Union included, had to make binding commitments not to conduct research on military uses of atomic energy, and to disclose their uranium sources. The United States would enjoy the right to withhold technical knowledge about atomic energy until a system of international inspection and control was working to American satisfaction. That was the "fatal defect," for as Wallace said, the United States was telling the Russians that if they were "good boys," according to unspecified standards Americans would set, they could join the atomic club after agreements by "easy stages" had prolonged the American monopoly. The Soviet Union could either accept a second-class status or reject the Baruch plan and proceed to develop its own bomb. The choice of the latter course, as Wallace concluded, was just the choice the United States would have made had the positions of the two nations been reversed.

The American military could enjoy their monopoly only until the Russians succeeded, a not distant eventuality. Accordingly, a few paranoid officers yearned for a pre-emptive strike against the Soviet Union. The prevailing strategy relied instead on that combination of bases, weapons, and technological expertness which could in theory balance any Soviet threat. Again Wallace saw the picture clearly. As he wrote Truman in July 1946:

> How do American actions since V-J day appear to other nations? I mean by actions the concrete things like $13 billion for the War and Navy Departments, the . . . tests of the atomic bomb and continued production of the bomb . . . production of B-29s and planned production of B-36s [super bombers], and the effort to secure air bases spread over half the globe from which the other half . . . can be bombed. . . . These facts rather make it appear either (1) that we

are preparing ourselves to win the war which we regard as inevitable or (2) that we are trying to build up a predominance of force to intimidate the rest of mankind.

So it did appear, though the basic intent of the federal government, in consonance with the aim of the American people, was to build up a predominance of force that would preclude another Pearl Harbor, that would assure the safety of the United States. In the emerging new world of atomic weapons, supersonic airplanes, and rockets, there would be no such assurance, but in the early postwar months Americans expected it, even considered it a just reward for the victory they had won. They were no more prepared to share their existing and potential military might than to share their unparalleled wealth or to diminish their inviolable sovereignty.

Yet Americans were not peculiar or malign. No other major power in 1945–46 had a broader, more generous foreign policy. No other people was less concerned with national safety or with building a new world first at home. The limits of American internationalism were neither historically accidental, for they emerged from the culture of the remote and the immediate past, nor comparatively selfish, for nationalism flourished in all the countries, existing or emerging, of the world. But limits there distinctly were, limits that national self-satisfaction and self-righteousness tended to obscure. They were the very limits that had guided American policy during the years of war and fixed American purpose on victory. Now that the war was over, they fixed American purpose on national security, a disguise for invincible national advantage.

3. Limits of American Liberalism

Conscious postwar policies to guarantee national welfare and prosperity, the general policies for which Walter Lippmann had called, had their particular origins in the domestic politics of the war years. In 1944 the voters had endorsed the composition of the Congress that had sat since 1942. Just as that Congress had abolished the National Resources Planning Board because of its liberal heresies, so did the new Congress turn away from the agenda for reform that American liberals sponsored in 1945–46. In his State of the Union message of January 1946, President Truman listed most of that agenda, but almost no one took him seriously. He seemed only to be going through the familiar routine of a Democratic chief executive, though less convincingly than had Roosevelt, and in the ensuing months he did not aggressively pursue the goals he had listed.

Those goals, the liberal inheritance of the war years, consisted in part of Keynesian economic doctrine and in part of compatible definitions of national purpose that emerged from the New Deal. In its wartime reports the NRPB had spelled out the particulars of that inheritance, as Roosevelt had in his 1944 campaign speeches. Those statements proposed postwar programs to initiate national health insurance, to expand social security and federal aid for housing and education, to develop the Arkansas and Columbia river valleys, to construct national electrical and transportation networks, and to reconstruct the American cities. The NRPB reports also designated the special beneficiaries of those programs as "the degraded and impoverished . . . the disinherited and despised." In his subdued candidacy for the Vice Presidency, Henry Wallace had stressed those themes and urged also equal rights for blacks and women. Yet had he inherited the Presidency, Wallace later admitted, Congress would have rebuffed him and the American people would have given him inadequate support.

Nevertheless, the agenda for reform remained alive in the hearts and minds of a sprightly minority—the liberals still in Washington or at the crusading journals or eastern universities, mostly veterans of New Deal or war agencies, many of them young economists filled with confidence in their theories and techniques. They were bright, proud, prolific, and articulate men, at once respectful and gently jealous of each other, disinclined to accept any individual as their spokesman. Not in that role, but as an expositor of a comprehensive statement of their essentially common program, Chester Bowles wrote *Tomorrow Without Fear,* a short illustrated tract, begun in 1945 and published in 1946.

The last of the wartime heads of the Office of Price Administration, Bowles had come to government from a successful career in advertising that he thereafter preferred not to discuss or perhaps even to recall. His administrative responsibilities for price control involved him with dozens of economists, many of them on his staff, which included some of the ablest in the country. So, too, the work of his office touched the daily lives of millions of Americans—businessmen, farmers, and wage earners—of whom thousands addressed questions to the OPA about rent control or price control or rationing, and as the war ended, about the outlook for prices, profits, taxes, and prosperity. In *Tomorrow Without Fear* Bowles attempted to answer those questions in language any layman could understand, within the context of the new economics, and in the tradition of liberal reform. While he wrote, he subdued his impatience with Truman's hesitations. Bowles, who had political amibitons of his own, intended not to criticize or offend men in office, but to persuade and inspire them. The experts whom he consulted directly or indirectly about his project applauded the result. His was, said John Kenneth Galbraith, then a young editor of *Fortune* and a former OPA officer, "an important and lively book." The dean of American Keynesians, Alvin H. Hansen, called it "a grand book." Others of similar mind found it "informed and enlightening," "a remarkable job." It was an outstanding primer of postwar liberal intentions.

Tomorrow Without Fear began with an explicit statement of the general purpose Bowles attributed to the people of the world, particularly to Americans: "Economic security, based on abundant production, fully shared, is our goal." He proceeded to take inventory of the United States in 1940, a time at which the country had "failed, and by a wide margin, to produce . . . the high standard of living for all . . . of which we were so clearly capable." Three of four Americans then needed a better balance in their diets, and throughout the country there existed a "great lack of dentists, doctors, nurses" and hospitals. Millions of Americans lived in overcrowded dwellings, and the majority of farmers "in dilapidated houses" without inside toilets, baths, or electricity. One of four children was not attending school. The "gnawing fear of tomorrow" gripped farmers, factory workers— of whom 8 million had no jobs—and stenographers, teachers, and clerks, especially women, who were "notably underpaid." Between 1900 and 1940, of 16 million small enterprises launched, only 2 million had survived, and their proprietors knew "hard work, long hours, sleepless nights." Corporations in 1940 earned less profit in the aggregate than they had earned in either 1929 or 1919. Though the United States was the richest nation in the world, there was then neither abundance, security, nor equity in American life.

In 1940, as Bowles knew, many eminent economists had concluded that stagnation had become a permanent condition of the national economy. As Bowles put it, the "baffling paradox" of that time had been "the inability of people on every hand to find markets for the goods that people on every hand so badly needed." That paradox provoked the theory of a "mature economy" that would no longer grow. But "in the war years . . . we again found our strength and confidence." By the end of 1943 industrial output had doubled that of the spring of 1940. "The more we produced for war, the more we had left for the rest of us," and with less than half of total output on V-E Day devoted to war purposes, "the standard of living of the *average* American family actually increased." The turnabout occurred because of "the government spending . . . made necessary by war" and the further impact of the billions of dollars that moved from industrial payrolls to the pockets of workers, who then spent their money to the degree that rationing and shortages permitted.

For Bowles the lesson was plain and the future bright: "There is a lot of government spending that is necessary in peacetime in order to get the educational and health services, the dam and power developments, the highways, schools, hospitals, and recreation facilities, that our government alone can provide." At times federal spending for national purposes would have to be increased just "to keep our economy running full blast." To avoid future depressions, the federal government had to "pledge itself to expand or contract its spending to whatever extent is necessary to balance any temporary slowing down or speeding up of business spending." The guarantee of that policy would lay the foundation for "full production and full employment." With them would come ample tax revenues within a balanced budget to finance social improvements, and there would be "above all else, good jobs for all our workers, good incomes for our farmers, and fair profits for our businessmen."

In Bowles's prediction, the government would not often have to spend beyond its income to stimulate the economy. Rather, in the spirit of the advertising industry he had left, he counted on "the increase of consumer spending on the things that consumers want and need" to provide "the basic and lasting solution of our economic future." To effect that solution, there would have to be a sufficient redistribution of income to permit the bottom third of the nation's families to earn and spend at several times their prewar capability. That redistribution would depend upon increases in minimum wages and extensions of social security. Further to stimulate consumer purchasing and business investment, Bowles recommended reductions in taxes on low and middle incomes, and on businesses taking risks by building new facilities for production. Taken together, those policies would assure prosperity. In the new elysium, Bowles continued, "we must all learn to live constantly better, a lot better. Our standard of living must rise steadily year by year to match the increase in our productive capacity." By the end of the 1960's, that process would "mean unlimited opportunity for health, recreation and good living . . . an end to poverty and insecurity."

There would then be, Bowles concluded, "not less for anybody, but *more* for everybody," enough good food, clothing, housing, provisions for good health, protection for the aged, and a college education for all who wanted one. "We Americans," he wrote, "are not going to accept another Great Depression as a substitute for prosperity and abundance." In an economy of full employment, he believed, "in such a nation, the tensions and conflicts, the inhumanity of man tᴏ man, which have poisoned our lives in the past, will disappear."

Even as Bowles described it in his utopian enthusiasm, the liberal agenda had surrendered much of the pugnacity of its recent past. In the hope for eternal prosperity, built on eternal consumer spending, the adversary toughness of the 1930's disappeared. It had begun to do so during the war. The distrust of big business that had so often marked American liberalism hardly survived the wartime accommodation between the service departments and the defense industries. During the New Deal and the war, the Treasury had argued continually for soaking the rich to help the poor, and for regulatory taxation to limit windfall profits and penalize monopolistic practices. After the defeat of its tax program in 1944, the department lost heart for further battle and looked instead for tax incentives for corporate investment. Bowles did not reverse the habitual antagonism of American liberalism toward big business; he merely accepted and recorded the change.

American corporate business accepted the resulting concessions as a sufficient capitulation. Leading business executives had also, in large numbers, subscribed to the gospel according to Keynes. There remained, especially in the National Association of Manufacturers, the United States Chamber of Commerce, and the Farm Bureau Federation, as well as among their allies in Congress, many ritualists who were still persuaded that federal planning and spending would pave the way to tyranny and bankruptcy. But during the war large corporations, with exceptions, to be sure, had come to look upon the federal government as a necessary partner in the preservation of prosperity. That was the view of the National Planning Association, whose membership included concerned businessmen, lawyers, journalists, and academicians. It was the opinion of those businessmen and economists who founded the Committee for Economic Development, an organization with continuing influence upon congressmen in both parties whose sympathies lay close to Bowles's. Just as Bowles had presented the program of liberal Keynesians, so did the CED represent the conservative Keynesians. The two groups disagreed about various elements in a proper economic policy but not about the intent of that policy itself. Beardsley Ruml, a force within the CED, remained, as he had been during the war, a familiar figure in liberal circles and a persuasive advocate of rapprochement between Washington and Wall Street.

Yet he and the CED did significantly shift the emphasis of Bowles's arguments toward the prejudices of the ritualists. In spite of his consumerism, Bowles had stressed government spending as essential to the maintenance of full employment—spending for social reform and for social improvements. So had Alvin Hansen in his NRPB reports and Paul Samuelson, then at the start of his brilliant career as an economist. "It is nothing less than pitiful," Samuelson had said, "to hear speakers, who admittedly reckon the backlog of deferred demand at $5 billion for the first six months after the war, claim . . . that this will take the place of a $35 billion deficit." Annual federal postwar spending, he estimated, would have to reach at least $7.5 billion and probably $25 billion. The CED did not necessarily reject those conclusions, but its program focused on tax reduction, not on spending. That shift gave to social considerations a secondary priority. It provided first for corporate investment and profits, and for private spending on goods and services that had at best an incidental social value.

The CED based its case on its reading of the experience of the war years. In those years government economists had constantly examined the problem of inflation that grew as aggregate personal income increasingly exceeded available consumer goods and services. The difference between the demand and the supply, "the inflationary gap," threatened to push prices through the levels set by OPA controls. The Treasury had regularly recommended closing the gap by increasing taxes, particularly on comfortable incomes and corporate profits, while other agencies advocated forced savings. Congress had raised taxes enough to buttress the system of controls, but the continuing debate about inflation underscored a fiscal lesson. With the economy at full employment, a sufficient increase in taxes rather than a decrease in government spending could contain inflation. The needs of the war precluded the decrease, but the end of the war permitted just that. As Keynesian theorists had known but New Dealers had rarely advertised, in the event of a decline in economic activity, a positive countercyclical fiscal policy could reduce taxes instead of raising spending. The stimulation of an economy operating below full employment depended on the resort to a deficit, however obtained.

Bowles had said as much but argued that constant changes in tax schedules would create intolerable business uncertainty. The CED went a step further. It proposed a substantial reduction in taxes instead of a program of federal spending, a reduction to a steady level on which business could rely in making expansionary plans. Bowles had noted that the yield from any schedule of taxes would rise as the economy approached full employment and full production. The resulting increase in revenue would permit the retirement of debts previously incurred and the funding of socially desirable programs. The CED, while accepting those points, stressed the utility of full-employment tax yields for balancing the budget, not annually but over the years of a business cycle. That prospect mollified even Robert Taft. In contrast to Bowles, then, the CED looked to corporate spending, encouraged by tax incentives, as the senior partner to government fiscal policy, whatever the social implications, in assuring the new prosperity.

The two schools within the university of the new economics had only small disagreements about theory. Both advocated a positive fiscal policy. Both, though in different measure, expected consumer purchasing to lift the economy, a prospect that delighted the American people who were so impatient to spend their way to comfort. Within the liberal camp there was an emphasis on the instrument of federal spending to complete the New Deal's agenda of reform. Still, there remained little of the New Deal's hostility to business. Within the conservative camp there was a stress on the instrument of reduced taxes to benefit business and increase disposable consumer income. Still, there remained little of the hatred of government of the 1930's, except of course among ritualists who rejected both Keynesian schools. But as Congress debated national economic policy, the conservatives and the ritualists, often in collaboration, left the liberals with only the minority dissent that had been their lot during the war years.

That weighting accounted for the crippling of the original Full Employment bill, introduced in January 1945 by Senator James Murray, of Montana, and his co-sponsors, all Democrats. The bill set out to establish in principle the obligation of the federal government to provide opportunities for employment for all those "able to work and seeking work." The President, on the basis of a regular analysis of the economy, was to devise and send to Congress a program to meet whatever needs were identified. When necessary, Congress would take steps to sustain employment, ultimately by federal deficit spending. Conservative Keynesians proceeded significantly to modify the measure. As the deletion of the word "full" from the title of the bill implied, they substituted high employment as the stipulated goal. The text of the amended measure referred to "maximum" employment, which could be interpreted to allow for 5 or 6 per cent unemployment, with resulting percentages for blacks and other minorities of several multiples of those figures. The change, as one critic put it, relieved the federal government of the responsibility which the sponsors had intended it to discharge: "to see that everybody . . . who desires to work has an opportunity to work."

Other amendments reflected the influence of the ritualists on the right, mostly Midwestern Republicans and Southern Democrats. Often in company with conservative Keynesians, they had particular success in the House of Representatives. Their amendments diluted the government's responsibility even for maximum employment. As described in the altered preface to the bill, that responsibility was confined to the use of practical means consistent with the "needs and obligations and other essential considerations of national policy"—a circumscription that encouraged the subordination of employment to other objectives. Another change in vocabulary specified "useful employment"—a euphemism that pleased conservatives determined to prevent a recurrence of New Deal work relief, which they despised. Moreover, the federal government was to exercise its obligations "with the assistance and cooperation of industry, agriculture, labor, and State and local governments"—a phrase that let Washington escape responsibility by seeming to share or to delegate it. Additionally, federal policy was to proceed in a manner calculated "to foster and promote free private enterprise" and maximum "production and consumption." That incantation disarmed the ritualists, satisfied the conversative Keynesians, and even suited the liberals except for the remnant still suspicious of business.

The conservative Keynesians won their major point by revising the vocabulary of the measure so that it called upon the government to use "all its plans, functions and resources" to foster maximum employment. There remained no direct reference to deficit spending. Though there was also no reference to tax reduction, the balance of power in Congress made that alternative the predictable preference on the Hill for at least the immediate future. The revised language also made more room for monetary policy as a device to temper the business cycle, an option that had fallen temporarily out of fashion in 1946. When its season again arrived, it provided, like decreased taxes, still another way in which opponents of federal spending for social purposes could rationalize their parsimony.

Even in its crippled form, the Employment Act of 1946 was hailed by the liberal Keynesians as a major step toward the elimination of depressions. That assessment blinked away the ambiguities inherent in the legislation, but liberals had little else to cheer about. They had spent their limited influence with Truman in getting him to use his declining prestige in behalf of the act. They had been unable to stop Congress from deferring action on an increase in minimum wages and an extension of social security. The rest of the agenda Bowles had described appeared less likely of adoption in 1946 than even in 1942 or 1944. So the liberals took their solace where they found it.

They had put too much hope in the Council of Economic Advisers, a new agency established by the Employment Act, which was supposed to provide professional, neutral, expert analyses of economic problems and their remedies. The council, to be appointed by the President and responsible to him, promised in the optimistic expectations of the New Dealers to constitute a continuing Brains Trust on economic policy. Roosevelt, their model President, had, after all, frequently consulted economists, ordinarily liberal economists, whom conservatives had constantly derided. But no more than his personal style did Roosevelt's method of governing set a precedent for Truman, much less for his successors. Though the existence of the Council of Economic Advisers guaranteed that professional advice would be available to Presidents who wanted it, it did not guarantee that any President would take that advice, or that Congress would agree with him if he did. Furthermore, the President had the authority to select candidates for the council whose biases about economic policy fit his own. Those men might be neither neutral politically nor expert professionally. Whatever their talent or bias, the President could defer instead to the recommendations of the Secretary of the Treasury, admittedly a political appointee, or the Federal Reserve Board, as ever a body that reflected in considerable measure the judgments of private bankers.

The Employment Act of 1946 acknowledged the concern of the Congress and of the American people with prosperity, their acceptance of the importance of the role of the federal government in national economic life, and their recognition of the utility of expert participation in the definition of economic policy. But the act guaranteed none of those conditions. It did not assert, and still less did it assure, a right to work. It did not even address itself to the question of "freedom from want." It advanced none of the objectives on the long agenda of liberal reform. No triumph of the liberals, who lacked the votes for triumphs, the act at most provided a working statement of the American consensus at the end of the war, a legislative proclamation of victory over that foremost public enemy, depression at home.

The great majority of Americans, embarking on the spending spree they had had to postpone, knew little about the Employment Act, though they had perhaps a vague sense of the heartening implications of what it proclaimed. For years the country had cared less about the tasks of social reform than about the chances for better living that the advertisements, including those for the Employment Act, had described. Now there were enough jobs, enough possibilities, enough income to make the accoutrements of contentment seem within reach. Weary of the memory of depression and of the demands of a foreign war, rather bored with politics, international or domestic, the American people accepted victory and prosperity as a sufficient achievement of the safety and security their hearts desired.

Epilogue: The Veterans

During each month from October 1945 through February 1946, three-quarters of a million persons were separated from the armed forces. By . . . June 1946, demobilization had been virtually completed. American . . . forces had returned about 12,807,000 veterans of World War II to civil life. . . . —*Annual Report*, 1946.

They were not much different from Americans who had always been civilians. They came in the same sizes and colors, and from the same kinds of backgrounds. They faced the same kinds of obstacles and shared many of the same kinds of opportunities. They had the same kinds of strengths and weaknesses, hopes and fears. But the veterans had had a different experience during the previous few years, and though the GI Bill offered them special help in their personal reconversions, they did not always change instantly or easily from fatigues to mufti.

The lieutenant reached home four months after the President died. He was dazzled by the gold braid at headquarters in New York, where they sent him to be mustered out. Except for her, he told his wife after they had spent a week alone together, what he most wanted was a new suit, any color but khaki or blue. So they went to Brooks Brothers. He had never been there before. He had never bought a suit that cost as much as those on the tables to which the salesman led them. He picked out a gray suit, single-breasted, with a waistcoat. His wife liked it. A week later, after it had been altered, he put it on as they set out to a dinner party. Catching his image in the mirror, he thought he looked odd. Most of the men his age on the bus they rode to the party were still in uniform. When he saw them, he knew he felt peculiar. The next day he packed his new suit and wore his uniform on the train to his wife's family home.

He would wear his khakis through three years of law school, until they were shreds. He was still not wholly comfortable in his three-piece gray suit when he went for his interview with the law firm he joined. But by that time he had stopped calling the floor the deck. In some ways, that had proved harder to learn than the law itself.

Many states had enacted legislation designed to cooperate with the Federal plan to aid veterans in acquiring homes, farms, and businesses. . . . —*Annual Report*, 1946.

The number of applications for guaranty or insurance of loans . . . increased sharply during . . . 1947. . . . A cumulative total of 823,548 loans had been reported closed, with a principal amount of $4,458,034,333. . . . Probably the most significant development . . . was the emergence of the VA loan program as a major factor in home financing activity. Approximately 93 per cent of the principal amount of . . . loans . . . was advanced to World War II veterans for home construction or purchase. . . . —*Annual Report*, 1947

"What about the *non*veteran," asked an old Irish-American politician sitting in the front row at the rally. "Yes, sir, the *non*veteran, too," the candidate replied.

Those two were talking about low-cost urban housing, but there was a commensurate demand for middle-class housing, particularly in the new suburbs then still in the planning stage. In California the booming suburban population included returning veterans, war workers and executives and their families who had gone west with industry, and just folks who wanted to live close to the sun.

Those prospects excited the major who left the Air Corps with ribbons on his chest and oak leaves on his shoulders. Those decorations balanced the absence of wings. For four years the major had interpreted the reports the pilots brought back from their missions. A meticulous intelligence officer, he was also a gregarious man. Pilots liked having a drink with him even though they knew he had received his first commission before taking any training.

The major had an eye for the main chance. He had barely kissed his wife at the airport before he told her that he was resigning from his father's insurance company. Real estate and development, he said, there lay their energetic future. He knew where to find attractive land to buy, how to negotiate a VA loan to set himself up in business, and how to instruct prospective purchasers to get the loans they needed.

Years later, when he gave his college $100,000, he was invited to select the costume for his classmates at their twenty-fifth reunion. He chose a replica of an Air Corps cap. After the reunion, he threw his cap away. After taxes, he scarcely missed his gift. He had already built more houses in more places than his wife could recall.

"Little boxes on the hillside, / Little boxes made of ticky tacky, / Little boxes on the hillside, Little boxes all the same; / There's a green one and a pink one / and a blue one and a yellow one / And they're all made out of ticky tacky and they all look just the same."

At the opening of fiscal year 1946, the Veterans' Administration had 2,300 doctors. . . . At the close of the year, more than 4,000. . . . In addition, almost 1,200 . . . physicians were assigned at VA hospitals.

The total number of patients . . . under hospital treatment at the end of the fiscal year was . . . World War II. 37,360. . . .

VA out-patient medical services examined 1,122,171 individuals.

VA social workers provided counsel . . . to more than 144,000 new cases during . . . 1946 . . . almost double the number . . . during 1944. . . . —*Annual Report*, 1946

Her husband, the lady said, had been in the hospital for thirteen months, and when he got out, he would walk with a limp. She didn't care—he was alive and that's what mattered. But he cared more than he admitted. He hated the artificial leg. It was clumsy and uncomfortable. It bothered him even when he drove the specially equipped car the VA had helped him finance. Sometimes after he left the hospital he drank a little too much before dinner, and sometimes he snapped at his wife and the children. But he had sometimes done those things before the war, his wife remembered, and she was glad he was alive. So was he when he discovered to his surprise that, with practice, he could even play golf again.

Grouped according to major type of disability . . . 57.6 per cent for neuropsychiatric conditions. . . . —*Annual Report*, 1946

The personal adjustment counseling program inaugurated in 1946 was under way by the end of the fiscal year 1947. This program is designed to assist veterans who are not well adjusted emotionally. . . . —*Annual Report*, 1947

He recognized Omaha. It looked much as it had always looked, not in the least like France. But he kept remembering France. One Sunday, at dinner with his girl and her parents, a Sunday only ten days before the wedding they were planning, an airplane flew over Omaha. The girl and her parents saw him dive under the table, curl up there, his eyes closed. For several months he dived under tables or ducked into doorways when airplanes flew over Omaha. For several months the Veterans Administration provided weekly psychiatric counseling. When he stopped diving he still remembered France. He talked about it now and then while he weighed packages at the Post Office where veterans' preference had assured him of a lifetime civil-service job.

> The number of new applications for training received during the fiscal year totaled 3,203,056. . . . The grand total received since the inception of the training program reached almost 6.6 million. . . . This is equivalent to about 46 per cent of the net veteran population of World War II. . . . —*Annual Report,* 1947

Many of the courses the VA supported involved on-the-job training. Many helped veterans convert skills acquired in service into trades useful in peacetime. The courses helped to instruct thousands of automobile mechanics, radio repair men, practical nurses, machine-tool operators, bookkeepers. With their new skills, veterans, once the postwar economy began to expand, earned incomes sufficient to permit them to raise families in the growing metropolitan areas of the country. But not all veterans made it, especially not black veterans. The VA saw to it that employers gave preference to hiring veterans, but the VA had no fair-employment-practice policy because Congress let the FEPC expire.

A former reporter for *Stars and Stripes* published a collection of articles about veterans. "This book isn't the final word," he wrote. "There are no statistics, no conclusions. I wondered what had happened to the men and women who had come home, and this is what I found out." One thing he found out about was Chester, a black veteran. "We had white officers in our outfit," Chester said. ". . . The way they talked, democracy and segregation meant the same thing." They still did in postwar Washington, which Chester reached on a Jim Crow bus. "I didn't expect any revolutionary changes in the attitude toward Negroes," Chester said. "All I wanted was . . . just the smallest sign. . . . If Congress had only passed an anti-lynch law or a permanent F.E.P.C., something, anything. . . ." So Chester was emigrating: ". . . as soon as I can finish this mortician's course. . . . I'll be the only mortician in Liberia."

She had been one of the few women trust officers at the Second National Bank. She liked the job, which absorbed her while her husband was overseas. She had not planned, or even wanted, to continue working after he got home, but she did resent the question she was asked daily during August 1945. When did she expect him? They needed her position for a returning lieutenant commander. They told her how well she had done, how valuable had been her help, but no one at the bank suggested that there might be a permanent place for a woman as a junior executive. She was so happy to see her husband again that she forgot her resentment for almost twenty years.

> Public Law 268 . . . provides that each veteran while in train-ing . . . shall be paid a subsistence allowance of $65 per month if without dependents and $90 per month if with dependent or depen-dents. . . .
> For the education . . . of veterans . .
> The Veterans' Administration uses . . . universities and col-leges, professional and technological institutions . . . trade and busi-ness schools. . . . The Veterans' Administration pays for tuition, books, and supplies, and grants subsistence allowances. . . . — *Annual Report*, 1946
> The number of veterans enrolled in institutions of higher learning reached a peak of 1,208,952 on April 30, 1947. . . . —*Annual Report*, 1947

He enlisted in 1939 because after high school he could find nothing else to do. He was a corporal in 1941 when the Army se-lected him for officer's training. For valor in battle he was promoted to captain in 1944. In 1945 he matriculated as a freshman at a small liberal-arts college with a beautiful campus and a dedicated faculty. He loved it. Because he graduated in three years, he had a remaining year of eligibility under the GI Bill at Chicago, where eventually he completed his Ph.D. program in Romance languages. Not every vet-eran was as lucky.

Some veterans quit college because the rents for the trailers or Quonset huts on the campuses exceeded their monthly allowance. Others found that two children in a trailer, with one or two more in each trailer alongside, made studying difficult, seemingly irrelevant, always frustrating. But most veterans did finish their courses, which graduated thousands of lawyers, doctors, engineers, teachers, man-agers. All those subsistence allowances and tuitions and books for which the VA paid spawned a new, young, eager middle class. Most of the veterans in it thought they were content. Many were. Some left the imposed uniformity of the Army for a voluntary uniformity of the spirit that made their adolescent children restless.

"And the people in the houses / All went to the university, /

Where they were put in boxes / and they all came out the same, / And there's doctors and there's lawyers, / And business executives, / And they're all made out of ticky tacky / And they all look just the same."

> Title V of the Serviceman's Readjustment Act . . . authorizes the payment of weekly and monthly allowances to qualified veterans of World War II to assist in their readjustment to civilian employment. . . .
>
> For each week of unemployment up to a maximum of 52, a totally unemployed veteran may draw a $20 allowance. . . . Eligibility requirements . . . include provisions that the claimant be available for and able to work. . . .
>
> Fiscal year 1947 began with a monthly total of 7,827,585 claims for unemployment, representing the peak . . . in program history . . . caused by industrial reconversion, labor disturbances, and material shortages.
>
> During August rolls began a decline. . . .
>
> Payments . . . from the beginning of the program totaled $2,483,564,448. . . . —*Annual Report*, 1947

By early morning the line in front of the VA office stretched two blocks. There mustered the 52-20 club, most of them in old uniforms stripped of insignia, many of them as worried as they had been in 1939, a few of them just looking for handouts, the rest trying to cope with their welcome home.

The sergeant recognized the trucker behind him and the salesman several places ahead. They had had a beer together a month ago. The youngest of the three, the sergeant, only twenty in 1939, had then had his first good role as a juvenile in a stock company near Buffalo. That was before the draft took him by stages to Burma for three years. Now no casting director remembered his name. Now he had only two more weeks of eligibility for standing in line at the VA.

He pocketed his check without looking around, walked downtown twelve blocks, and called his agent from a drugstore near her office. They needed an entertainer, she reported, at the longshoremen's rally at the Garden the next night. Fifty dollars. He would be there, he said. Back at his flat he worked out ten minutes of patter. At the Garden the longshoremen clapped and whistled.

Later the parts began to come again. Several busy seasons of repertory theater. One short run on Broadway. Then television, regular work, favorable reviews, regular paychecks. But the producer discovered after some years that the sergeant had once entertained a Communist union. So he fired the sergeant. There was, he told him, no place for fellow travelers on the national networks.

> As of June 30, 1947, living World War II veterans had attained an average of 29.1 years. Three-fifths were married and two-thirds resided in urban areas. World War II veterans and their immediate families compromised almost one-fourth of the total United States population. . . . —*Annual Report*, 1947

Those veterans and their families voted in patterns typical for their age groups, national origins, levels of education, and occupations. They responded like nonveterans to national issues and national candidates. After Harry Truman retired, in six successive quadrennial contests the veterans and their families and their neighbors—the American people—elected World War II veterans to the Presidency. The first two so elected had been heroes.

General Eisenhower's D-Day orders had launched the greatest invasion in history:

> You are about to embark upon the Great Crusade, toward which we have striven these many months. The eyes of the world are upon you. The hopes and prayers of liberty-loving people everywhere march with you. . . . You will bring about the destruction of the German war-machine, the elimination of Nazi tyranny over the oppressed peoples of Europe, and security for ourselves in a free world. . . .

Americans liked Ike, who never wholly stopped crusading. Perhaps as a civilian he forgot what he had learned as a soldier. "Veterans," he had written, ". . . acquire a steadiness that is not shaken. . . . But when kept too long in the fight they . . . become subject to physical and mental weariness. . . ." It was just after Ike's first inauguration that the producer, a friend of the President, fired the sergeant from his weekly role on TV.

He was asked how he became a hero. "It was involuntary," Jack Kennedy replied. "They sank my boat." In 1946 he addressed a group of gold-star mothers. "I think I know how all you mothers feel," he said, "because my mother is a gold-star mother, too." Kennedy's older brother, Joe, had died in the war. As Kennedy's first campaign got under way, "Red Fay, his fellow PT-boat officer in the Pacific, was summoned from San Francisco"; Torbert MacDonald,

"Jack's Harvard roommate," also a veteran, worked in Cambridge and Somerville, Massachusetts; and Bobby Kennedy, "a youngster just out of the service as an enlisted man in the Navy, pitched in . . . for three months knocking on doors." Mark Dalton was in the 52-20 club when he joined the campaign. There were others, too, veterans not of World War II but of Massachusetts politics, and, assisted by volunteers, they passed out dozens of pamphlets telling the story of *PT-109* and its courageous skipper. In winning his first victory in 1946 he did not hesitate voluntarily to talk about the war and its veterans:

> Most of the courage shown in the war came from men's understanding of their interdependence on each other. Men were saving other men's lives at the risk of their own simply because they realized that perhaps the next day their lives would be saved in turn. . . .
> Now . . . they miss the close comradeship, the feeling of interdependence, that sense of working together for a common cause. In civilian life, they feel alone. . . .
> In fact, if we only realized it, we are in time of peace as interdependent as the soldiers were in time of war.
> The institutions and principles for which we fought will be under a growing fire in the years ahead. . . .
> We must work together. We must recognize that we face great dangers. . . . We must have the same unity that we had during the war. . . .

That speech expressed in 1946 a point of view that the candidate and his fellow veterans would bring to Washington in 1961. Kennedy in that later year became the most senior of the former junior officers of World War II who then reached positions of influence and authority. Their rhetoric, though refined, carried the same message it had earlier carried: "Ask not what your country can do for you; ask what you can do for your country." As during the years of war, as in the interceding fifteen years, the Commander in Chief and his aides and their countrymen found some answers to that question that were better than others. And after 1961, as in the period since the victory of 1945, there were in the policies Americans debated, now rejected, now pursued, echoes of the policies that had provoked debate during World War II, echoes of the politics and culture of that earlier time, echoes of varying intensity that sounded through the republic for a generation.

Acknowledgments

The writing of this book has consumed much of the last ten years, during which I have accumulated more debts than I can properly repay here. To put first what ordinarily appears last, my wife in that decade has been engaged in several scholarly projects of her own. I hope that I have rendered to her the indispensable support she has given me. Several friends have criticized part or all of the manuscript of this volume. Though they will recognize that I have not always taken their advice, I am nonetheless grateful to William B. Goodman, John Hersey, Archibald MacLeish, James T. Patterson, Allan M. Winkler, and C. Vann Woodward. I owe special thanks to the staffs of the Franklin D. Roosevelt Library and the National Archives, of which so many, particularly the late Herman Kahn, helped me find elusive materials. Three grants—a fellowship from the National Endowment for the Humanities, a leave of absence from Yale University, and a subvention from the Concilium for International Studies at Yale University—facilitated my work. So did a number of seminar papers by various students in the graduate program in American history at Yale, papers I have cited at appropriate places, Priscilla Hunt translated Russian newspapers for me. The late Henry Morgenthau, Jr., and Mrs. Henry A. Wallace gave me access respectively to the Morgenthau Diary and the Wallace Diary before those rich sources were regularly open to scholars, as they now are. My largest intellectual obligation is to other historians who published important studies about subjects of interest to me before I completed my own labors. Those studies are cited in the notes, but two on which I have depended continually deserve special mention: James M. Burns, *Roosevelt: The Soldier of Freedom* (New York, 1970), and Richard Polenberg, *War and Society: The United States 1941–1945* (Philadelphia, 1972). I am no less indebted, as my notes also reveal, to the shrewd reporting of *Fortune* and the *Wall Street Journal*.

Notes

Materials at the Franklin D. Roosevelt Library in Hyde Park, New York, are cited to FDRL, and are in the Roosevelt Papers unless otherwise indicated. Materials from the National Archives in Washington, D.C., or from its branch in Suitland, Maryland, are cited to NA. In both cases, I have minimized information about the exact location of documents in order to save space and costs, but interested scholars will have no trouble retrieving the documents I have noted if they consult those indispensable finding aids "Historical Materials in the Franklin D. Roosevelt Library," undated, copies available from the Library, and *Federal Records of World War II*, I, *Civilian Agencies* (Washington, 1950). The unpublished research papers cited in the notes are in the Yale Miscellaneous Manuscripts, Historical Manuscripts, Yale University Library, New Haven, Connecticut.

Prologue

The two opening quotations are from Ben D. Zevin, ed., *Franklin D. Roosevelt: Nothing to Fear* (Popular Library, New York, 1961), pp. 312–314, 462–465. My impressions of Roosevelt derive from works about him and his associates cited in the chapters that follow. The episodes I have described are true, though I have ordinarily avoided the use of proper names in order to protect the privacy of those discussed. For the same reason I have altered a few insignificant details. I either observed each of those episodes or learned about them from close friends whose memory I trust.

Chapter 1. The Selling of the War

1. Miles from the Battlefields

The opening quotation from the *Wall Street Journal* (hereafter *WSJ*) is from April 15, 1943. The English visitor was H. G. Nicholas; see his "The Comparability of American History: The Second World War—Comments on Paper of John M. Blum," ms., Yale University Library (hereafter YUL).

2. War Bonds and War Aims

Except as noted below, this section is adapted from John M. Blum, *From the Morgenthau Diaries*, II (Boston, 1965), pp. 289–296, and *op. cit.*, III (Boston, 1967), pp. 16–22. For Roosevelt to Barton, see *WSJ*, June 23, 1942; the same issue contains the Joneses' dialogue. The Textron, Community, and Arden advertisements are in *Glamour*, respectively, October 1943, September 1943, and April 1944.

3. Truth as a Handicap

On the establishment of the OFF and the organization of its records, which are in the National Archives (hereafter NA) with the OWI records, see *Federal Records of World War II*, I, *Civilian Records* (Washington, 1950), pp. 284–288; and "Preliminary Inventory of the Records of the Office of War Information (Record Group 208)," NA, March 1967. MacLeish's essay is reprinted in Chester E. Eisinger, ed., *The 1940s: Profile of a Nation in Crisis* (Garden City, 1969), pp. 213–217. On the strategy of truth and later challenges to it, see Sherman H. Dryer, *Radio in Wartime* (New York, 1942). MacLeish's recollections of Pearl Harbor and of Lasswell are from statements to the author. On the War and Navy communiqués, see Ulric Bell to MacLeish, February 6, 1942, OWI Records, NA. The views of MacLeish's subordinate are in "Remarks by Philleo Nash to Mrs. Sharp," Memos, January 9, 11, 1952, Official Files, Franklin D. Roosevelt Library (hereafter OF, FDRL). On the movies, see Hollywood Writers Mobilization for Defense, *Communique*, March 6, 1942, OF, FDRL; and "Hollywood in Uniform," *Fortune*, April 1942; "Walt Disney, Great Teacher," *ibid.*, August 1942. The material on radio, the tactic of vividness, and the OFF radio dramas is from Dryer, *op. cit.* Moley's criti-

cisms are from *WSJ*, February 27, 1942; for other responses, see Leo
C. Rosten to W. B. Lewis, March 10, 1942, OWI Records, NA. On Oboler
and MacLeish, see Liam O'Connor to MacLeish, May 5, 1942, and Mac-
Leish to Michael J. Ready, May 27, 1942, OWI Records, NA. The end
of this section is based upon MacLeish to Roosevelt, May 16, 1942, May
18, 1942, OF, FDRL, and statements by MacLeish to the author.

4. Prescriptions for War Information

On the establishment of OWI and on its records, see *Federal Records of
World War II*, pp. 867–919; also Memo, Davis to Branch Directors, Depu-
ties, Bureau Chiefs, October 8, 1942, OF, FDRL. The best full study of
OWI is Allan M. Winkler, "Politics and Propaganda: The Office of War
Information, 1942–1945," Ph.D. dissertation, Yale University, 1974. On
the debate over the mission of OWI, see MacLeish to Davis, November
30, 1942; M. S. Eisenhower to MacLeish, December 1, 1942; MacLeish to
Eisenhower, December 2, 1942; Eisenhower to Davis, December 3, 1942;
Eisenhower to MacLeish, December 3, 1942; MacLeish to Eisenhower,
December 5, 1942; MacLeish to Davis, December 7, 1942; MacLeish to
Welles, December 9, 1942, OFF Records, Director's File, NA. On the
Navy, see New York *Times* (hereafter *NYT*), July 3, 10, October 29,
November 23, 28, 1942. On the OSS, see Stimson to Roosevelt, February
17, 1943, OF, FDRL. On Jeffers and Ickes, see *NYT*, April 19, 20, 21,
1943, August 22, 29, September 2, 1943; Davis to Ickes, August 27, 1943,
OF, FDRL. On the content of films and comic strips, see *WSJ*, November
2, 3, 1942. Bureau of Intelligence reports are filed serially under the name
of the division in OWI Records, NA. The account of the controversy over
advertising, culminating in Pringle's resignation, rests upon the *NYT*,
February 6, 19, April 11, 13, 14, 15, 18, 1943; and "Remarks of Philleo
Nash . . . ," *loc. cit.* The latter contains the version, one of several, of the
Coca-Cola story that I have used.

5. The Politics of Propaganda

The account of Congress and the OWI is derived from *NYT*, February 9,
10, 11, 13, March 5, 7, 14, 18, 22, 25, April 20, 22, May 14, June 19, 20,
22, 27, 30, July 1, 2, 3, 6, 8, 14, 16, 1943. On the episode of the Italian
King and Sherwood's role, see Winkler, "Politics and Propaganda"; also
NYT, July 27, 28, 29, 1943; *WSJ*, July 28, 29, 1943. The President's new
directive is explained in M. S. Eisenhower to Davis, August 2, 1943, and
Roosevelt to Secretaries of War, Navy, September 1, 1943, OF, FDRL;
also *NYT*, September 8, 1943. On Davis and Sherwood, see *NYT*, Febru-

ary 8, July 26, September 26, 1944. The British expert was Nicholas, ". . . Comments . . . ," *loc. cit.*

6. Pictures of the Enemy

The data on public opinion are from Lawrence S. Wittner, *Rebels Against War: The American Peace Movement, 1941–1960* (New York, 1969), pp. 120, 122. The analysis of attitudes toward the Japanese is based upon Harold R. Isaacs, *Images of Asia: American Views of China and India* (Harper Torchbook, New York, 1972), especially pp. xix, 64, 172. *Time* described its friends on December 22, 1941; on the WPB ad, see Wittner, *op. cit.*, p. 105. The balance of the section, except for the comment by film critic John O'Connor, *NYT*, May 8, 1975, consists of my readings of the following books, in the order they are discussed: Vincent Sheean, *Not Peace but a Sword* (New York, 1939); Lillian Hellman, *The Collected Plays of Lillian Hellman* (Boston, 1971), "Watch on the Rhine," pp. 201–266; "The Searching Wind," pp. 267–324; John Steinbeck, *The Moon Is Down* (Bantam, New York, 1964); Helen MacInnes, *Above Suspicion* (Boston, 1941) and *Assignment in Brittany* (Boston, 1942); Glenway Wescott, *Apartment in Athens* (Greenwood Press, Westport, 1972); Nevil Shute, *Pied Piper* (New York, 1942) and *Most Secret* (London, 1945); Upton Sinclair, *Presidential Agent* (New York, 1944).

Chapter 2. Homely Heroes

1. The GI's: American Boys

On Kate Smith, see Robert K. Merton, *Mass Persuasion: The Social Psychology of a War Bond Drive* (New York, 1946), Ch. I; on Franklin, Edward L. Bernays, *Biography of an Idea: Memoirs of Public Relations Counsel* (New York, 1965), Ch. 59. Sherrod's comment is in Robert Sherrod, *Tarawa: The Story of a Battle* (New York, 1944), pp. 150–151, the book from which I have also later quoted his descriptions of the marines in the battle. Mauldin's comment is from Bill Mauldin, *Up Front* (New York, 1944), pp. 19–21, a book I have also later quoted in all references to Mauldin. See, too, W. L. White, *They Were Expendable* (New York, 1942), on MTB Squadron 3; on Wake Island, S. I. Smith, ed., *The United States Marine Corps in World War II* (New York, 1969), p. 26. On "Biffle Manor," see Memo, Natalie Davis to David Frederick, June 15, 1944, OWI Records, NA. The *Time* references to athletes are from the issues of August 31 and October 19, 1942. The articles I used from "These Are the Generals" ran in *The Saturday Evening Post* on October 3, 1942 and weekly from the issue of January 30, 1943 through that of April 10, 1943. The material in this chapter from Pyle is in Ernie Pyle, *Here Is Your War* (New York, 1943) and *Brave Men* (New York, 1944). On Guadalcanal, see John Hersey, *Into the Valley: A Skirmish of the Marines* (Bantam, New York, 1966), to which I have also referred often in later sections, and Ira Wolfert, *Battle for the Solomons* (Boston, 1943). The Eisenhower profile is in the earliest article of "These Are the Generals," *loc. cit.* The discussion of aviators and its immediate sequel is based upon a comparison of the imagery applied to foot soldiers and aviators in Mara Nacht Mayor, "The Heroic Image During World War II," ms., YUL. On the ethnic make-up of platoons, see Arthur Knight, *The Liveliest Art* (New York, 1957), p. 264. On OWI initiatives, see Leo C. Rosten to Nicholas Roosevelt, December 8, 1942, and to Henry Luce, March 19, 1943, as well as April 1944 Memo and August 31, 1944 Memo, Entries 76, 84, OWI Records, NA.

2. In Foreign Foxholes

See citations to Pyle, Hersey, and Mauldin in Section 1. On reporting about blacks, including that by Duckett, see *This Is Our War* (Baltimore, 1945). On the marine lieutenant colonel, see W. C. Leinly, letter written at sea, May 14, 1943, OF 4351, FDRL. The article, "What I Am Fighting For," is from *The Saturday Evening Post*, July 17, 1943. On the Redistribution Center, see Robert A. Lovett, Memo for the Secretary of War, May 25, 1944, President's Personal File (hereafter PPF) 1820, FDRL. I have quoted too from Bob Hope, *I Never Left Home* (New York, 1944), and Marion Hargrove, *See Here, Private Hargrove* (New York, 1942).

3. Cossacks and Comrades

The generalizations about the British derive from samplings of the *Observer* of London and the Manchester *Guardian* and, as do the remarks about the Germans, from "A Study of War Communiqués: Methods and Results," ms., New School for Social Research, New York, 1942, copy in National Resources Planning Board Records, Entry 14, NA. Himmler is quoted in Press Release, June 19, 1942, OF 5015, FDRL. The quotations about the Russians are from *Pravda* and *Izvestiia*, all of them freely translated, in sequence as follows: about Terekhin, *Pravda*, June 8, 1941; about Mel'nikov, *ibid.*, July 4, 1941; about Mamedov, *Izvestiia*, December 19, 1941; the tankmen, *ibid.*, July 20, 1941; Tokarev, *ibid.*, August 5, 1942;

the Cossacks, *Pravda,* September 19, 1941; Toknev, *ibid.,* October 25, 1941; the Armenians, *ibid.,* December 4, 1943; the Ukrainian, *ibid.,* December 18, 1943, May 8, 1944; Kabanov, *ibid.,* August 25, 1941; the brothers, *Izvestiia,* July 1, 1942; the march to Germany, *Pravda,* October 7, 1944; other, lesser references, *Pravda,* July 9, 1941, December 23, 1943; *Izvestiia,* August 23, September 24, November 13, 1942. For Kennan's comments, see George F. Kennan, *Memoirs, 1925–1950* (Boston, 1967), pp. 190–195.

4. The War in Literature

On art, see "The Painters Interpret the War," *Fortune,* March 1943. Edward Newhouse, "Time Out," is in Jack Salzman, ed., *The Survival Years* (New York, 1969), pp. 101–110; William L. Worden, "Mist from Attu," in *The Saturday Evening Post Stories, 1942–1945* (New York, 1946), pp. 31–45; Salinger in *ibid.,* pp. 314–320. The novels I have discussed, in order of sequence, are James Jones, *From Here to Eternity* (New York, 1951), James Gould Cozzens, *Guard of Honor* (New York, 1948), John Hersey, *Bell for Adano* (New York, 1944), Irwin Shaw, *The Young Lions* (New York, 1948), Norman Mailer, *The Naked and the Dead* (New York, 1948), Thomas Heggen, *Mister Roberts* (Boston, 1946), John Horne Burns, *The Gallery* (London, 1948), Alfred Hayes, *All Thy Conquests* (New York, 1946), John Hersey, *The Wall* (New York, 1950). For Randall Jarrell's poem, see "The Death of the Ball Turret Gunner," Salzman, ed., *Survival Years,* p. 88; for Marianne Moore's, see "In Distrust of Merits," *ibid.,* pp. 68–70. See also "Army Orientation," *Fortune,* March 1944. This entire section owes much to Stuart Schoffman, "Social Protest in American World War II Novels," ms., YUL.

Chapter 3. Getting and Spending

1. The Return of Prosperity

The impact of war on the economy and economic thought is described admirably in Robert Lekachman, *The Age of Keynes* (New York, 1966), Ch. 6; see also Bureau of the Budget, *The United States at War* (Washington, 1946).

2. The Wartime Consumer

Average family income was reported in *WSJ*, November 17, 1942. On Washington, see John Dos Passos, "Washington Sketches," *Fortune*, December 1943. On transportation, see *WSJ*, July 1, 7, 14, 1943; on the impact of war on cities, *ibid.*, April 24, November 14, 1942; on vitamins, *ibid.*, March 19, 1943; on pawnshops, *ibid.*, October 1, 1942. Two books that contain excellent accounts of similar phenomena are Richard R. Lingeman, *Don't You Know There's a War On: The American Home Front, 1941–1945* (New York, 1970), and Geoffrey Perrett, *Days of Sadness, Years of Triumph: The American People, 1939–1945* (New York, 1973). The study of the cities and of hoarding was reported in *WSJ*, January 29, 1942; of men's suits, *ibid.*, March 7, April 1, 1942; women's wear, *ibid.*, April 3, 1943. The best account of women during the war is in William H. Chafe, *The American Woman* (New York, 1972), Part Two. On the motion-picture industry, see *WSJ*, July 3, 1942; on fun in the city, *ibid.*, November 30, 1942, January 20, 1943; on golf and resorts, *ibid.*, May 6, 10, 1943; on music and books, *ibid.*, August 6, October 18, November 23, 1943, March 19, 1945; also "The Boom in Books," *Fortune*, November 1943. On race tracks see *WSJ*, September 22, 1943; "A Billion Across the Board," *Fortune*, September 1944. On black-market prices, see *WSJ*, August 23, 1943; on pharmaceuticals, *ibid.*, October 2, 1943; on jewelry, *ibid.*, November 24, 1942, August 17, November 10, 1943, December 7, 1944; on holiday buying, *ibid.*, November 29, December 23, 1943, March 30, August 8, November 2, 1944, January 6, 1945; "more pie" is in *ibid.*, July 3, 1944; "ox tongues" in *ibid.*, May 25, 1945. The discussion about *Gourmet* derives from the issues of January 1941, February, May, June, July, October 1942, January, April 1944. Naomi Torrence located those and other *Gourmet* tidbits. For Wylie, see Philip Wylie, *Generation of Vipers* (New York, 1942), p. 223. The advertisement for Miami appeared in *Time*, January 5, 1942; for playing cards, in *The Saturday Evening Post*, August 15, 1942; for Formfit, in *ibid.*, October 17, 1942. On advertising budgets and markets, see *WSJ*, May 13, 15, June 2, 1944. On Yamamoto, see *The Saturday Evening Post*, August 29, 1942; on Johnston and Murphy, *Time*, October 18, 1943; on Nash Kelvinator, *The Saturday Evening Post*, February 27, April 17, 1943; on Pan American, *ibid.*, January 5, 1943; on General Electric, *Time*, June 26, 1944. The material about housing is from a manuscript study by Stephen Foster. For the OWI findings, see "Excerpts from Confidential Report to the President . . . by Special Housing Mission to Great Britain," ca. November 1943, OWI Records, Entry 27, NA.

3. Enterprisers at War

Except as otherwise indicated, this entire section is based upon articles in *Fortune,* sequentially as follows: "Bernard Gimbel: Top Merchant," July 1945; "Beardsley Ruml," March 1945; "Pushing the Pens," October 1944; "Bob Woodruff of Coca-Cola," September 1945; "Chewing Gum: War Material," January 1943; "Five That Were Small," June 1943; "The Earth Movers," August, September, October 1943 and also on Kaiser, *WSJ,* September 12, 15, 1942, and Christopher K. Seglem, "Henry J. Kaiser: Wartime Entrepreneur," ms., YUL. For the "free enterprise" advertisements, see, *i.a., The Saturday Evening Post,* August 15, 29, 1942, January 29, February 12, 1944; and *Time,* January 4, 1943, January 24, 1944, April 30, 1945.

Chapter 4. War Lords and Vassals

1. Industry and Government

For Roosevelt on monopolistic power, see Zevin, ed., *Franklin D. Roosevelt,* pp. 139–145. The discussion of Stimson and his associates is based heavily upon Part Three of Elting E. Morison, *Turmoil and Tradition: A Study of the Life and Times of Henry L. Stimson* (Boston, 1960), a splendid book from which my interpretations sometimes deviate; *cf.* Blum, *Morgenthau Diaries,* II, Ch. IV–V. For Stimson on profits, see Richard Polenberg, *War and Society* (Philadelphia, 1972), Ch. 1. That chapter speaks also to Nelson and Eberstadt, about whom see also Bruce Catton, *The War Lords of Washington* (New York, 1948), and Donald M. Nelson *Arsenal of Democracy* (New York, 1946). On the proclivity to favor large companies, see Polenberg, *op. cit.,* and Chief Counsel to Senator Tom Connally, August 7, 1941, Truman Committee, Entry 49, NA; also John M. Blair and Harrison Houghton, "The Concentration of American Business and Finance in Peace and War," undated ms., Records of Smaller War Plants Corporation, Record Group 240, NA.

2. Pains for Small Business

This section rests primarily upon the two documents cited immediately above, as well as two others from the same Record Group: C. W. Fowler,

"The History of the Administrative Policies and Achievements of the Smaller War Plants Corporation," undated ms., and, probably also by Fowler, "The Mature Program of the Smaller War Plants Corporation As Developed by Chairman Maury Maverick," undated ms. On Taft, see James T. Patterson, *Mr. Republican: A Biography of Robert A. Taft* (Boston, 1972). Catton's judgment is from his *War Lords of Washington,* which discusses, as does Polenberg, *op. cit.,* the reconversion question. On that question, see also William D. Leahy to Donald Nelson, July 7, 1944; Nelson to Roosevelt, July 10, 1944; Nelson to Leahy, July 10, 1944; Chit, Roosevelt to James F. Byrnes, July 11, 1944; Memo, Byrnes to the President, July 29, 1944; C. E. Wilson to Roosevelt, August 23, 1944; Roosevelt to Wilson, August 24, 1944; all FDRL. For objections to Nelson's departure, see letters and telegrams about that subject in OF 4735, FDRL.

3. Holiday for the Trusts

Arnold's views, early and later, are in Thurman Arnold, *The Folklore of Capitalism* (New Haven, 1937) and *Fair Fights and Foul: A Dissenting Lawyer's Life* (New York, 1965), especially Ch. 15. I also used Douglas Ayer, "In Quest of Efficiency: The Ideological Journey of Thurman Arnold in the Interwar Period," *Stanford Law Review,* June 1971, pp. 1049–1086; Mark H. Lytle, "Thurman Arnold and the Wartime Cartels," ms., YUL; and Thurman Arnold et al., "Anti-Trust Enforcement Now and in the Post-War," *American Forum of the Air,* September 1943. The agreement of March 20, 1942, was released to the press on March 28 for publication the next day, see, *e.g.,* Press Release, March 28, 1942, FDRL. On Berge, see "Mr. Berge's Antitrust," *Fortune,* August 1944, and Wendell Berge, *Cartels: Challenge to a Free World* (Washington, 1944). The latter discusses the cases I have discussed, but those cases receive richer treatment in correspondence at FDRL among the principals concerned with them; the most important of the relevant letters and memoranda are: Forrestal to Biddle, April 25, 1944; Halifax to Hull, April 27, 1944; Biddle to Stimson, May 4, 1944; Forrestal to Biddle, May 8, 1944; Patterson to Byrnes, May 10, 1944; Biddle to Hull, May 12, 1944; Biddle to Roosevelt, May 12, 1944; Byrnes to Roosevelt, May 16, 1944; Biddle to Roosevelt, June 23, 1944; Patterson to Roosevelt, July 6, 1944; Biddle to Roosevelt, July 12, 1944; Forrestal to Biddle, July 12, 1944; Roosevelt to Stimson, July 13, 1944; Forrestal to Roosevelt, July 17, 1944; Biddle to Roosevelt, July 24, 1944; Roosevelt to Forrestal, July 26, 1944; Biddle to Roosevelt, August 4, 1944; Roosevelt to Biddle, August 14, 1944. The comment at the end of the section is from an oral statement of Ward S. Bowman, Jr.

4. Of Size and Abundance

On general developments concerning labor and agriculture, see Polenberg, *op. cit.* On academic morale, see Odell to Gulick, October 6, 1942, NRPB Records, Entry 14, NA. On eastern colleges, see *WSJ*, May 4, 1942. On ASTP and V-12, see Patterson to Rosenman, January 27, 1944, OF 25-NN, FDRL, and "Navy V-12 Curricula Schedules, Course Descriptions," November 1, 1943, Bureau of Personnel, Washington, D.C. On Conant and Rosenman, see *WSJ*, January 18, 1943, and Bard to Rosenman, January 18, 19, 1944; Knox to Roosevelt, March 21, 1944, OF 18-HH, FDRL. The best general study of OSRD is James P. Baxter, 3rd, *Scientists Against Time* (Boston, 1946). I have also relied upon Michael S. Sherry, "Preparing for the Next War: American Plans for Postwar Defense, 1941–1945," Ph.D. dissertation, Yale University, 1975. On size and related questions, see J. Kenneth Galbraith, *American Capitalism* (Boston, 1952).

Chapter 5. Outsiders

1. Little Italy

On "Roberto," see "Steam from the Melting Pot," *Fortune*, September 1942, an article that deals, not always sympathetically, with sundry European nationality groups, as the quotations I have used about them suggest. See also "Summary of a Report Submitted by the Nationalities Section, November 25, 1940," OWI Records, Nationalities Section, NA. The contrasting plea for courage is from John H. Tolan to MacLeish, March 8, 1942, OWI Records, NA. My discussion of the Italian-Americans rests in large part upon David M. Kennedy, "Italian-Americans, 1935–1942: A Study in Assimilation," ms., YUL, a monograph which contains translations of the Italian-language newspapers I have quoted, as well as other useful data. See also John P. Diggins, "The Italo-American Anti-Fascist Opposition," *Journal of American History*, December 1967, pp. 579–598. Stimson is quoted in Allan R. Bosworth, *America's Concentration Camps* (New York, 1967), pp. 68–69. On MacLeish's proposal and related developments, see MacLeish to Hull, May 9, 1942; Cranston to MacLeish, May 14, 1942; Wilkinson to MacLeish, May 26, 1942; Bureau of Intelligence, OFF to MacLeish, May 25, 1942; Cranston to MacLeish, June 6,

1942, OWI Records, NA. On Biddle, see Kennedy, *op. cit.;* on *Seventeen,* see issue of November 1944.

2. "A Jap's a Jap"

On Roosevelt, Ickes, and Eisenhower, see Ickes to Roosevelt, April 13, 1943; Roosevelt to Eisenhower, April 19, 1943; Roosevelt to Ickes, April 22, 1943; Eisenhower to Roosevelt, April 22, 1943; and Eisenhower to M. H. McIntyre, May 15, 1942, FDRL. My general discussion about the Japanese-Americans and the treatment accorded them derives from four studies: Bosworth, *America's Concentration Camps;* Audrie Girdner and Anne Loftis, *The Great Betrayal* (Toronto and London, 1969), an especially helpful account; and Dorothy S. Thomas *et al., The Spoilage* (Berkeley, 1946) and *The Salvage* (Berkeley, 1952), the earlier of which contains an excellent analysis of life in the camps, of the loyalty program, and of Tule Lake not the least. The quotations I have used, except as otherwise indicated, are from those sources and the good synthesis in Polenberg, *War and Society,* Ch. 2. Protests against the government's policy are quoted from *The Nation,* February 14, 1942; *The New Republic,* March 2, April 6, 1942; *The Christian Century,* March 25, April 1, June 10, 1942; *Fortune,* April 1944. On the Senate in 1943, see D. S. Myer to Roosevelt, July 14, 1943, FDRL. The "hard-liners" spoke out in George E. Outland *et al.,* to Roosevelt, January 28, 1944, FDRL. I have also used that classic article by Eugene V. Rostow, "The Japanese American Cases— A Disaster," *Yale Law Journal,* June 1945.

3. Jews in Great Britain

There are two brief references to the issues discussed in this section in Henry Pelling, *Britain and the Second World War* (Glasgow, 1970), pp. 87, 315. Otherwise the section rests upon reports in English newspapers and magazines, of which some are cited fully in the text. The others are cited here in sequence: On developments during the phony war, *Economist,* January 27, March 30, 1940; rebuttal to the *Times, Economist,* May 18, 1940; "measure of precaution," *Times,* May 17, 1940; "inexplicable" prohibitions, *Economist,* August 10, 1940; Rugby master, *Times,* July 16, 1940; Isle of Man, *ibid.,* July 23, 24, August 6, November 6, 1940; Trevelyan, *ibid.,* August 24, 1940; Murray, *ibid.,* July 8, 1940, *New Statesman,* July 20, 1940; Home Office "muddle" and excuses, *New Statesman,* August 3, 10, 1940, October 5, 1940, *Economist,* August 31, 1940; Liversidge, Atkin, and Laski, *New Statesman,* November 15, 1941; Commons debate, *ibid.*, December 6, 1941; citizenship and Roosevelt, *ibid.*, February 22, 1941, *Times,* January 18, 1943.

4. American Jews

On wartime opinion about Jews, see "Memorandum on Discrimination Against Jews in Defense Industries," February 16, 1942; Detroit Hearings, May 24, 25, 1943; FEPC Records, NA; Isaac Frank to OFF, and enclosures, April 24, 1942, OWI Records, NA; Wittner, *Rebels Against War.* On unconscious anti-Semitism, see "Jews in America," *Fortune,* February 1936. On Mrs. Roosevelt, see Polenberg, *War and Society,* and Joseph P. Lash, *Eleanor and Franklin* (New York, 1971). On suburban moods, see Laura Z. Hobson, *Gentleman's Agreement* (New York, 1947). The study on which I have relied for the material on Long and on the development of German policy and American responses thereto is Henry L. Feingold, *The Politics of Rescue: The Roosevelt Administration and the Holocaust, 1938–1945* (New Brunswick, 1970). I have also, particularly for the period after the Treasury became involved, adapted my account in *Morgenthau Diaries,* III, pp. 207–227. On Ibn Saud, see John M. Blum, ed., *The Price of Vision: The Diary of Henry A. Wallace, 1942–1946* (Boston, 1973), pp. 254–255, 300, 420.

Chapter 6. Black America: The Rising Wind

1. Jim Crow

On the loyalty of blacks, see John R. Carlson, *Under Cover* (New York, 1943), pp. 154–163; and "Survey of Intelligence Materials" #14, March 16, 1942; T. N. Berry to George Barnes, March 18, 1942, OWI Records, NA. On USES, see William J. Schuck, "Eliminating Employer Discriminatory Hiring Practices," January 31, 1946, Historical Section, War Manpower Commission Records, NA; see also Jervis Anderson, *A. Philip Randolph: A Biographical Portrait* (New York, 1973). God as a white person is from Charles Jenkins to Roosevelt, December 3, 1941, OF 93, FDRL. On Army policy, including testing and the views of Stimson and others, see the authoritative Ulysses Lee, *The Employment of Negro Troops* (Washington, 1966), a volume in the extraordinary series *United States Army in World War II.* "Grave apprehensions" is from Anderson, *Randolph,* p. 247. Stimson and Marshall are quoted in Polenberg, *War and Society,* pp. 123–124. Polenberg has written an excellent account of all the issues treated in this chapter. On Hastie and Davis and *The Crisis,* see Lee, *Negro Troops,* pp. 79–80, 83. My account of Randolph and of the march, and of Roosevelt's negotiations, is derived from Anderson, *op. cit.* I have also consulted and quoted from Randolph to Roosevelt, May 29, 1941; Robert P. Patterson Memo to General Watson, June 3, 1941; Stephen Early to Wayne Coy, June 6, 1941; Watson Memo to Roosevelt, June 14, 1941; advertisement, undated, "Join Negro March for Jobs and Justice," all in FDRL.

2. The Persistence of Prejudice

The OFF material is in "Survey of Intelligence Materials," #21, May 19, 1942; #25, May 27, 1942; "The Negro Looks at the War," May 19, 1942, all in OWI Records, NA. On ODT, see Walter White to MacLeish, August 3, 1942, in *ibid.;* on the "intolerable situation" in Arlington, see Edgar G. Brown to Roosevelt, May 20, 1942, FDRL. On Walla Walla, see Roy Wilkins to MacLeish, March 27, 1942; Theodore M. Berry to Lt. Col. David P. Page, April 20, 1942; and the reply of May 2, 1942, OWI Records, NA. The account of Salina is from Lloyd L. Brown, "Brown v. Salina, Kansas," *NYT,* February 26, 1973. On congestion, see "Leland Tension Area Study," undated, FEPC Record Group 228, NA. The "dumb" Army is from Roosevelt to Davis, June 17, 1942, and reply June 19, 1942, OF 93, FDRL. On Hastie, see *Time,* February 8, 1943. The Lynchburg ladies wrote Roosevelt February 22, 1944; the Philadelphia matron wrote Grace Tully, June 4, 1942, OF 93, FDRL. On the rural South, see Arthur Raper to Jonathan Daniels, September 25, 1943 and enclosure: "Race Tension and Farm Wages in the Rural South," FDRL. On "Bull" Connor, see Connor to Roosevelt, August 7, 1942 and attachments, President's Secretary's File (hereafter PSF) 4245-G, FDRL. On the dissatisfactions of blacks, see Ulric Bell to MacLeish, March 16, 1942, OWI Records, NA. There are copies of Chandler Owen's *Negroes and the War* there and in FDRL. The *Courier* and *Crisis* are quoted in Memo, Paul Richman to John Wildring, April 14, 1942, OWI Records, NA. On the failure of the pamphlet, see Theodore M. Berry to Elmer Davis, July 24, 1942, and Lester P. Granger to Davis, March 27, 1943, in *ibid.* MacLeish wrote Mrs. Roosevelt on April 29, 1942, in *ibid.* The definition of principles is in Robert Huse to Gardner Cowles, Jr., February 4, 1943, *ibid.* On the FEPC, see "Final Report, Fair Employment Practice Committee," June 29, 1946, FEPC Records Group 228, NA. On the WMC see Schuck, ". . . Discriminatory Hiring Practices," *loc. cit.* The material on segregated education is from PSF 4245-G, FDRL, especially the following documents: Cramer to Roosevelt, July 3, 1942, and enclosure, ". . . Hearing . . . on Discrimination in Defense Training," April 13, 1942;

Chit, Roosevelt to Patterson, July 9, 1942; Patterson to Roosevelt, July 14, 1942; Chit, Roosevelt to McIntyre, July 17, 1942; McIntyre to Cramer, July 21, 1942. On the Mexican-Americans, see the file cited above, especially the following documents: Welles to Roosevelt, June 20, 1942; Chit, Roosevelt to McIntyre, June 23, 1942; Cramer to Roosevelt, July 10, 1942; Welles to McIntyre, July 24, 1942; Chit, Rosenman to McIntyre, July 29, 1942; McIntyre to Cramer, July 30, 1942; Chit, McIntyre to Roosevelt, July 31, 1942.

3. Race Riot

The housing problem is reviewed in "The Negro's War," *Fortune*, June 1942. On "Sojourner Truth," see Arthur W. Mitchell to Roosevelt, January 10, 1942; H. A. White to McIntyre, January 16, 1942; C. F. Palmer to McIntyre, January 20, 30, 1942, all OF 93, FDRL; Walter White to Roosevelt, undated, OF 2538, FDRL; Herbert Emmerich to Jonathan Daniels, November 24, 1943, PSF 4245-G; FDRL; Nelson Foote, Special Report, March 5, 1942, OWI Records, NA. On the background to the Detroit riot and the riot itself, see Francis Biddle to Corrington Gill, July 24, 1943 and enclosed report, "Memo for Attorney General Re: Detroit Race Riot, June 20–21," Department of Justice Records, Committee on Congested Areas, NA; on post-riot opinion, see "Opinions in Detroit Thirty-six Hours After the Race Riot," June 30, 1943, Special Memo #64, Surveys Division, OWI Records, NA; see also, with respect to both of the foregoing subjects, John H. Witherspoon to the Common Council, Detroit, June 28, 1943, OF 93C, FDRL, and Polenberg, *War and Society*, Ch. 4; Robert Shogan and Tom Craig, *The Detroit Race Riot: A Study in Violence* (Philadelphia, 1964). In my account of the riot itself, I have leaned more to the Justice Department's report than to the narrative in Shogan and Craig. The pleas for a statement from Roosevelt are in Walter White to Roosevelt, June 21, 1943; Fiorello La Guardia to Roosevelt, June 27, 1943, OF 93C, FDRL; Grace Tully (quoting Wallace) to Roosevelt, July 7, 1943; Daniels to Roosevelt, June 22, 1943; Roosevelt to Rosenman, June 23, 1943, all PPF, FDRL. On Los Angeles, see *Time*, June 21, 1943; Polenberg, *op. cit.*, Ch. 4, which also discusses Harlem; White to Roosevelt, June 11, 1943, which is also in OWI Directors File, NA; Philip Murray to Roosevelt, June 18, 1943; Chit, Roosevelt to Perkins, June 23, 1943; Roosevelt to Murray, July 14, 1943, FDRL. On the end of this section, see Biddle to Daniels, August 27, 1943, enclosing Granger to Biddle, August 27, 1943, OF 4245-G, FDRL; Pauli Murray's poem is quoted by Shogan and Craig.

4. Protest and Prognosis

The early part of this section is based upon Carol O'Connor, "The Double V, American Negroes and World War II," ms., YUL, and Polenberg, *op. cit.*, pp. 105 ff. The black reporters are anthologized in *This Is Our War* (Baltimore, 1945). Walter White, "Observations in the North African and Middle Eastern Theatres of Operation," April 22, 1944, is in OF 93, FDRL. The various protests about recreational facilities for black soldiers are in OF 93B, FDRL, which also contains Stimson to Roosevelt, September 20, 1944; White to Roosevelt, October 3, 1944; Chit, Roosevelt to Daniels, October 9, 1944; Roosevelt to White, October 11, 1944. On the VA, see White to Hines, September 8, 1944; Hines to White, September 21, 1944; White to Roosevelt, October 5, 1944; Roosevelt to White, October 14, 1944; White to Roosevelt, October 23, 1944; Daniels to White, October 27, 1944, FDRL. On black voting rights, see Daniels to Roosevelt, September 28, 1944, OF 93, FDRL. On Ross, see Polenberg, *op. cit.*, p. 120; "Final Report F.E.P.C.," *loc. cit.; Time*, December 27, 1943. On the cases and hearings, see "Final Report"; *The Nation*, December 30, 1944; "Summary, Findings and Directives Relating to Kaiser Company, Inc., and the Oregon Shipbuilding Corporation, International Brotherhood of Boiler Makers . . . Subordinate Lodge No. 72 . . . 401, Auxiliary Lodge No. A–32," December 9, 1943, FEPC ms., FEPC Records, NA; Allan M. Winkler, "The Philadelphia Transit Strike of 1944," *Journal of American History*, June 1972. My accounts of Muste and CORE and of the demonstrations are based upon August Meier and Elliott Rudwick, *CORE* (New York, 1973), Ch. 1, and, especially, Eleanor Holmes (Norton), "World War II and the Beginning of Non-Violent Action in Civil Rights," ms., YUL. Norton had extensive conversations with Pauli Murray, who wrote McIntyre to reach Roosevelt on June 18, 1943, FDRL. For the end of the section, see Walter White, *A Rising Wind* (Garden City, 1945).

Chapter 7. Congress and the Politics of Comfort

1. Resurgent Republicans

This chapter and this section of it cover questions that have received thoughtful treatment from a number of authors whose works have been of large importance to me. I have especially relied upon Polenberg, *War and Society*, particularly Chapters 3, 7, and 8; James M. Burns, *Roosevelt: The Soldier of Freedom* (New York, 1970), particularly Chapters VIII, IX, and XVII; Roland Young, *Congressional Politics in the Second World War* (New York, 1956), particularly Appendices A and B; and Richard N. Chapman, "Contours of Public Policy, 1939–1945," Ph.D. dissertation, Yale University, 1976. On municipal power plants, see *WSJ*, December 26, 1941; on "no blueprint," *ibid.*, January 9, 1942; on Hansen's *After the War—Full Employment*, *ibid.*, January 29, February 9, March 7, April 9, 1942. On Roosevelt and the NRPB, see Delano to Roosevelt, July 1, 1942; Roosevelt to Delano, July 6, 1942, OF 1092D, FDRL, and Roosevelt to Edwin R. Embree, March 16, 1942, OF 93, FDRL. Taber and Flynn are quoted in Polenberg, *op. cit.*, pp. 186–187. Roosevelt on Norris is quoted in Burns, *op. cit.*, p. 279. The discussion of revenue legislation is adapted from Blum, *Morgenthau*, III, Ch. II. The composite picture of the Hartford voters is derived from the significant study by John W. Jeffries, "Testing the Roosevelt Coalition: Connecticut Society and Politics, 1940–1946," Ph.D. dissertation, Yale University, 1974. On Cantril, see Burns, *op. cit.*, pp. 280–281. Post-election comments are from *WSJ*, November 6, 19, 1942; *Fortune*, December 1942; Ickes to Roosevelt, November 16, 1942, PSF 12, FDRL; Oscar Ewing to Roosevelt, March 16, 1943, OF 5015-3, FDRL; Pauley to Roosevelt, December 14, 1942, PPF 1820, FDRL.

2. The New Deal at Bay

The analysis of Halleck as a typical Republican is based upon Lewis L. Gould, "Congressional Republicans in the Second World War: The Example of Halleck and Brown," ms., YUL. On the CCC and WPA, see Polenberg, *op. cit.*, pp. 80 ff. On NYA, see Aubrey Williams to Ralph Turner, February 9, 1943 and attachment, Post-War File, FDRL; Owen D. Young to Roosevelt, December 23, 1941; Roosevelt to Young, January 3, 1942; Williams to Roosevelt, April 15, 1942; Roosevelt to Williams, April 18, 1942; Williams to Eleanor Roosevelt, October 13, 1942; Williams to Roosevelt, December 28, 1942; Willard S. Townsend to Frank Knox, April 19, 1943; Williams to Jonathan Daniels, August 31, 1943, all OF 444-D, FDRL; also *WSJ*, March 25, 1942. On the NRPB, see *WSJ*, January 5, February 18, March 12, 1943; Delano to Roosevelt, February 10, 1943; Roosevelt to Delano, February 22, 1943; Roosevelt to Carter Glass, March 24, 1943; Delano to Roosevelt, April 15, 1943, May 28, 1943, OF 1092, FDRL. On the balance of the discussion of developments in 1943, see the first three works cited in Section 1, above; on Jones and Wallace, see Blum, ed., *Price of Vision*, Introduction and Part I; and concerning taxation, Blum, *Morgenthau Diaries*, III, Ch. II.

3. Two Bills of Rights

On FDR's health, see Howard G. Bruenn, "Clinical Notes on the Illness and Death of President Franklin D. Roosevelt," *Annals of Internal Medicine*, April 1970, the article to which I have turned for all my references to the subject in this chapter and the next. My discussion of social security and related issues rests upon the reports of the NRPB cited in the text and Watson Memo to Roosevelt, December 16, 1942, with attachments including A. J. Altmeyer to Watson, December 14, 1942 and its enclosed analysis; also Fred K. Huehler to Roosevelt, December 4, 1942; Altmeyer to Roosevelt, December 4, 1942 and its attached analysis; Edward S. Greenbaum to Watson, December 12, 1942 and attached chits; S. I. Rosenman to Roosevelt, June 30, 1943; all FDRL. See, too, in the same depository, "Preliminary Report to the President of the United States from the Armed Forces Committee on Post War Educational Opportunities for Service Personnel," July 30, 1943, ms. In addition, there are excellent analyses of both the GI Bill and the economic bill of rights, and also of the soldiers'-vote question, in Burns, *op. cit.* and Polenberg, *op. cit.* On the Southern fear of radicals, see *WSJ*, May 26, 1944. On national service, I have followed the accounts in Burns and Polenberg, though either or both authors may disagree with my analysis of the balance between the right hand and the left.

Chapter 8. Referendum on Roosevelt

1. Mise en Scène

The Cantril report is attached to Rosenman to Roosevelt, November 24, 1943, Post-War File, FDRL. Jerome S. Bruner's conclusions are in his perceptive *Mandate from the People* (New York, 1944), p. 5; that book is the source for my other quotations from Bruner. On Roosevelt and the defeat of "our common enemies," see Oscar Cox to Roosevelt, May 31, 1944, PPF 1820, FDRL. The press conference of May 26 and the June statement are reported as part of the excellent analysis of Roosevelt's intentions in Robert A. Divine, *Second Chance: The Triumph of Internationalism in America During World War II* (New York, 1967), p. 204. My judgments about Roosevelt's purposes, and about those of other nations and of his domestic critics, were formed while I worked on the Morgenthau and Wallace diaries for books cited elsewhere in the notes.

On Taft, I have relied upon Patterson, *Mr. Republican;* on Ball, Burton *et al.,* upon Divine, *op. cit.*

2. Willkie: Republican Heretic

The narrative of Willkie's career, particularly during the war years, and the quotations from his speeches and about him, except as otherwise noted, are derived from that first-rate biography, Ellsworth Barnard, *Wendell Willkie: Fighter for Freedom* (Marquette, Michigan, 1966). My analysis of Wendell L. Willkie's *One World* (New York, 1943) is of the soft-back edition of that work. I have quoted the April 30, 1943 issue of *The Commonweal;* the other reviews quoted are from *Foreign Affairs,* October 1943; *The New Yorker,* April 17, 1943; *The Atlantic,* May 1943; *The Nation,* April 24, 1943; *The New Republic,* April 19, 1943. The quotations of Taft are from Patterson, *Mr. Republican,* pp. 285–286 and, later, p. 271. Vandenberg is quoted in Divine, *Second Chance,* pp. 106, 113. That book also provided the basis for my account of the Mackinac meeting. On Dewey and Gallup, see Polenberg, *War and Society,* pp. 210, 271. On Dulles, see John Foster Dulles, "Peace Without Platitudes," *Fortune,* January 1942.

3. Wallace: Democratic Pariah

This section is adapted from Blum, ed., *Price of Vision,* Introduction and Part II, especially, insofar as the maneuverings just before and during the convention are concerned, pp. 360–374. On "the American Century," see also Divine, *op. cit.,* pp. 40, 65–66. I have quoted, too, *WSJ,* April 2, 1943.

4. And Still the Champion

On Wallace and Porter, see Blum ed., *op. cit.,* pp. 382, 390. On the campaign in general I have relied continually on Burns, *Roosevelt* and Polenberg, *War and Society;* also Bruenn, "Clinical Notes on . . . Roosevelt," *loc. cit.* Roosevelt's speeches are quoted by Burns. The "goat and dog" is from Paul Porter to Roosevelt, September 28, 1944, PPF 1820, FDRL. There is a good study, which includes quotations from Bricker, Dewey, and Roosevelt that I have used, by Peter Hayes, "The Communist Issue in the Election of 1944," ms., YUL. Polenberg, *op. cit.,* contains a crisp analysis of the election returns. For Wallace to Pepper, see Blum, ed., *op. cit.,* p. 403. On the Jefferson Day draft, see Burns, *op. cit.,* p. 596.

Chapter 9. Victory

1. Fundamental Issues

See Walter Lippmann, *United States War Aims* (Boston, 1944), Ch. XV.

2. Limits of American Internationalism

This section is adapted from John M. Blum, "Limits of American Internationalism, 1941–1945," in Leonard Krieger and Fritz Stern, eds., *The Responsibility of Power* (Garden City, 1967). The essays cited therein by F. G. Morain and D. W. Crofts are in YUL. See also Eugene V. Rostow, "The Great Transition," *Fortune*, January 1945, on economic policy, and on lend-lease, *WSJ*, March 13, April 7, 12, 1943, April 11, 1944, March 7, 1945; and undated, unsigned memo, "Subject: Russian Lend-Lease Program," OWMR Records, Deputy Director for Reconversion, NA. I have also used the account on the UN Charter in Divine, *Second Chance*. On weapons, I have relied upon Baxter, *Scientists Against Time;* Blum, ed., *Price of Vision*, p. 591; and especially the invaluable history of the development of the atomic bomb and of early efforts to control atomic energy in Richard G. Hewlett and Oscar E. Anderson, Jr., *The New World* (University Park, Pennsylvania, 1962). See, too, the stimulating study, published after my own went to press, Marvin J. Sherwin, *A World Destroyed* (New York, 1975).

3. Limits of American Liberalism

On Wallace, see Blum, ed., *Price of Vision*. For Bowles, see Chester Bowles, *Tomorrow Without Fear* (New York, 1946); the endorsements are in the book. On NRPB, see Polenberg, *War and Society*, p. 84. Samuelson is quoted in an important essay from which I profited, Byrd L. Jones, "The Role of Keynesians in Wartime Policy and Postwar Planning, 1940–1946," ms., American Economic Association, 1972, copies in the possession of Professor Jones. On the CED, see Herbert Stein, *The Fiscal Revolution in America* (Chicago, 1969), Chs. 8–9. My account of the Employment Act of 1946 derives from those chapters, from Lekachman, *Age of Keynes*, Chs. 6–7, and especially from Stephen K. Bailey, *Congress Makes A Law* (New York, 1950), which I find more persuasive than I do Stein.

Epilogue: The Veterans

The annual reports cited are those of the *Administrator for Veterans Affairs, Annual Report* (Washington, D.C., 1946) and *ibid.*, 1947. The episodes described are in the pattern explained in the note to the Prologue, except in the case of the old Irish-American, which is from James M. Burns, *John Kennedy: A Political Profile* (New York, 1960), and the case of Chester, which is from Ralph G. Martin, *The Best Is None Too Good* (New York, 1948). The verses are from Malvina Reynolds, "Little Boxes," *Little Boxes and Other Handmade Songs* (New York, 1964). The quotations from Eisenhower are from Dwight D. Eisenhower, *Crusade in Europe* (Garden City, 1948). For Kennedy in 1946, see Kenneth P. O'Donnell and David F. Powers, *"Johnny, We Hardly Knew Ye"* (Boston, 1972).

Index

22801926R00225

Made in the USA
Lexington, KY
14 May 2013